The Human Impact
on the Natural Environment

D1100560

B

The Human Impact
on the Natural Environment

Fourth Edition

ANDREW GOUDIE

BLACKWELL
Oxford UK & Cambridge USA

Copyright © Andrew Goudie 1981, 1986, 1990, 1993

First published 1981
Reprinted 1981, 1982, 1984
Second edition 1986
Reprinted 1987, 1988
Third edition 1990
Reprinted 1992.

Fourth edition 1993
Reprinted 1994, 1995

Blackwell Publishers Ltd
108 Cowley Road, Oxford, OX4 1JF, UK

British Library Cataloguing in Publication Data

A CIP catalogue record for this book is available from the British Library.

ISBN 0–631–18483–X
ISBN 0–631–18484–8 Pbk

Typeset by TecSet Ltd, Wallington, Surrey
Printed in Great Britain by Hartnolls Ltd, Bodmin, Cornwall

This book is printed on acid-free paper

Contents

SIX   Human Agency in Geomorphology

SEVEN   The Human Impact on Climate and the Atmosphere

EIGHT   The Future

NINE   Conclusion

Preface to the Fourth Edition

In this edition I have retained the structure of the three previous editions. However, I have updated substantial portions of the text to take account of the ever-increasing amount of material appearing on this theme, most especially those parts dealing with deforestation, desertification, ecological invasions, climatic change and the climatic future. I have also revised the guide to further reading, augmented and updated the references, and included some new figures.

A.S.G.

Acknowledgements

In the Oxford School of Geography there is thankfully (though others might say unfashionably) a concern with the unity of geography and an abomination of fissiparist tendencies. Such an atmosphere makes the writing of a book such as this a real possibility, for one receives many ideas and much information from colleagues which is germane to the theme of the volume. Some of the ideas come from the lecture courses of others, some ideas come from discourses in a house in Holywell, Oxford, and some ideas come from the overall atmosphere of the School. Mrs Pam Berry, Dr Alayne Perrott, Professor Bruce Atkinson (of Queen Mary College), Miss Ruth Goodwin, Mr Steve Lewitt, and Mr C. L. M. H. Gibbons have made comments on parts of an early draft, and the office, secretarial and technical staff of the School have assisted in many ways – especially Julie. Christine Zwart has assisted me greatly in obtaining illustrations for the book, and I am grateful to all those authors who have permitted me to use their figures and plates. I am also in the debt of the Faculty of Anthropology and Geography and of Hertford College for granting me a spell of sabbatical leave to facilitate the completion of this book. My main debt is to several generations of hapless undergraduates who, unknowingly, have provided a critical appraisal of some of this material in tutorials and lectures.

The publisher and author are grateful to the following for permission to use and for their help in supplying photographs: Alexander Turnbull Library, New Zealand (2.1); Australian Information Service (2.13); British Geological Survey (6.10); British Museum, photograph Michael Holford (3.12); British Museum (Natural History) (1.6); Janet and Colin Bord (2.12, 6.4 top); Jean Bottin (2.6); Cambridge University Collection, copyright reserved (3.5, 4.2, 9.1); Camera Press, photograph Colin Jones (2.5): J. Allan Cash (2.10, 6.1); Daily Telegraph Photograph Library (6.13, 7.4); Louise Davies (6.19 bottom); Michael Dent (5.1); © Earth Satellite Corporation/Geopic (2.7, 6.9, 6.20); Ecoscene (2.3 [Winkley], 2.11 [Harwood]); Forestry Commission (6.15); P. J. Green/

Ardea (1.4); Sonia Halliday (6.4 bottom); Grant Heilman, Lititz, PA (5.6, 5.7, 6.3); Hoa Qui, Paris (2.2); Hong Kong Government (6.5); Peter Kain © Richard Leakey (1.5); Eric Kay (5.3, 6.16); Frank Lane (3.1, 6.11); Peter Loughran (7.1); D. McGregor (3.3); Magma Copper Co., Arizona (6.2); Middle East Photographic Archive (6.14); Jeremy Moeran (6.19 top and centre); Tony Morrison (2.4); Maggie Murray/Format (2.9); NASA (7.10); National Meteorological Office Library, Bracknell (7.5); National Portait Gallery, London (1.1); National Publicity Studios, Wellington, New Zealand (3.7); National Resources Defence Council Inc., Washington (3.8); National Society for Clean Air (7.7); Nigel Press Associates (5.4); Novosti Photographic Library (3.10, 5.2, 5.5); Oxfam © Badal (2.8, 8.1); Panos Pictures, photograph Martin Adler (7.2); Press Association (3.2, photograph Chris Bacon, inset photograph David Giles); Brenda Prince/Format (1.8); © Mirella Ricciardi/Colorific (3.6); G. R. Roberts, Nelson, New Zealand (2.16); Royal Geographical Society (6.15 top); Thomas Photos, Oxford (7.8); Gianni Tortoli/Colorific (6.7); Transafrica Pix (2.14); United Press International/Bettman Archive (8.3); © US Soil Conservation Servies (4.4); USWB–G. L. Gillespie (Frank Lane) (7.5); Western Americana/Peter Newark (6.6); Colin Wheeler (5.8); WWF (1.8 [photograph Paul Forster], 2.15 [Mauri Rautkari], 3.9 [Miroslav Levy], 3.13 [Rick Weyerhaeuser], 7.9 [anon], 8.2 [Michele Depraz]); George Wright (6.18); Xinhua News Agency (4.6). All other photographs are by the author.

ONE

Introduction

The development of ideas

In the history of Western thought, men have persistently asked three questions concerning the habitable earth and their relationship to it. Is the earth, which is obviously a fit environment for man and other organic life, a purposefully made creation? Have its climates, its relief, the configuration of continents influenced the moral and social nature of individuals, and have they had an influence in molding the character and nature of human culture? In his long tenure of the earth, in what manner has man changed it from its hypothetical pristine condition?

C. J. Glacken, 1967, p. vii

Glacken's third question forms the focus of this book. Since he wrote, its importance has been increasingly recognized, for as Yi-fu Tuan (1971) put it: 'The fact of diminishing nature and of human ubiquity is now obvious.'

This last question received relatively little systematic attention until the eighteenth century, and overall it has been less central in the history of geographical thought than the second – the one which asks whether environment has an influence on people (Glacken, 1967). Indeed, in the late seventeenth century, when the Creation was deemed to have occurred in 4004 BC, a grand conception of nature as a divinely created order for the well-being of all life was formulated (Glacken, 1956). In other words it was the first question to which people addressed themselves. Men and women were seen to be caretakers, stewards of God, and their task was thought to improve the primeval aspect of the earth through tillage and in other ways. Many thinkers at this time, such as John Ray, author of *The wisdom of God manifested in the works of the Creation* (1691), believed people were living in a world – which they could change and improve – where the order and beauty of nature, manifestations of the Creator's skill and wisdom, could be seen everywhere. Indeed the Genesis notion which set humans above and against nature, entitled to rule like absolute despots (Passmore, 1974: 9), was formulated into a philosophy of science and progress by men like Francis Bacon, Galileo and Descartes (Swift, 1974). Bacon foresaw the need to combine knowledge with invention to give humankind control of nature. Also, for a time in the nineteenth century, biological interpretations fortified people's belief in the right to deal with nature however they chose. Darwin's theory of natural selection was transformed by Herbert Spencer into the doctrine of 'the survival of the fittest'.

Count Buffon can be regarded as the first Western scientist to be concerned directly and intimately with the human impact on the natural environment (Glacken, 1963). He contrasts the appearance of inhabited and uninhabited lands: the anciently inhabited countries have few woods, lakes or marshes, but they have many heaths and scrub; their mountains are bare, and their soils are less fertile because they lack the organic matter which woods, felled in inhabited countries, supply, and the herbs are browsed. Buffon was also much involved with the domestication of plants and animals – one of the major transformations in nature brought about by human actions.

Studies of the torrents of the French and Austrian Alps, undertaken in the late eighteenth and early nineteenth centuries, deepened immeasurably the realization of human capacity to change the environment. Fabre and Surell studied the flooding, the siltation, the erosion and the division of watercourses brought about by deforestation in the Alps. Similarly de Saussure showed that Alpine lakes had suffered a lowering of water levels in recent times because of deforestation. In Venezuela, in the valley of Aragua, von Humboldt concluded that the lake level of Lake Valencia in 1800 (the year of his visit) was lower than it had been in previous times, and that deforestation, the clearing of plains, and the cultivation of indigo, were among the causes of the gradual drying up of the basin.

Comparable observations were made by the French rural economist, Boussingault (1845). He returned to Valencia some 25 years after Humboldt and noted that the lake was actually rising. He ascribed this reversal to political and social upheavals following the granting of independence to the colonies of the erstwhile Spanish Empire. The freeing of slaves had led to a decline in agriculture, a reduction in the application of irrigation water, and the re-establishment of forest.

Boussingault also reported some pertinent hydrological observations that had been made on Ascension Island (p. 685):

In the Island of Ascension there was an excellent spring situated at the foot of a mountain originally covered with wood; the spring became scanty and dried up after the trees which covered the mountain had been felled. The loss of the spring was rightly ascribed to the cutting down of the timber. The mountain was therefore planted anew. A few years afterwards the spring reappeared by degrees, and by and by followed with its former abundance.

Charles Lyell, in his *Principles of geology*, one of the most influential of all scientific works, referred to the human impact and recognized that tree-felling and drainage of lakes and marshes tended 'greatly to vary the state of the habitable surface'. Overall, however, he believed that the forces exerted by people were insignificant in comparison with those exerted by nature:

If all the nations of the earth should attempt to quarry away the lava which flowed from one eruption of the Icelandic volcanos in 1783, and the two following years, and should attempt to consign it to the deepest abysses of the ocean they might toil for thousands of years before their task was accomplished. Yet the matter borne down by the Ganges and Burrampooter, in a single year, probably very much exceeds, in weight and volume, the mass of Icelandic lava produced by that great eruption. (Lyell, 1835: 197)

Lyell somewhat modified his views in later editions of the *Principles* (see, for example, Lyell, 1875), largely as a result of his experiences in the United States, where recent deforestation in Georgia and Alabama had produced numerous ravines of impressive size.

One of the most important physical geographers to show concern with our theme was Mary Somerville (1858) (plate 1.1), who clearly appreciated the unexpected results that occurred as man 'dexterously avails himself of the powers of nature to subdue nature':

Plate 1.1
Mary Somerville (1780–1872)

Man's necessities and enjoyments have been the cause of great changes in the animal creation, and his destructive propensity of still greater. Animals are intended for our use, and field-sports are advantageous by encouraging a daring and active spirit in young men; but the utter destruction of some races in order to protect those destined for his pleasure, is too selfish, and cruelty is unpardonable: but the ignorant are often cruel. A farmer sees the rook pecking a little of his grain, or digging at the roots of the springing corn, and poisons all his neighbourhood. A few years after he is surprised to find his crop destroyed by grubs. The works of the Creator are nicely balanced, and man cannot infringe his Laws with impunity. (Somerville, 1858, p. 493)

This is in effect a statement of one of the basic laws of ecology: that everything is connected to everything else and that one cannot change just one thing in nature.

Considerable interest in conservation, climatic change and extinctions arose amongst European colonialists who witnessed some of the consequences of western-style economic development in tropical lands (Grove, 1990). However, the extent of human influence on the environment was not explored in detail and on the basis of sound data until George Perkins Marsh (plate 1.2) published *Man and nature* (1864), in which he dealt with human influence on the woods, the waters and the sands. The following extract illustrates the breadth of his interests and the ramifying connections he identified between human actions and environmental changes:

Plate 1.2
George Perkins Marsh in 1861. Marsh was the author of *Man and nature* – probably the most important landmark in the history of the study of the role of humans in changing the face of the earth

Vast forests have disappeared from mountain spurs and ridges; the vegetable earth accumulated beneath the trees by the decay of leaves and fallen trunks, the soil of the alpine pastures which skirted and indented the woods, and the mould of the upland fields, are washed away; meadows, once fertilized by irrigation, are waste and unproductive, because the cisterns and reservoirs that supplied the ancient canals are broken, or the springs that fed them dried up; rivers famous in history and song have shrunk to humble brooklets; the willows that ornamented and protected the banks of lesser watercourses are gone, and the rivulets have ceased to exist as perennial currents, because the little water that finds its way into their old channels is evaporated by the droughts of summer, or absorbed by the parched earth, before it reaches the lowlands; the beds of the brooks have widened into broad expanses of pebbles and gravel, over which, though in the hot season passed dryshod, in winter sealike torrents thunder, the entrances of navigable streams are obstructed by sandbars, and harbors, once marts of an extensive commerce, are shoaled by the deposits of the rivers at whose mouths they lie; the elevation of the beds of estuaries, and the consequently diminished velocity of the streams which flow into them, have converted thousands of leagues of shallow sea and fertile lowland into unproductive and miasmatic morasses. (Marsh, 1965:9)

More than a third of the book is concerned with 'the woods'; Marsh does not touch upon important themes like the modifications of mid-latitude grasslands, and he is much concerned with Western civilization. Nevertheless, employing an eloquent style and copious footnotes, Marsh, the versatile Vermonter, stands as a landmark in the study of environment (Thomas, 1956).

Marsh, however, was not totally pessimistic about the future role of humankind or entirely unimpressed by positive human achievements (1965: 43–4):

New forests have been planted; inundations of flowing streams restrained by heavy walls of masonry and other constructions; torrents compelled to aid, by depositing the slime with which they are charged, in filling up lowlands, and raising the level of morasses which their own overflows had created; ground submerged by the encroachment of the ocean, or exposed to be covered by its tides, has been rescued from its dominion by diking; swamps and even lakes have been drained, and their beds brought within the domain of agricultural industry; drifting coast dunes have been checked and made productive by plantation; seas and inland waters have been repeopled with fish, and even the sands of the Sahara have been fertilized by artesian fountains. These achievements are far more glorious than the proudest triumphs of war . . .

Reclus (1873), one of the most prominent French geographers of his generation, and an important influence in the USA, also recognized that the 'action of man may embellish the earth, but it may also disfigure it; according to the manner and social condition of any nation, it contributes either to the degradation or glorification of nature' (p. 522). He warned rather darkly (p. 523) that 'in a spot where the country is disfigured, and where all the grace of poetry has disappeared from the landscape, imagination dies out, and the mind is impoverished; a spirit of routine and servility takes possession of the soul, and leads it on to torpor and death.' Reclus (1871) also displayed a concern with the relationship between forests, torrents (p. 335) and sedimentation (p. 338):

It is, however, the improvidence of the inhabitants, and not so much the geological constitution of the soil, which is the principal cause of the devastating action of the streams. In the mountains of Dauphimy and Provence . . . during the course of centuries, the trees have been cut down by greedy speculators, and by senseless farmers who wished to add some little strips of land to the fields in the valleys and to the pastures on the summits; but when they destroyed the forest they also destroyed the very land it stood on.

Although the torrents lower the mounts, on the other hand they elevate the plains; but their deposits not being pulverised into clays and sands, are often the means of bringing another disaster on the inhabitants, who find their fertile land covered beneath enormous masses of rocks and pebbles.

Reclus obtained many of his ideas from Marsh, and another important contributor, Woeikof (1842–1914) of St Petersburg, obtained some of his ideas from Reclus. Woeikof believed that the earth's surface was composed of mobile bodies (*Corps meubles*), such as soil, subsoil and gravels, and that humans influenced them through an intermediary, vegetation. He refers to the effects of fire, the climatic effects of forest removal, the destruction of chernozems by ploughing, gullying, the siltation of streams and the barrenness of karst.

In 1904 Friedrich coined the term 'Raubwirtschaft' which can be translated as economic plunder, robber economy or, more simply, devastation. This concept has been extremely influential but is open to criticism. He believed that destructive exploitation of resources leads of necessity to foresight and to improvements, and that after an initial phase of ruthless exploitation and resulting deprivation human measures would, as in the old countries of Europe, result in conservation and improvement:

I can conceive the Raubwirtschaft of our time as a depletion of land, wild plants, and animals, which will disappear quickly as the culture of northwest Europe is substituted for that of colonial and primitive peoples. Such a substitution will bring an intensive use of earth resources which aims at a firmer and firmer establishment of man on the earth. (Friedrich, cited in Whitaker, 1940: 157)

This idea was opposed by Sauer (1938) and Whitaker (1940), the latter pointing out that some soil erosion could well be irreversible (p. 157):

It is surely impossible for anyone who is familiar with the eroded loessial lands of northwestern Mississippi, or the burned and scarred rock hills of north central Ontario, to accept so complacently the damage to resources involved in the process of colonization, or to be so certain that resource depletion is but the forerunner of conservation.

None the less Friedrich's concept of robber economy was adopted and modified by the great French geographer, Jean Brunhes, in his

Human geography (1920). He recognized the interrelationships involved in anthropogenic environmental change (p. 332): 'Devastation always brings about, not a catastrophe, but a series of catastrophes, for in nature things are dependent one upon the other.' Moreover, Brunhes acknowledged that the 'essential facts' of human geography included 'Facts of Plant and Animal Conquest' and 'Facts of Destructive Exploitation'. At much the same time other significant studies were made of the same theme. Shaler of Harvard (*Man and the earth*, 1912) was very much concerned with the destruction of mineral resources (a topic largely neglected by Marsh).

In spite of this work, however, the subject of the human role in changing the face of the earth received relatively little attention from geographers in the nineteenth century and early decades of the twentieth century. One of the main features of the 'classical' view of geography put forward by men like Ratzel (Wrigley, 1965:5) was the investigation of the effects of the physical environment on the functioning and development of societies, and environmentalism persisted as a potent force, notably in American geography, through the works of Davis, Brigham, Semple and Huntington.

Very definite shifts of emphasis appeared, however, in the 1920s and 1930s when, under the influence of Harlan Barrows and Carl Sauer, there was an eager search for an alternative to so-called 'naive environmental determinism'. Sauer (plate 1.3) led an effective campaign against destructive exploitation (Speth, 1977), reintroduced Marsh to a wide public, recognized the ecological virtues of some primitive peoples, concerned himself with the great theme of domestication, concentrated on the landscape changes that resulted from human action, and gave clear and far-sighted warnings about the need for conservation (Sauer, 1938: 494):

We have accustomed ourselves to think of ever expanding productive capacity, of ever fresh spaces of the world to be filled with people, of ever new discoveries of kinds and sources of raw materials, of continuous technical progress operating indefinitely to solve problems of supply. We have lived so long in what we have regarded as an expanding world, that we reject in our contemporary theories of economics and of population the realities which contradict such views. Yet our modern expansion has been effected in large measure at the cost of an actual and permanent impoverishment of the world.

Likewise, in the former USSR, in spite of the fact that in *Das Kapital* Marx took an environmental line (Matley, 1966), the era of Stalin produced a strong attack on environmentalist views. Grandiose 'practical' schemes were developed for the alteration of the natural environment, such as the 1948 'Stalin Plan for the Transformation of Nature'. These views in turn, however, were thought by some geographers to be excessive. Indeed, much recent Soviet geography has been concerned with the sometimes undesirable transformation of the natural environment by human action (see Komarov, 1978).

Plate 1.3
Carl Ortwin Sauer
(1889–1975), a cultural and historical geographer, whose studies examined such central themes as domestication, the effects of resource exploitation, conservation and the evolution of cultural landscapes

Gerasimov and others (1971) point to some of the effects produced by the exploitation of natural resources in the erstwhile USSR, including erosion and pollution, and Gerasimov (1976: 77) has placed the human impact on nature in its historical context:

By taking energy and matter from the environment and returning them in converted – industrial, domestic and other – forms society interferes with the dynamically balanced cyles of natural processes. However, as a result of its long evolution, nature has acquired an ability to restore disrupted natural processes . . . Thus the natural environment taken as a whole was able, up to a point, to withstand anthropogenic disturbances, although there were also local irreversible changes. Since the industrial revolution, the general intensity of human impact on the environment has exceeded its potential for restoration in many large areas of the earth's surface, leading to irreversible changes not only on a local but also on a regional scale.

He also indicated that some systems had been more disrupted than others (1976: 79):

Almost all European capitalist countries find themselves 'pressed' between the two jaws of a vice – the complex historical inheritance of a considerably disrupted natural environment and the increasing negative effects on the environment of industrialization and urbanization.

Gerasimov wrote in the same work (p. 78) of a subdiscipline of '*constructive geography*' ('a science of the planned transformation and management of the natural environment for the sake of the future of mankind'). This concept has been criticized for being insufficiently geographical and for lacking a distinctive object of study by Mil'kov and his co-workers (Nesterov, 1974), who propose an alternative subdiscipline, '*anthropogenic landscape science*', concerned with the human construction of entire landscapes. Isachenko (1974, 1975) in turn attacks the Mil'kov school with vigour and urges the need to see natural and human influences working together in the landscape:

An oasis created in desert remains part of the desert just as a clearing and plot of cropped land in the tayga belongs to the tayga and will continue to belong to it as long as zonal and azonal external conditions will be preserved or until man acquires the ability to control incoming solar radiation, the circulation of air masses and tectonic processes. (1974: 474)

To the non-Russian geographer the wonder is not so much that their Russian colleagues seem to hold such firmly divergent viewpoints, but that Gerasimov, Mil'kov and Isachenko are all so greatly concerned with the human impact.

The theme of the human impact on the environment has, however, been central to some Western historical geographers studying the evolution of the cultural landscape. The clearing of woodland (Darby, 1956; Williams, 1989), the domestication process (Sauer, 1952), the draining of marshlands (Williams, 1970), the introduction of alien plants and animals (McKnight, 1959), and the reclamation of heathland are among some of the recurrent themes of a fine tradition of historical geography.

In 1956, some of these themes were explored in detail in a major symposium volume, *Man's role in changing the face of the earth* (Thomas, 1956). Its impact is difficult to assess, for although it contained many brilliant individual contributions it did not really herald a new generation of environmental research. Kates et al. (1991: 4) write of it:

Man's Role seems at least to have anticipated the ecological movement of the 1960s, although direct links between the two have not been demonstrated. Its dispassionate, academic approach was certainly foreign to the style of that movement. . . . Rather, *Man's Role* appears to have exerted a much more subtle, and perhaps more lasting, influence as a reflective, broad-ranging and multidimensional work.

In the last twenty years too many Anglo-Saxon geographers have largely retreated from examining relationships between human agency and the physical environment. Much geography has focused on spatial and distributional analysis and has found little space for consideration of the natural environment (see, for example, Abler et al., 1971). None the less, this retreat has not been total, though it is salutary to reflect that at a time of increasing anxiety about environmental issues at public and political levels, one of the main traditional concerns of geography, human–land relationships, has been effectively carried out by other disciplines under the guise of ecology, environmental studies and the like.

There have, however, been notable exceptions to this tendency, and many geographers have contributed to, and been affected by, the phenomenon which is often called the environmental revolution or the ecological movement. The subject of the human impact on the environment, dealing as it does with such matters as environmental degradation, pollution and desertification, has close links with these developments, and is once again a theme in many textbooks and research monographs in geography (see, for example, Detwyler, 1971; Berry, 1974; Manners and Mikesell, 1974; Wagner, 1974; Cooke and Reeves, 1976; Gregory and Walling, 1979; Simmons, 1979; Tivy and O'Hare, 1981; Mannion, 1991; Turner et al., 1990; Bell and Walker, 1992).

The development of human population and stages of cultural development

During the early part of the Ice Age, some three million years ago, human life probably first appeared on earth. The oldest remains have been found in sediments from the Rift Valleys of East Africa, notably in Tanzania, Kenya and Ethiopia. Since that time the human population has spread over virtually the entire land surface of the planet (figure 1.1). Table 1.1 gives data on recent views of the dates for the arrival of humans in selected areas.

Estimates of population levels in the early stages of human development are difficult to make with any degree of certainty

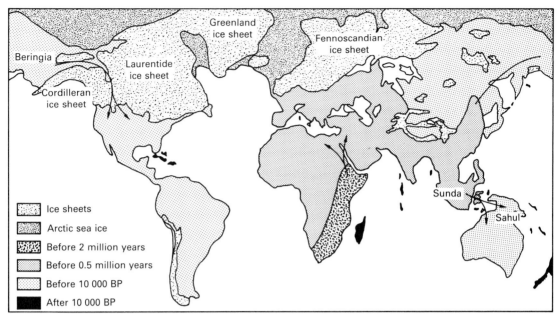

Ice sheets
Arctic sea ice
Before 2 million years
Before 0.5 million years
Before 10 000 BP
After 10 000 BP

Figure 1.1
The human colonization of
Ice-Age earth (after Roberts,
1989, figure 3.7)

(figure 1.2). Before the agricultural 'revolution' some 10 000 years ago, human groups lived by hunting and gathering in parts of the world where this was possible. At that time the world population *may* have been of the order of five million people (Ehrlich et al., 1977: 182) and large areas would only recently have witnessed human migration. The Americas and Australia, for example, were virtually uninhabited until about 30 000 years ago (or possibly even later).

The agricultural revolution probably enabled an expansion of the total human population to about 200 million by the time of Christ, and to 500 million by AD 1650. It is since that time, helped by the medical and industrial revolutions and developments in agriculture

Table 1.1 Dates of human arrivals

Area	Source	Date (years BP)
Africa	Klein (1983)	2 700 000–2 900 000
Europe	Champion et al. (1984)	c.1 600 000 but most post-350 000
Britain	Green (1981)	c.200 000
Japan	Ikawa-Smith (1982)	c.50 000
New Guinea	Bulmer (1982)	c.50 000
Australia	Allen et al. (1977)	c.40 000
North America	Irving (1985)	15 000–40 000
South America	Guidon and Delibrias (1986)	32 000
Ireland	Edwards (1985)	9 000
Caribbean	Morgan and Woods (1986)	4500
Polynesia	Kirch (1982)	2000
Madagascar	Battistini and Verin (1972)	c. AD 500
New Zealand	Green (1975)	AD 700–800

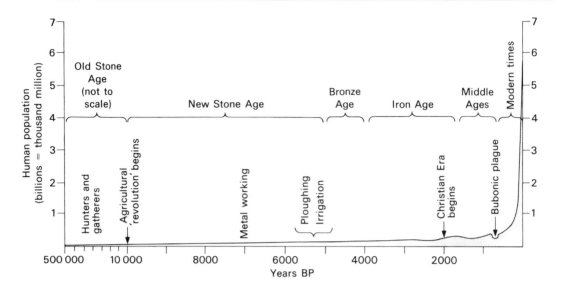

Figure 1.2
The growth of human numbers
for the past half-million years
(after Ehrlich et al., 1977,
figure 5.2)

and colonization of new lands, that human population has exploded, reaching about 1000 million by AD 1850, 2000 million by AD 1930 and 4000 million by AD 1975. The figure had reached 5150 million by 1987. Victory over malaria, smallpox, cholera and other diseases has been responsible for marked decreases in death-rates throughout the non-industrial world, but death-rate control has not in general been matched by birth control. Thus the annual population growth rate in the late 1980s in South Asia was 2.64 per cent, Africa 2.66 per cent and Latin America (where population increased sixfold between 1850 and 1950), 2.73 per cent.

Clearly, this growth of the human population of the earth is in itself a highly important cause of the transformation of nature. Of no lesser importance, however, has been the growth and develop-ment of culture and technology. Sears (1957: 51) has put the power of humankind into the context of other species:

Man's unique power to manipulate things and accumulate experience presently enabled him to break through the barriers of temperature, aridity, space, seas and mountains that have always restricted other species to specific habitats within a limited range. With the cultural devices of fire, clothing, shelter, and tools he was able to do what no other organism could do without changing its original character. Cultural change was, for the first time, substituted for biological evolution as a means of adapting an organism to new habitats in a widening range that eventually came to include the whole earth.

We now turn to a consideration of the major cultural and technical developments that have taken place during the past two to three million years. Three main phases will form the basis of this analysis: the phase of hunting and gathering; the phase of plant cultivation, animal keeping and metal working; and the phase of modern urban and industrial society. These developments are treated in much greater depth by Simmons (1989).

Hunting and gathering

The definition of 'human' is something of a problem, not least because, as is the case with all existing organisms, new forms tend to emerge by perceptible degrees from antecedent ones. Moreover, the fossil evidence is scarce, fragmentary and can rarely be dated with precision. Although it is probably justifiable to separate the hominids from the great apes on the basis of their assumption of an upright posture, it is much less justifiable or possible to distinguish on purely zoological grounds between those hominids that remained pre-human and those that attained human status. To qualify as a human, a hominid must demonstrate cultural development: the systematic manufacture of implements as an aid to manipulating the environment.

The oldest records of human activity and technology, pebble tools (crude stone tools which consist of a pebble with one end chipped into a rough cutting edge), have been found with human bone remains in various parts of Africa. For example, at Lake Turkana in northern Kenya, and the Omo Valley in southern Ethiopia, a tool-bearing bed of volcanic material called tuff has been dated by isotopic means at about 2.6 million years old, while another bed at the Olduvai Gorge in Tanzania, made famous by the researches of the Leakey family, has been dated by similar means at 1.75 million years.

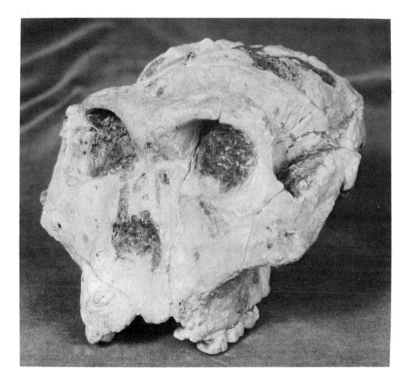

Plate 1.4
Some of the earliest hominids, which were more like humans than apes, belonged to the genus *Australopithecus*. This skull of *Australopithecus robustus* probably dates back to Plio-Pleistocene times. These creatures probably used stones and bones as tools

Figure 1.3 (opposite)
Developments in stone tool technology:
(a) a primitive pebble tool from early Pleistocene strata – mode 1
(b) a hand axe of Abbeville type from a younger level – mode 2
(c) another Lower Palaeolithic hand axe, from Rajasthan, India – mode 2
(d) a Middle Palaeolithic keeled scraper and core from Rajasthan – mode 3
(e) Upper Palaeolithic burin and blade from Rajasthan – mode 4
(f) Mesolithic artefacts from Rajasthan – mode 5

As the Stone Age progressed the tools became more sophisticated, more varied and more effective. Greater exploitation of plant and animal resources became feasible. It is possible to observe a clear progression during the Palaeolithic Age in the technology of working stone (figure 1.3), although it needs to be stressed that in reality the stages were not synchronous in different areas nor among different human groups in the same area. In addition, particular industries are often seen to combine techniques from more than one stage of development.

The lithic technology in the Old Stone Age in Europe has been divided into five main modes: mode 1 (figure 1.3a) consists of chopper tools of Lower Palaeolithic type; mode 2 (figure 1.3b and c) comprises bifacially flaked hand axes of the Lower Palaeolithic; mode 3 (figure 1.3d) Middle Palaeolithic flake tools from prepared cores; mode 4 (figure 1.3e) Upper Palaeolithic punch-struck blades

Plate 1.5
A skull of one of the first humans, *Homo habilis*, from near Lake Turkana in Kenya. This skull probably dates back to about two and a half million years ago. These were probably the first creatures to make tools

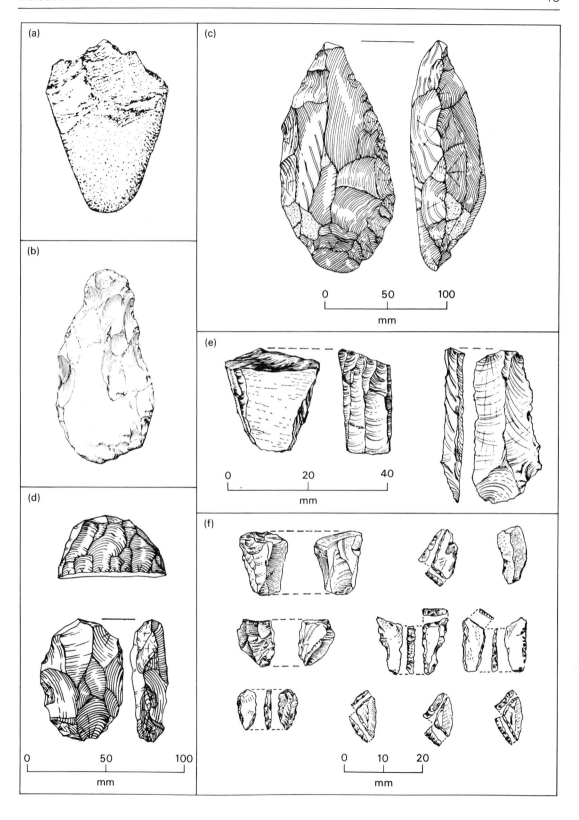

(a)

(b)

(c)

```
0        50      100
|_____|_____|
        mm
```

(e)

```
0          20        40
|_____|_____|
          mm
```

(d)

```
0        50      100
|_____|_____|
        mm
```

(f)

```
0      10     20
|_____|_____|
       mm
```

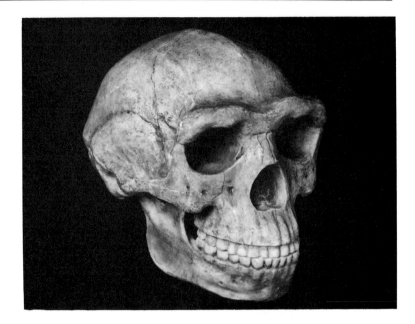

Plate 1.6
Homo erectus remains have
been found outside Africa: in
China, South-East Asia and in
Europe. These early humans
used fire, but their skull form
was different from that of
modern humans, *Homo
sapiens*

with steep retouch; and mode 5 (figure 1.3f) microlithic components of composite artefacts from the Mesolithic Age.

Stone may not, however, have been the only material used by early civilizations. Sticks and animal bones, the preservation of which is less likely than stone, are among the first objects that may have been used as implements, although the sophisticated utilization of antler and bone as materials for weapons and implements appears to have developed surprisingly late in pre-history. There is certainly a great deal of evidence for the use of wood throughout the Palaeolithic Age, for building, ladders, fire, pigment (charcoal), the drying of wood and digging sticks. Tyldesley and Bahn (1983: 59) go so far as to suggest that 'The Palaeolithic might more accurately be termed the "Palaeoxylic" or "Old Wood Age".'

The building of shelters and the use of clothing became a permanent feature of human life as the Palaeolithic period progressed, and permitted habitation in areas where the climate was otherwise not congenial. European sites from the Mousterian of the Middle Palaeolithic have revealed the presence of purposefully made dwellings as well as caves, and by the Upper Palaeolithic more complex shelters were in use, allowing people to live in the tundra lands of Central Europe and Russia.

Another feature of early society which seems to have distinguished humans from the surviving non-human primates was their seemingly omnivorous diet. Biological materials recovered from settlements in many different parts of the world indicate that in the Palaeolithic Age humans secured a wide range of animal meats, whereas the great apes, though not averse to an occasional taste of animal food, are predominantly vegetarian. One consequence of enlarging the range of their diet, was that, in the long run, humans were able to explore a much wider range of environment (G. Clark, 1977b: 19).

Moreover, at an early stage humans discovered the use of fire, and as Sparks and West (1972) have argued, 'the regularity with which hearths are found associated with the Middle and Upper Palaeolithic sites leaves little doubt that Neanderthal Man and his successors were capable of fire production.' Fire, as we shall see (chapter 2), is a major agent by which humans have influenced their environment. It has been found in association with Peking Man in deposits at Choukoutien dating back to perhaps 600 000–700 000 years, and may have been employed even earlier in East Africa, where Gowlett et al. (1981) have claimed to find evidence for deliberate manipulation of fire from over 1.4 million years ago.

Another major difference that set humankind above the beasts was the development of communicative skills such as speech. Until hominids had developed words as symbols, the possibility of transmitting, and so accumulating, culture hardly existed. Animals can express and communicate emotions, never designate or describe objects.

Overall, compared with later stages of cultural development, early hunters and gatherers had neither the numbers nor the technological skills to have a very substantial effect on the environment. Wymer (1982) believes that the first clear evidence of human influence on vegetation may be found at Hoxne in eastern England where pollen analysis indicates an increase in grassland at the same time as Acheulian societies occupied the edge of a small lake. Specks of charcoal have been found in the lake muds and the inference is that hunters had set light to the forest, intentionally or accidentally, on such a scale that the local vegetational succession was disrupted.

Besides the effects of fire, early civilizations may have caused some diffusion of seeds and nuts, and through hunting activities (see chapter 3) may have had some effects on animal extinctions (the so-called 'Pleistocene overkill'). Locally some eutrophication may have occurred, and around some archaeological sites phosphate and nitrate levels may be sufficiently raised to make them an indicator of habitation to archaeologists today (Holdgate, 1979).

Humans as cultivators, keepers and metal workers

Around the beginning of the Holocene, about 10 000 years ago, humans started in various parts of the world to domesticate rather than to gather food plants and to keep, rather than just hunt, animals. This phase of human cultural development is well reviewed in Roberts (1989). By domesticating food plants, they reduced enormously the space required for sustaining each individual by a factor of the order of 500 at least (Sears, 1957: 54). As a consequence we see shortly thereafter, notably in the Middle East, the establishment of the first major settlements – towns. So long as man had

to subsist on the game animals, birds and fish he could catch and trap, the insects and eggs he could collect and the foliage, roots, fruits and seeds he could gather, he was limited in the kind of social life he could develop; as a rule he could only live in small groups, which gave small scope for specialization and the subdivision of labour, and in the course of a year he would have to move over extensive tracts of country, shifting his habitation so that he could tap the natural resources of successive areas. It is hardly to be wondered at that among communities whose energies were almost entirely absorbed by the mere business of keeping alive, technology remained at a low ebb. (Clark, 1962: 76)

Although it is now recognized that some subsistence hunters and gathers had considerable leisure, there is no doubt that through the controlled breeding of animals and plants humans were able, by hard work, to develop a more reliable and readily expandable source of food and thereby create a solid and secure basis for cultural advance, an advance which included civilization and the 'urban revolution' of Childe (1936) and others. Indeed, Isaac (1970) has termed domestication 'the single most important intervention man had made in his environment'; while G. Clark (1977b) has remarked that 'the milk-stool and the mattock were forerunners of the conveyor-belt and the punch-card.'

A distinction can be drawn between cultivation and domestication. Whereas cultivation involves deliberate sowing or other management, and entails plants which do not necessarily differ genetically from wild populations of the same species, domestication results in genetic change brought about through conscious or unconscious human selection. This creates plants that differ morphologically from their wild relatives and which may be dependent on humans for their survival. Domesticated plants are thus necessarily cultivated plants, but cultivated plants may or may not be domesticated. For example, the first plantations of *Hevea* rubber and quinine in the Far East were established from seed which had been collected from the wild in South America. Thus at this stage in their history these crops were cultivated but not yet domesticated.

The whole question of the origin of agriculture remains controversial (Bender, 1975). Some early workers saw agriculture as a divine gift to humankind, while others thought that animals were domesticated for religious reasons. They argued that it would have been improbable that humans could have predicted the usefulness of domestic cattle before they were actually domesticated. Wild cattle are large and fierce beasts, and no one could have foreseen their utility for labour or milk until they were tamed – tamed for ritual sacrifice in connection with lunar goddess cults (the great curved horns being the reason for the association). Another major theory was that domstication was produced by crowding, possibly brought on by a combination of climatic deterioration (post-Glacial progressive desiccation) and population growth. Such pressure, according to Childe, forced communities to intensify their methods of food production. Current palaeo-climatological research tends not to support this interpretation. Sauer (1952), a geographer, believed that the domestication of plants was initiated in South-East Asia by fishing folk, who found that lacustrine and riverine resources would underwrite a stable economy and a sedentary or semi-sedentary life style. He surmises, in an almost totally theoretical model, that the initial domesticates would be multi-purpose plants set around small fishing villages to provide such items as starch foods, substances for toughening nets and lines and making them water-resistant, and drugs and poisons. He suggested that 'food production was one and perhaps not the most important reason for bringing plants under cultivation.' Yet another model was advanced by Jacobs (1969) which turned certain more traditional models upside down. Instead of following the classic pattern whereby farming leads to village which leads to town which leads to civilization, she proposed that one could be a hunter-gatherer and live in a city, and that agriculture originated in and around such cities rather than in the countryside. Her argument suggests that even in primitive hunter-gatherer societies particularly valuable commodities such as fine stones, pigments and shells could create and sustain a trading centre which would possibly become large and stable. Food would be exchanged for goods, but natural produce brought any distance would have to be durable, so meat would be transported on the hoof, for example, but not all the animals would be consumed immediately; some would be herded together and might breed. This might be the start of domestication.

The process of domestication and cultivation was also once considered a revolutionary system of land procurement that had

Plate 1.7
The domestication of plants such as wheat involved the deliberate selection of particular genetic characteristics thought to be advantageous

evolved in one or two hearths and diffused over the face of the earth, replacing the older hunter-gathering systems by stimulus diffusion. It was felt that the deliberate rearing of plants and animals for food was a discovery or invention so radical and complex that it could have developed only once (or possibly twice) – the so-called 'Eureka model'.

The first domesticated plants occur at approximately the same time in widely separated areas (figure 1.4). This might be construed to suggest that developments in one area triggered experiments with local plant materials in others. However, if plant domestication were proved to begin at different times in different parts of the world this could also be used to argue that diffusion took place from one single early centre. Yet the balance of botanical and archaeological evidence seems to suggest that humans started experimenting with domestication and cultivation at different times in different parts of the world.

The current belief (Harlan, 1976) is that agriculture evolved through an extension and intensification of longstanding practices, it evolved over wide areas in different parts of the world, and the innovative pattern was complex and diffuse. Harlan (1957b: 57) has gone so far as to write:

I am inclined to develop a no-model which leaves room for whole arrays of motives, actions, practices and evolutionary processes. What applies in Southeast Asia may not apply in Southwest Asia . . . A search for a single

Figure 1.4
Dates for the domestication of crops and animals (from J. R. Harlan, 'The plants and animals that nourish man', *Scientific American*, 1976 Copyright © 1976 Scientific American, Inc. All rights reserved)

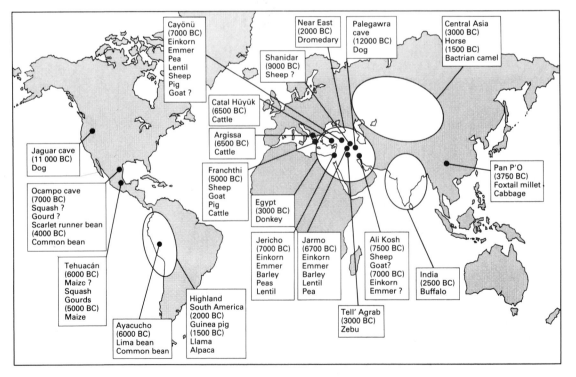

overriding cause for human behaviour is likely to be frustrating and fruitless.

His interpretation of this spatial and temporal complexity is shown in figure 1.4. Certainly the pre-eminence once accorded the Near East is now lessened (Simoons, 1974); while Jarman (1977) has stressed the continuity of human–animal relationships and cast doubt on any clear-cut wild–domestic dichotomy. Indeed, Jarman proposes six stages in the process of human exploitation of animals:

1 *Random predation*, during which the human group makes no effort to control or benefit from the regularity of the animal behaivour pattern, simply exploiting the animals when they happen to be available.
2 *Controlled predation*, which would not necessarily amount to a year-round association of humans and their prey. The control of animal movement in game drives and corralling are examples.
3 *Herd following*, in which the human group, or part of it, echoes the animal movements, maintaining a degree of contact, so that a given human population tends to become associated with a given animal population.
4 *Loose herding*, where seasonal control is exerted.
5 *Close herding*, in which a high degree of control of herds is maintained throughout the year.
6 *Factory farming*, whereby animals are kept immobile throughout their lives, and maintained in a wholly artificial environment.

One highly important development in agriculture, because of its rapid and early effects on environment, was irrigation. This came rather later than domestication. Amongst the earliest evidence of artificial irrigation is the mace-head of the Egyptian Scorpion King which shows one of the last pre-dynastic kings ceremonially cutting an irrigation ditch around 5050 years ago (Butzer, 1976) although it is possible that irrigation in Iraq started even earlier.

A major difference has existed in the development of agriculture in the Old and New Worlds: in the New World there were few counterparts to the range of domesticated animals which were an integral part of Old World systems (Sherratt, 1981). A further critical difference was that in the Old World the secondary applications of domesticated animals were explored. The plough was particularly important in this process – the first application of animal power to the mechanization of agriculture. Closely connected to this was the use of the cart, which both permitted more intensive farming and enabled the transportation of its products. Furthermore, the development of textiles from animal fibres afforded, for the first time, a commodity which could be produced for exchange in areas where arable farming was not the optimal form of land use. Finally, the use of animal milk provided a means

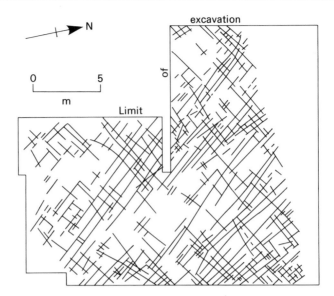

excavation

N

0   5

m

Limit

of

Figure 1.5
Plan of Neolithic plough marks:
some of the first tangible
evidence of the first
agricultural revolution in the
British Isles (after Taylor, 1975,
figure 1a)

whereby large herds could use marginal or exhausted land, encouraging the development of the pastoral sector with trans-
humance or nomadism.

This secondary utilization of animals therefore had radical
effects, and the change took place over quite a short period. The
plough was invented some 5000 years ago, and was used in
Mesopotamia, Assyria and Egypt. The remains of plough marks
have also been found beneath a burial mound at South Street,
Avebury in England, dated at around 3000 BC (see figure 1.5), and
ever since that time have been a dominant feature of the English
landscape (Taylor, 1975). The wheeled cart was first produced in
the Near East in the fourth millennium BC, and rapidly spread from
there to both Europe and India during the course of the third
millennium. The diffusion of the plough is shown in figure 1.6 and
that of the cart in figure 1.7.

The development of other means of transport preceded the
wheel. Sledge-runners found in Scandinavian bogs have been dated
to the Mesolithic period (Cole, 1970: 42), while by the Neolithic
era humans had developed boats, floats and rafts that were able to
cross to Mediterranean islands and sail the Irish sea. Dugout canoes
could hardly have been common before polished stone axes and
adzes came into general use during Neolithic times, although some
paddle and canoe remains are recorded from Mesolithic sites in
northern Europe. The middens of the hunter-fishers of the Danish
Neolithic contain bones of deep-sea fish such as cod, showing that
these people certainly had seaworthy craft with which to exploit
ocean resources.

Both the domestication of animals and the cultivation of plants
have been among the most significant causes of the human impact.
Pastoralists have had many major effects – for example, on soil
erosion – though Passmore (1974: 12) believes that nomadic

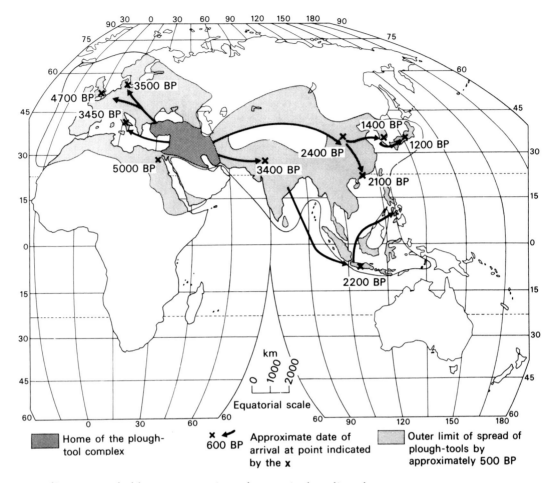

Figure 1.6
The spread of plough
agriculture in the Old World
(after Spencer and Thomas,
1978, figure 4.2)

Legend:
- Home of the plough-tool complex
- × ← 600 BP : Approximate date of arrival at point indicated by the **x**
- Outer limit of spread of plough-tools by approximately 500 BP

pastoralists are probably more conscious than agriculturalists that
they share the earth with other living things. Agriculturalists, on the
other hand, deliberately transform nature in a sense which nomadic
pastoralists do not. Their main role has been to simplify the world's
ecosystems. Thus in the prairies of North America, by ploughing
and seeding the grasslands, farmers have eliminated a hundred
species of native prairie herbs and grasses, which they replace with
pure stands of wheat, corn or alfalfa. This simplication may reduce
stability in the ecosystem (but see chapter 9, section on 'The
susceptibility to change'). Indeed, on a world basis (see Harlan,
1976) such simplification is evident. Whereas people once enjoyed
a highly varied diet, and have used for food several thousand
species of plants and several hundred species of animals, with
domestication their sources are greatly reduced. For example, today
four crops (wheat, rice, maize and potatoes) at the head of the list
of food supplies contribute more tonnage to the world total than
the next twenty-six crops combined. Simmonds (1976) provides an
excellent account of the history of most of the major crops
produced by human society.

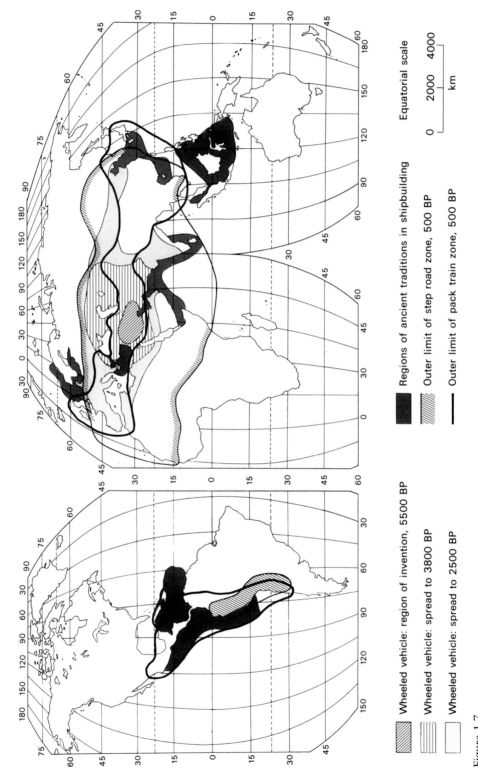

Figure 1.7
The early development of transport (after Spencer and Thomas, 1978, figure 4.5)

Wheeled vehicle: region of invention, 5500 BP

Wheeled vehicle: spread to 3800 BP

Wheeled vehicle: spread to 2500 BP

Regions of ancient traditions in shipbuilding

Outer limit of step road zone, 500 BP

Outer limit of pack train zone, 500 BP

Equatorial scale

0   2000   4000

km

It is becoming increasingly clear that Neolithic and Bronze Age peoples were able to achieve very notable changes in the soils, plants and animals of large areas of Europe and the Near East (see, for example, Dimbleby, 1974).

One further development in human cultural and technological life which was to increase human power was the mining of ores and the smelting of metals. This may have started around 5700 years ago, possibly in what is today north-west Iran, with the smelting of copper-oxide ores into metallic copper. Recent finds of copper and bronze implements in north-east Thailand, however, raise questions about the dating and the place of origin, since tentative dates of between 7000 and 6000 years ago have been given to the Thailand finds. The spread of metal working into other areas was rapid (figure 1.8), and by 2500 BC bronze products were in use from

Figure 1.8
The diffusion of mining and smelting in the Old World (after Spencer and Thomas, 1978, figure 4.4)

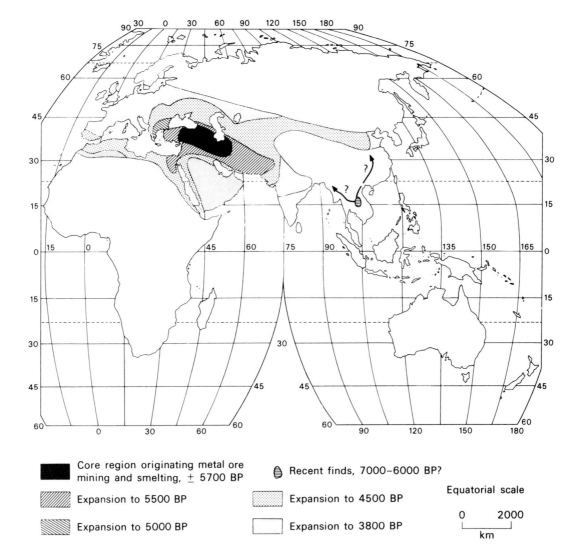

Core region originating metal ore mining and smelting, ± 5700 BP

Recent finds, 7000–6000 BP?

Expansion to 5500 BP

Expansion to 4500 BP

Expansion to 5000 BP

Expansion to 3800 BP

Equatorial scale

0   2000
km

Britain in the West to northern China in the East. The smelting of iron ores may date back to as late as 1500 BC (Spencer and Thomas, 1978). Metal working required enormous amounts of wood, and Sir William Flinders Petrie in his investigation of the third millennium BC copper industry at Wadi Nasb in western Sinai found a bed of wood ashes 30 m long, 15 m wide and 0.5 m deep.

In recent decades fossil-fuelled machinery has allowed mining activity to expand to such a degree that in terms of the amount of material moved its effects are reputed to rival the natural processes of erosion. Taking overburden into account, the total amount of material moved by the mining industry globally is probably at least 28 billion tonnes – about 1.7 times the estimated amount of sediment carried each year by the world's rivers (Young, 1992). The environmental impacts of mineral extraction are diverse but extensive, and relate not only to the process of excavation and removal, but also to the processes of mineral concentration, smelting and refining (table 1.2).

Modern industrial and urban civilizations

In ancient times, certain cities had evolved which had considerable human populations. It has been estimated that Nineveh may have

Table 1.2 Environmental impacts of minerals extraction

Activity	Potential impacts
Excavation and ore removal	• Destruction of plant and animal habitat, human settlements, and other surface features (surface mining) • Land subsidence (underground mining) • Increased erosion; silting of lakes and streams • Waste generation (overburden) • Acid drainage (if ore or overburden contain sulphur compounds) and metal contamination of lakes, streams and groundwater
Ore concentration	• Waste generation (tailings) • Organic chemical contamination (tailings often contain residues of chemicals used in concentrators) • Acid drainage (if ore contains sulphur compounds) and metal contamination of lakes, streams, and groundwater
Smelting/refining	• Air pollution (substances emitted can include sulphur dioxide, arsenic, lead, cadmium and other toxic substances) • Waste generation (slag) • Impacts of producing energy (most of the energy used in extracting minerals goes into smelting and refining)

Source: Young, 1992, table 5

Plate 1.8
Urbanization (and, in particular, the growth of huge conurbations) is an increasingly important phenomenon, not least in some less developed countries. Urbanization causes and accelerates a whole suite of environmental problems. These include air pollution caused by high traffic levels in cities like Jakarta, Indonesia

had a population of 700 000, that Augustan Rome may have had a population of around one million, and that Carthage, at its fall in 146 BC, had 700 000 (Thirgood, 1981). Such cities would have already exercised a considerable influence on their environs, but this influence was never as extensive as that of the last few centuries; for the modern era, especially since the late seventeenth century, has witnessed the transformation of, or revolution in, culture and technology – the development of major industries (Pawson, 1978). This, like domestication, has reduced the space required to sustain each individual and has increased the intensity with which resources are utilized. Modern science and modern medicine have compounded these effects, leading to accelerating population increase even in non-industrial societies. Urbanization has gone on apace, and it is now recognized that large cities have their own environmental problems (Cooke et al., 1982), and a multitude of environmental effects (Douglas, 1983). If present trends continue, many cities in the less developed countries will become inconceivably large and crowded. For instance, by the year 2000, it is projected that Mexico City will have more than 30 million people (roughly three times the present population of the New York metropolitan area), and Calcutta, Greater Bombay, Greater Cairo, Jakarta and Seoul are each expected to be in the 15–20 million range. In all, around 400 cities will have passed the million mark (Council on Environmental Quality and the Department of State, 1982: 12), and UN estimates indicate that by the end of the century over 3000 million people will live in cities, compared with around 1400 million people in 1970 (figure 1.9).

The perfecting of sea-going ships in the sixteenth and seventeenth centuries was part of this industrial and economic transformation, and this was the time when mainly self-contained but developing regions of the world coalesced so that the ecumene became to all intents and purposes continuous. The invention of the steam engine

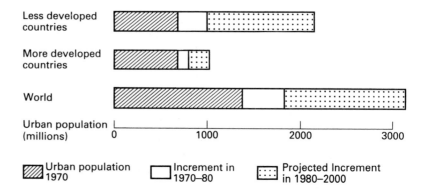

Figure 1.9
Trends in urbanization (after
Holdgate et al., 1982, figure
9–1)

in the late eighteenth century, and the internal combustion engine
in the late nineteenth century, massively increased human access to
energy and lessened dependence on animals, wind and water.

Modern science, technology and industry have also been applied
to agriculture, and in recent decades some spectacular progress has
been made through, for example, the use of fertilizers and the
selective breeding of plants and animals.

To conclude, we can recognize certain trends in human manipulation of the environment which have taken place in the modern
era. The first of these is that the ways in which humans are affecting
the environment are proliferating. For example, nearly all the
powerful pesticides post-date the Second World War, and the same
applies to the increasing construction of nuclear reactors. Secondly,
environmental issues that were once locally confined have become
regional or even global problems. An instance of this is the way in
which substances such as DDT, lead and sulphates are found at the
poles, far removed from the industrial societies that produced them.
Thirdly, the complexity, magnitude and frequency of impacts are
probably increasing; for instance, a massive modern dam like that

Table 1.3 Some indicators of changes in the global economy from 1950 to
1991

World indicator	1950	1991	Change (xn)
Grain production (million tonnes)	631	1696	2.7
Soybean production (million tonnes)	18	106	5.9
Meat production (million tonnes)	46	173	3.8
Fish catch (million tonnes)	22	97	4.4
Irrigated area (million ha)	94	235	2.5
Fertilizer use (million tonnes)	14	136	9.7
Oil production (million b/d)	10.4	59.0	5.7
Natural gas production (trillion ft^3)	6.7	77.0	11.5
Car production (million)	8	35	4.4
Bike production (million)	20	95	4.8
Human population (million)	2565	5409	2.1

Source: data processed by author from Brown et al. 1992

at Aswan has a very different impact from a small Roman dam. Finally, compounding the effects of rapidly expanding populations is a general increase in *per capita* consumption and environmental impact (table 1.3). Energy resources are being developed at an ever-increasing rate, giving humans enormous power to transform the environment. One index of this is world commercial energy consumption which trebled in size between the 1950s and 1980.

Likewise, we can see stages in the pollution history of the Earth. Mieck (1990), for instance, has identified a sequence of changes in the nature and causes of pollution: *pollution microbienne* or *pollution bactérielle*, caused by bacteria living and developing in decaying and putrefying materials and stagnant water associated with settlements of growing size; *pollution artisanale*, associated with small-scale craft industries such as tanneries, potteries and other workshops carrying out various rather disagreeable tasks, including soap manufacture, bone burning and glue-making; *pollution industrielle*, involving large-scale and pervasive pollution over major centres of industrial activity, particularly from the early nineteenth century in areas like the Ruhr or the English 'Black Country'; *pollution fondamentale*, in which whole regions are affected by pollution, as with the desiccation and subsequent salination of the Aral Sea area; *pollution foncière*, in which vast quantities of chemicals are deliberately applied to the land as fertilizers, and biocides; and finally, *pollution accidentale*, in which major accidents can cause pollution which is neither foreseen nor calculable (e.g. the Chernobyl disaster).

Above all, as a result of the escalating trajectory of environmental transformation it is now possible to talk about *global* environmental change. There are two components to this (Turner et al., 1990): systemic global change and cumulative global change. In the systemic meaning, 'global' refers to the spatial scale of operation and comprises such issues as global changes in climate brought about by atmospheric pollution. This is a topic discussed at length in chapters 7 and 8. In the cumulative meaning, 'global' refers to the areal or substantive accumulation of localized change, and a change is seen to be 'global' if it occurs on a worldwide scale, or represents a significant fraction of the total environmental phenomenon or global resource. Both types of change are closely intertwined. For example, the burning of vegetation can lead to systemic change through such mechanisms as carbon dioxide release and albedo change, and to cumulative change through its impact on soil and biotic diversity (table 1.4).

We can conclude this introductory chapter by quoting from Kates et al. (1991: 1):

Most of the change of the past 300 years has been at the hands of humankind, intentionally or otherwise. Our ever-growing role in this continuing metamorphosis has itself essentially changed. Transformation has escalated through time, and in some instances the scales of change have

Table 1.4 Types of global environmental change

Type	Characteristic	Examples
Systemic	Direct impact on globally functioning system	(a) Industrial and land use emissions of 'greenhouse' gases (b) Industrial and consumer emissions of ozone-depleting gases (c) Land cover changes in albedo
Cumulative	Impact through worldwide distribution of change	(a) Groundwater pollution and depletion (b) Species depletion/genetic alteration (biodiversity)
	Impact through magnitude of change (share of global resource)	(a) Deforestation (b) Industrial toxic pollutants (c) Soil depletion on prime agricultural lands

Source: from B. L. Turner et al., 1990, table 1

shifted from the locale and region to the earth as a whole. Whereas humankind once acted primarily upon the visible 'faces' or 'states' of the earth, such as forest cover, we are now also altering the fundamental flows of chemicals and energy that sustain life on the only inhabited planet we know.

TWO

The Human Impact on Vegetation

Introduction

In any consideration of the human impact on the environment it is probably appropriate to start with vegetation, for humankind has possibly had a greater influence on plant life than on any of the other components of the environment. Through the changes humans have brought about in plant cover they have modified soils (see chapter 4), influenced climates (see chapter 7), affected geomorphic processes (see chapter 6), and changed the quality (see chapter 5) and quantity of some natural waters. Indeed, the nature of whole landscapes has been transformed by human-induced vegetation change (figure 2.1). Following Hamel and Dansereau (1949, cited by Frenkel, 1970) we can recognize five principal degrees of interference – each one increasingly remote from pristine conditions. These are:

Figure 2.1
Some ramifications of human-induced vegetation change

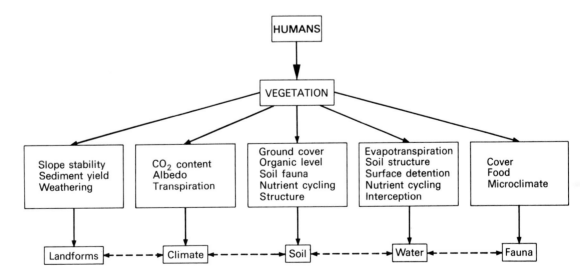

1 *Natural habitats*: those that develop in the absence of human activities.
2 *Degraded habitats*: those produced by sporadic, yet incomplete, disturbances; for example, the cutting of a forest, burning and the non-intensive grazing of natural grassland.
3 *Ruderal habitats*: where disturbance is sustained but where there is no intentional substitution of vegetation. Roadsides are an example of a ruderal habitat.
4. *Cultivated habitats*: when constant disturbance is accompanied by the intentional introduction of plants.
5 *Artificial habitats*: which are developed when humans modify the ambient climate and soil, as in greenhouse cultivation.

An alternative model for classifying the extent of human influence on vegetation is provided by Westhoff (1983), who adopts a four-part scheme:

1 *Natural*: a landscape or an ecosystem not influenced by human activity.
2 *Subnatural*: a landscape or ecosystem partly influenced by humans, but still belonging to the same (structural) formation type as the natural system from which it derives (for example, a wood remaining a wood).
3 *Semi-natural*: a landscape or ecosystem in which flora and fauna are largely spontaneous, but the vegetation structure is altered so that it belongs to another formation type (for example, a pasture, moorland or heath deriving from a wood).
4 *Cultural*: a landscape or ecosystem in which flora and fauna have been essentially affected by human agency in such a way that the dominant species may have been replaced by other species (for example, arable land).

In this chapter we shall be concerned mainly with degraded and ruderal habitats, or sub-natural and semi-natural habitats; but first we need to consider some of the processes that human societies employ; notably fire, grazing and the physical removal of forest.

The use of fire

Humans are known to have used fire since Palaeolithic times (see pp. 14–15). As Sauer (1969), one of the great proponents of the role of fire in environmental change, has put it (pp. 10–11):

Through all ages the use of fire has perhaps been the most important skill to which man has applied his mind. Fire gave to man, a diurnal creature, security by night from other predators . . . The fireside was the beginning of social living, the place of communication and reflection.

People have utilized fire for a great variety of reasons (Bartlett, 1956; Stewart, 1956; Wertine, 1973): to clear forest for agriculture

Plate 2.1
Large areas of the world's forests are being destroyed by fires started by humans, such as this one in eastern New Zealand. On steep slopes serious erosion may follow such fires

(plate 2.1), to improve grazing land for domestic animals or to attract game; to deprive game of cover; to drive game from cover in hunting; to kill or drive away predatory animals, ticks, mosquitoes and other pests; to repel the attacks of enemies, or to burn them out of their refuges; for cooking; to expedite travel; to burn the dead and raise ornamental scars on the living; to provide light; to transmit messages via smoke signalling; to break up stone for tool-making: to protect settlements or encampments from great fires by controlled burning; to satisfy the sheer love of fires as spectacles; to make pottery; to smelt ores; to harden spears; to provide warmth; to make charcoal; and to assist in the collection of insects such as crickets for eating. Given this remarkable utility it would be surprising if it had not been turned to account. Indeed it is still much used, especially by pastoralists such as the cattle-keepers of Africa (Lemon, 1968), and by the practitioners of shifting agriculture (plate 2.2). For example, the Malaysian and Indonesian *ladang* and the *milpa* system of the Maya in Latin America involved the preparation of land for planting by felling or deadening forest, letting the debris dry in the hot season, and burning it before the commencement of the rainy season. With the first rains, holes were dibbled in the soft ash-covered earth with a planting stick. This system was suited to areas of low population density with sufficiently extensive forest to enable long intervals of 'forest fallow' between burnings. Land which was burned too frequently became overgrown with perennial grasses, which tended to make it difficult to farm with primitive tools. Land cultivated for too long rapidly suffered a deterioration in fertility, while land recently burned was temporarily rich in nutrients (see figure 9.1).

The use of fire, however, has not been restricted to primitive peoples in the tropics. Remains of charcoal are found in Neolithic soil profiles in highland Britain; large parts of North America

appear to have suffered fires at regular intervals prior to European settlement; and in the case of South America the 'great number of fires' observed by Magellan during the historical passage of the Strait that bears his name resulted in the toponym, 'Tierra del Fuego'. Indeed, says Sternberg (1968: 718): 'for thousands of years, man has been putting the New World to the torch, and making it a "land of fire".'

Fire was also central to the way of life of the Australian aboriginals, including those of Tasmania, and the carrying of fire sticks was a common phenomenon. As Blainey (1975: 76) has put it, 'Perhaps never in the history of mankind was there a people who could answer with such unanimity the question: "have you got a light, mate?". There can have been few if any races who for so long were able to practise the delights of incendiarism.'

In neighbouring New Zealand, Polynesians carried out extensive firing of vegetation in pre-European settlement times, and hunters used fire to facilitate travel and to frighten and trap a major food source – the flightless moa (Cochrane, 1977). The changes in vegetation that resulted were substantial (figure 2.2). The forest cover was reduced from about 79 per cent to 53 per cent, and fires were especially effective in the drier forests of central and eastern South Island in the rain shadow of the Southern Alps. The fires continued over a period of about a thousand years up to the period of European settlement (Mark and McSweeney, 1990).

Figure 2.2 (opposite)
The changing state of the
vegetation cover in New
Zealand (from Cochrane,
1977):
(a) early Polynesian
    vegetation, c.AD 700
(b) Pre-Classical Maori
    vegetation, c.AD 1200
(c) Pre-European vegetation,
    c.AD 1800
(d) present-day vegetation

Fires: natural and anthropogenic

Although people have used fire for all the reasons that have been mentioned, before one can assess the role of this facet of the human impact, one must ascertain how important fires started by human action are in comparison with those caused naturally, especially by

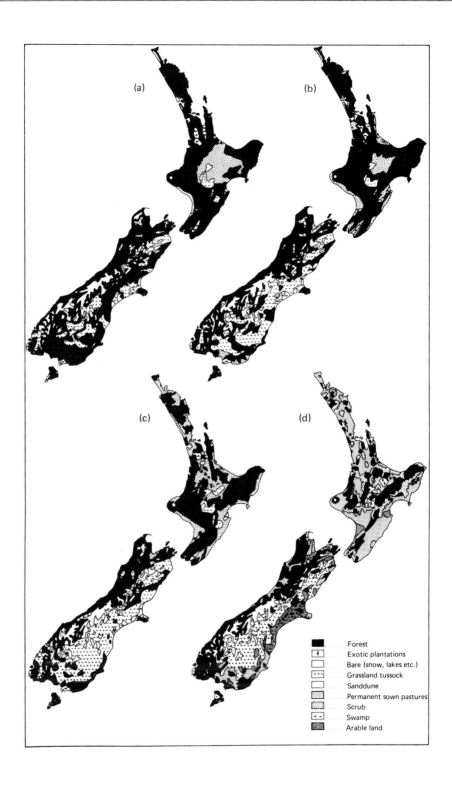

Forest
Exotic plantations
Bare (snow, lakes etc.)
Grassland tussock
Sanddune
Permanent sown pastures
Scrub
Swamp
Arable land

Table 2.1 Fire frequencies and their causes for selected areas

1   Western United States

State	Total average yearly number of fires	% caused by lightning
Arizona	1486	84
California	3608	26
Colorado	413	36
Idaho	1458	69
Montana	852	71
Nevada	86	34
New Mexico	614	79
Oregon	1860	52
South Dakota	173	62
Utah	236	35
Washington	1807	28
Wyoming	157	62
TOTAL	12 750	49

Source: Brown and Davis, 1973

2   Montane savanna, Belize

Year	Total number of fires	% caused by lightning
1963	7	43
1964	11	55
1965	7	43
1966	2	100
1967	5	80
1968	8	100
1969	3	67
1970	7	71
1971	4	25
1972	12	42
TOTAL	66	59

Source: Kellman, 1975

3   Bush fires in Australia

Cause	% of total
Deliberate burning off	35
Cigarettes and matches	7
Camp fires	6
Lightning	8
Trains	6
Tractors	5
Motor vehicles	6
Domestic	9
Miscellaneous	18

Source: Luke, 1962

4   Mediterranean France (Bouches du Rhône), 1962

Cause	% of total
Children playing with fire	39
Cigarettes	17
Agricultural accidents	12.5
Railways	11.8
Rubbish tips	6.8
Spontaneous fires	6.4
Others	6.0

Source: Nicod, cited by Wright and Wanstall, 1977

5   Fires in the fynbos heathlands of the Southern Cape, South Africa

Cause	% of total
Lightning	27.4
Trains	23.8
Unknown	15.5
Prescribed	11.9
Camp fires	10.7
Arson	7.1
Smokers	2.4
Money hunters	1.2

Source: Booysen and Tainton, 1984

lightning, which on average strikes the land surface of the globe 100 000 times each day (Yi-fu Tuan, 1971). Some natural fires may result from spontaneous combustion (Vogl, 1974), for in certain ecosystems heavy vegetal accumulations may become compacted, rotted and fermented, thus generating heat. Other natural fires can result from sparks produced by falling boulders and by landslides (Booysen and Tainton, 1984). Some of the available data on the causes of fire are summarized in table 2.1. In the forest lands of the western United States about half the fires are caused by lightning. Lightning starts over half the fires in the pine savanna of Belize, Central America, more than a quarter of the fires in the fynbos of South Africa, about 8 per cent of the fires in the bush of Australia; while in the south of France, nearly all the fires are caused by people.

One method of gauging the long-term frequency of fires is to look at tree rings and lake sediments, for these are affected by them. Studies in the USA indicate that over wide areas fires occurred with sufficient frequency to have effects on annual tree rings and lake cores every 7 to 80 years (table 2.2) in pre-European times.

The frequency of fires in different environments tends to show some variation. Rotation periods may be in excess of a century for tundra, 60 years for boreal pine forest, 100 years for spruce-dominated ecosystems, 5 to 15 years for savanna, 10 to 15 years for chaparral, and less than 5 years for semi-arid grasslands (Wein and Maclean, 1983).

Table 2.2 Long-term forest burning frequencies in North America determined from tree-ring analyses and lake core studies

Tree or forest type	Location	Interval (year)	Source
Ponderosa pine	Eastern Oregon	18	H. Weaver (1974)
	Southern Nevada	8–10	H. Weaver (1974)
	Eastern Washington	8	H. Weaver (1974)
Sequoias	South-west USA	7–9	H. Weaver (1974)
Coniferous forest	Minnesota	80	Wright (1974)
Coniferous forest	Ontario	80	Cwynar (1978)
Sequoia + mixed conifer forest	California	9–16	Kilgore and Taylor (1979)

Fires can sometimes extend over vast areas. In 1963 in Parana, Brazil, no less than 2 million hectares of forest were consumed in 3 weeks.

The temperatures attained in fires

The effects which fires have on the environment depend very much on their size, duration and intensity. Some fires are relatively quick and cool, and only destroy ground vegetation. Other fires, crown fires, affect whole forests up to crown level and generate very high temperatures (table 2.3). In general forest fires are hotter than grassland fires. Perhaps more significantly in terms of forest management, fires which occur with great frequency do not attain too high a temperature because there is inadequate inflammable material to feed them. However, when humans deliberately suppress fire, as has frequently been normal policy in forest areas (see figure 2.3), large quantities of inflammable materials accumulate, so that when a fire does break out it is of the hot, crown type. Such fires can be ecologically disastrous and there is now much debate about the wisdom of fire suppression given that, in many forests, fires under so-called 'natural' conditions appear, as we have seen, to have been a relatively frequent and regular phenonemon.

Table 2.3 Temperatures attained in fires (maxima)

Vegetation	Source	Temperature (°C)
Minnesota jack pine	Ahlgren (1974)	800
Chaparral scrub	Ahlgren (1974)	538
British heath	Whittaker (1961)	840
Senegal savanna	Daubenmire (1968)	715
Japanese grassland	Daubenmire (1968)	887
Chaparral scrub	Mooney and Parsons (1973)	1100
Nigerian savanna	Hopkins (1965)	>538–640
Sudanese savanna	Hopkins (1965)	850

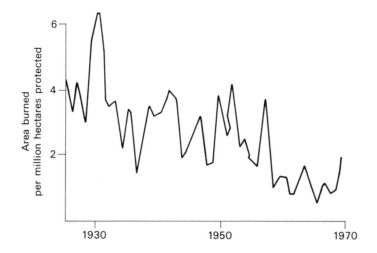

Figure 2.3
The reduction in number per million hectares burned per million hectares protected for the USA between 1926 and 1969 as a result of fire-suppression policies (after Brown and Davis, 1973, figure 2.1)

Some consequences of fire suppression

Given that fire has long been a feature of many ecosystems, and irrespective of whether the fires were or were not caused by people, it is clear that any deliberate policy of fire suppression will have important consequences for vegetation.

Fire suppression, as has already been suggested, can magnify the adverse effects of fire. The position has been well stated by Sauer (1969: 14):

The great fires we have come to fear are effects of our civilization. These are the crown fires of great depths and heat, notorious aftermaths of the pyres of slash left by lumbering. We also increase fire hazard by the very giving of fire protection which permits the indefinite accumulation of inflammable litter. Under the natural and primitive order, such holocausts, that leave a barren waste, even to the destruction of the organic soil, were not common.

Recent studies have indicated that rigid fire-protection policies have often had undesirable results and as a consequence many foresters stress the need for 'environmental restoration burning' (Vankat, 1977). So, for example, in the coniferous forests of the middle and upper elevations in the Sierra Nevada mountains of California fire protection since 1890 has made the stands denser, shadier and less park-like, and sequoia seedlings have decreased in number. Likewise, at lower elevations the Mediterranean semi-arid shrubland, called chaparral, has had its character changed. The vegetation has increased in density, the amount of combustible fuel has risen, fire-intolerant species have encroached, and vegetation diversity has decreased, resulting in a monotony of old-age stands instead of a mosaic of different successional stages. Unfavourable consequences of fire suppression have also been noted in Alaska (Oberle, 1969). It has been found that when fire is excluded from many lowland sites an insulating carpet of moss tends to accumulate and

Plate 2.3
Fires are a highly important
ecological process in areas
like the Yellowstone National
Park, USA. Fire suppression
can result ultimately in forest
devastation as increasingly
large amounts of combustible
matter accumulate through
time

raise the permafrost level. Permafrost close to the surface encourages the growth of black spruce, a low-growing species with little
timber or food value. In the Kruger National Park, in South Africa,
fires have occurred less frequently after the establishment of the
game reserve, when it became uninhabited by natives and hunters.
As a result, bush encroachment has taken place in areas that were
formerly grassland, and the carrying capacity for grazing animals
has declined. Controlled burning has been reinstituted as a necessary game-management operation.

Some effects of fire on vegetation

There is evidence that fire has played an important role in the
formation of various major types of vegetation. This applies, for
instance, to some savannas, mid-latitude grasslands and shrublands
(like the garrigue and maquis of the Mediterranean lands and the
chaparral of the south-west United States). Before examining these,
however, it is worth looking at some of the general consequences of
burning.

Fire may assist in seed germination. For example, the abundant
germination of dormant seeds on recently burned chaparral sites
has been reported by many investigators, and it seems that some
seeds of chaparral species require scarification by fire. The better
germination of those not requiring scarification may be related to
the removal by fire of competition, litter and some substances in the
soils which are toxic to plants (Hanes, 1971). Fire alters seedbeds.
If litter and humus removal are substantial, large areas of rich ash,
bare soil, or thin humus may be created. Some trees, such as
Douglas fire and the giant sequoia benefit from such seedbeds
(Heinselman and Wright, 1973). Fire sometimes triggers the release

of seeds (as with the Jack pine, *Pinus bankdiana*) and seems to stimulate the vegetative reproduction of many woody and herbaceous species. Fire can control forest insects, parasites and fungi – a process termed 'sanitization'. It also seems to stimulate the flowering and fruiting of many shrubs and herbs, and to modify the physicochemical environment of the plants. Mineral elements are released both as ash and through increased decomposition rates of organic layers. Above all, areas subject to fire often show greater species diversity, which is a factor that tends to favour stability.

One can conclude by quoting at length from Pyne (1982: 3), who provides a detailed and scholarly analysis of the history of cultural fires in America:

Hardly any plant community in the temperate zone has escaped fire's selective action, and, thanks to the radiation of *Homo sapiens* throughout the world, fire has been introduced to nearly every landscape on earth. Many biotas have consequently so adapted themselves to fire that, as with biotas frequented by floods and hurricanes, adaption has become symbiosis. Such ecosystems do not merely tolerate fire, but often encourage it and even require it. In many environments fire is the most effective form of decomposition, the dominant selective force for determining the relative distribution of certain species, and the means for effective nutrient recycling and even the recycling of whole communities.

The role of grazing

Many of the world's grasslands have long been grazed by wild animals like the bison of North America or the large game of East Africa, but the introduction of pastoral economies also affects their nature and productivity (Coupland, 1979).

Light grazing may increase the productivity of wild pastures (Warren and Maizels, 1976). Nibbling, for example, can encourage the vigour and growth of plants, and in some species such as the valuable African grass, *Themeda triandra*, the removal of coarse, dead stems permits succulent sprouts to shoot. Likewise the seeds of some plant species are spread efficiently by being carried in cattle guts, and then placed in favourable seedbeds of dung or trampled into the soil surface. Moreover, the passage of herbage through the gut and out as faeces modifies the nitrogen cycle, so that grazed pastures tend to be richer in nitrogen than ungrazed ones. Also, like fire, grazing can increase species diversity by opening out the community and creating more niches (plate 2.4).

On the other hand, heavy grazing may be detrimental. Excessive trampling when conditions are dry will reduce the size of soil aggregates and break up plant litter to a point where they are subject to aeolian deflational processes. Trampling, by puddling the soil surface, can accelerate soil deterioration and erosion as infiltration capacity is reduced. Heavy grazing can kill plants or lead to a marked reduction in their level of photosynthesis. In addition, when relieved of competition from palatable plants or plants liable

Plate 2.4
Grazing by domestic animals
has many environmental
consequences and assists in
the maintenance of grasslands
such as the Pampas at Entre
Rios in Argentina. Excessive
grazing can compact the soil
and contribute to both wind
and water erosion

to trampling damage, resistant and usually unpalatable species
expand their cover. Thus in the western USA poisonous burroweed
(*Happlopappus* spp.) has become dangerously common, and many
woody species have intruded. These include the mesquite (*Prosopis
juliflora*), the big sagebrush (*Artemisia tridentata*) (Vale, 1974), the
one-seed juniper (*Juniperus monosperma*) (Harris, 1966), and the
Pinyon Pine (Blackburn and Tueller, 1970).

Grover and Musick (1990) see shrubland encroachment by
creosote bush (*Larrea tridentata*) and mesquite as part and parcel
of desertification and indicate that in southern New Mexico the
area dominated by them has increased several fold over the last
century. This has been as a result of a corresponding decrease in the
areal coverage of productive grasslands. They attribute both tendencies to excessive livestock overgrazing at the end of the nineteenth century, but point out that this was compounded by a phase
of rainfall regimes that were unfavourable for perennial grass
growth.

Some of the plants that have invaded grasslands in California are
aliens, and these have been studied in detail by Burcham (1970). It
is clear from his long-term analysis of their progress that the aliens
have moved in in four distinct stages:

> Stage 1 (1845 →)
> Wild oats (*Avena fatua* and *A. barbata*) and black mustard
> (*Brassica nigra*).
> Stage 2 (1855 →)
> Filarees (*Erodium* spp.), wild barleys (*Hordeum* spp.),
> nitgrass (*Gastridium ventricosum*) and native annuals.
> Stage 3 (1870 →)
> Mouse barley (*Hordeum leporinum*), red brome (*Bromus*

rubens), silver hairgrass (*Aira caryophyllea*), Chile tarweed (*Madia sativa*) and star thistles (*Centaurea*).

Stage 4 (1900 →)

Medusa-head (*Taeniatherum asperum*), barb goatgrass (*Aegilops triuncialis*), dogtail grass (*Cynosaurus echinatus*) and annual false-brome (*Brachypodium distachyon*).

Each stage involves progressively less desirable species and indicates increased intensities of grazing use. In general Burcham demonstrates that perennials have tended to be replaced through time by greater numbers of annuals, which have the ability to germinate quickly, grow rapidly and produce large quantities of viable seed.

In Australia (R. M. Moore, 1959) the widespread adoption of sheep grazing led to significant changes in the nature of grasslands over extensive areas. In particular, the introduction of sheep led to the removal of kangaroo grass (*Themeda australis*) – a predominantly summer-growing species – and its replacement by essentially winter-growing species such as *Danthonia* and *Stipa*. Also in Australia, not least in the areas of tropical savanna in the north, large herds of introduced feral animals (e.g. *Bos taurus*, *Equus caballus*, *Camelus dromedarius*, *Bos banteng* and *Cervus unicolor*) have resulted in overgrazing and alteration of native habitats. As they appear to lack significant control by predators and pathogens, their densities, and thus their effects, became very high (Freeland, 1990).

Similarly in Britain many plants are avoided by grazing animals because they are distasteful, hairy, prickly or even poisonous (Tivy, 1971). The persistence and continued spread of bracken (*Pteridium aquilinium*) on heavily grazed rough pasture in Scotland is aided by the fact that it is slightly poisonous, especially to young stock. The success of bracken is furthered by its reaction to burning, for with its extensive system of underground stems (rhizomes) it tends to be little damaged by fire.

The survival and prevalence of shrubs such as elder (*Sambucus nigra*), gorse (*Ulex* spp.), broom (*Sarothamus scoparius*) and the common weeds ragwort (*Senecio jacobaea*) and creeping thistle (*Cirsium arvense*), in the face of grazing, can be attributed to their lack of palatability.

The role of grazing in causing marked deterioration of habitat has been the subject of further discussion in the context of upland Britain. Darling (1956: 780–1) has written:

It may be said in general that man's ecological dominance by pastoralism of domesticated animals over wild lands has resulted in marked deterioration of habitat. Vegetational climaxes have been broken insidiously rather than by some grand traumatic act, and, just as cultivation of food plants involves setting back ecological succession to a primary stage, pastoralism deflects succession to the xeric, a profound and dangerous change.

In particular, Darling stresses that while trees bring up nutrients from rocks and keep minerals in circulation, pastoralism means the export of calcium phosphate and nitrogenous organic matter. The vegetation gradually deteriorates, the calcicoles disappear and the herbage becomes deficient in both minerals and protein. Progressively more xerophytic plants come in: *Nardus stricta, Molinia caerulaea, Erica tetralix* and then *Scirpus caespitosa.*

However, this is not a view that now receives universal support (Mather, 1983). Studies in several parts of upland Britain have shown that mineral inputs from precipitation are very much greater than the nutrient losses in wool and sheep carcasses. During the 1970s the idea of upland deterioration was gradually undermined.

In general terms it is clear that in many parts of the world the grass family is well equipped to withstand grazing. Many plants have their growing points located on the apex of leaves and shoots, but grasses reproduce the bulk of fresh tissue at the base of their leaves. This part is least likely to be damaged by grazing and allows regrowth to continue at the same time that material is being removed.

Communities severely affected by the treading of animals (and indeed people) tend to have certain distinctive characteristics. These include diminutiveness (since the smaller the plant is the more protection it will get from soil surface irregularites); strong ramification (the plant stems and leaves spread close to the ground); small leaves (which are less easily damaged by treading); tissue firmness (cell-wall strength and thickness to limit mechanical damage); a bending ability, strong vegetative increase and dispersal (for example, by stolons); small hard seeds which can be easily dispersed, and the production of a large number of seeds per plant (which is particularly important because the mortality of seedlings is high under treading and trampling conditions).

Deforestation

Controversy surrounds the meaning of the word 'deforestation' and this causes problems when it comes to assessing rates of change and causes of the phenomenon. It is best defined (Grainger, 1992) as 'the temporary or permanent clearance of forest for agriculture or other purposes'. According to this definition, if clearance does not take place then deforestation does not occur. Thus, much logging in the tropics, which is selective in that only a certain proportion of trees and only certain species are removed, does not involve clear-felling and so cannot be said to constitute deforestation. However, how clear is clear? Many shifting cultivators in the humid tropics leave a small proportion of trees standing (perhaps because they have special utility). At what point does the proportion of trees left standing permit one to say that deforestation has taken place?

The deliberate removal of forest is one of the most longstanding and significant ways in which humans have modified the environment, whether achieved by fire or cutting. Pollen analysis shows that temperate forests were removed in Mesolithic and Neolithic times and at an accelerating rate thereafter. Since preagricultural times world forests have declined approximately one fifth, from five to four billion hectares. The highest loss has occurred in temperate forests (32–5 per cent), followed by subtropical woody savannas and deciduous forests (24–5 per cent). The lowest losses have been in tropical evergreen forests (4–6 per cent) because many of them have for much of their history been inaccessible and sparsely populated (*World Resources*, 1990–91: 107).

In Britain, Birks (1988) has analysed dated pollen sequences to ascertain when and where tree pollen values in sediments from upland areas drop to 50 per cent of their Holocene maximum percentages, and takes this as a working definition of 'deforestation'. From this work he identified four phases:

1 *3700–3900 years BP*: north-west Scotland and the eastern Isle of Skye
2 *2100–2600 years BP (the pre-Roman iron age)*: Wales, England (except the Lake District), northern Skye, and northern Sutherland
3 *1400–1700 years BP (post-Roman)*: Lake District, southern Skye, Galloway, and Knapdale–Ardnamurchan
4 *300–400 years BP*: the Grampians and the Cairngorms.

Sometimes forests are cleared to allow agriculture; at other times to provide fuel for domestic purposes, or to provide charcoal or wood for construction; sometimes to fuel locomotives, or to smoke fish; and sometimes to smelt metals. The Phoenicians were exporting cedars as early as 4600 years ago (Mikesell, 1969) both to the Pharoahs and to Mesopotamia. Attica in Greece was laid bare by the fifth century BC and classical writers allude to the effects of fire, cutting and the destructive nibble of the goat. The great phase of deforestation in central and western Europe, described by Darby (1956: 194) as 'the great heroic period of reclamation' occurred properly from AD 1050 onwards for about 200 years (figure 2.4). In particular, the Germans moved eastward: 'What the new west meant to young America in the nineteenth century, the new east meant to Germany in the Middle Ages' (Darby, 1956: 196).

The landscape of Europe was transformed, just as that of North America, Australia, New Zealand and South Africa was to be as a result of the European expansions, especially in the nineteenth century.

Temperate North America underwent particularly brutal deforestation (Williams, 1989), and lost more woodland in 200 years than Europe did in 2000. The first colonialists arriving in the *Mayflower* found a continent that was wooded from the Atlantic seaboard as far as the Mississippi River (figure 2.5). The forest originally occupied some 170 million hectares. Today only about 10 million hectares remain.

Fears have often been expressed that the mountains of High Asia (for example in Nepal) have been suffering from a wave of

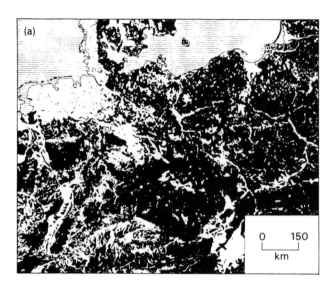

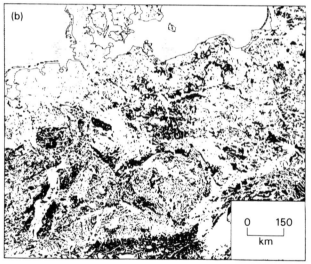

Figure 2.4
The changing distribution of forest in Central Europe between (a) AD 900 and (b) AD 1900 (reprinted from Darby, 1956, pp. 202, 203, in *Man's role in changing the face of the Earth*, ed. W. L. Thomas, by permission of The University of Chicago Press © The University of Chicago 1956)

deforestation that has led to a whole suite of environmental consequences, which include accelerated landsliding, flooding in the Ganges Plain and sedimentation in the deltaic areas of Bengal. Ives and Messerli (1989) doubt that this alarmist viewpoint is soundly based and argue (p. 67) that 'the popular claims about catastrophic post-1950 deforestation of the Middle Mountain belt and area of the high mountains of the Himalaya are much exaggerated, if not inaccurate'.

With regard to the equatorial rain forests, the researches of Flenley and others (Flenley, 1979) have indicated that forest clearance for agriculture has been going on since at least 3000 BP in Africa, 7000 BP in South and Central America, and possibly since 9000 BP or earlier in India and New Guinea.

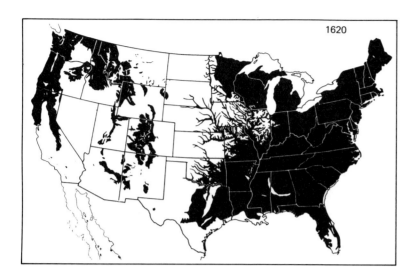

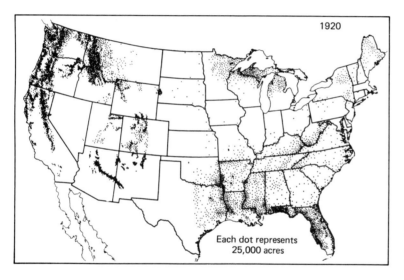

Figure 2.5
The distribution of American
natural forest in 1620 and 1920
(modified from Williams, 1989)

The causes of this particular spasm of deforestation are complex
and multifarious and are summarized in Grainger (1992) and table
2.4. Grainger also provides an excellent review of the problems of
measuring and defining loss of tropical forest area. Indeed, there
are very considerable difficulties in estimating rates of deforestation, in part because different groups of workers use different
definitions of what constitutes a forest (Allen and Barnes, 1985)
and what distinguishes rain forest from other types of forest. There
is thus some variability in views as to the present rate of rain forest
removal. Recent FAO estimates (Lanly et al., 1991) show that the
total annual deforestation in 1990 for 62 countries (representing
some 78 per cent of the tropical forest area of the world) was 16.8
million hectares, a figure significantly higher than the one obtained

Table 2.4 The causes of deforestation

A. Immediate causes – land use

---

1 Shifting agriculture

   (a) Traditional long-rotation shifting cultivation
   (b) Short-rotation shifting cultivation
   (c) Encroaching cultivation
   (d) Pastoralism

2 Permanent agriculture

   (a) Permanent staple crop cultivation
   (b) Fish farming
   (c) Government sponsored resettlement schemes
   (d) Cattle ranching
   (e) Tree crop and other cash crop plantations

3 Mining

4 Hydro-electric schemes

5 Cultivation of illegal narcotics

B. Underlying causes

---

1 Socio-economic mechanisms

   (a) Population growth
   (b) Economic development

2 Physical factors

   (a) Distribution of forests
   (b) Proximity of rivers
   (c) Proximity of roads
   (d) Distance from urban centres
   (e) Topography
   (f) Soil fertility

3 Government policies

   (a) Agriculture policies
   (b) Forestry policies
   (c) Other policies

Source: from Grainger, 1992

for these same countries for the period 1976–80 (9.2 million hectares per year). Myers (1992) suggests that there has been an 89 per cent increase in the tropical deforestation rate during the 1980s (compared with an FAO estimate of a 59 per cent increase). He believes that the annual rate of loss in 1991 amounted to about 2 per cent of the total forest expanse.

There is, however, a considerable variation in the rate of forest regression in different areas, with some areas under relatively modest threat (e.g. western Amazonia, the forests of Guyana, Surinam and French Guiana, and much of the Zaire Basin in central

Africa (Myers, 1983, 1984). Some other areas are being exploited so fast that minimal areas will be left by the year 2000, for example: the Philippines, peninsular Malaya, Thailand, Australia, Indonesia, Vietnam, Bangladesh, Sri Lanka, Central America, Madagascar, West Africa and eastern Amazonia (plate 2.5). Myers (1992) refers to particular 'hot spots' where the rates of deforestation are especially threatening, and presents data for certain locations where the percentage loss of forest is more than three times the global figure of 2 per cent: southern Mexico (10 per cent), Madagascar (10 per cent), northern and eastern Thailand (9.6 per cent), Vietnam (6.6 per cent) and the Philippines (6.7 per cent).

The rapid loss of rain forest is potentially extremely serious, because as Poore (1976: 138) has stated, these forests are a source-book of potential foods, drinks, medicines, contraceptives, abortifacients, gums, resins, scents, colourants, specific pesticides, and so on, of which we have scarcely turned the pages. Their removal may contribute to crucial global environmental concerns (e.g. climatic change and loss of biodiversity), besides causing regional and local problems, including lateritization, increased rates of erosion and accelerated mass movements. The great range of potential impacts of tropical deforestation is summarized in table 2.5.

Plate 2.5
One of the most serious environmental problems in tropical areas is the removal of the rain forest. The felling of trees in Amazonia on slopes as steep as these will cause accelerated erosion, loss of soil nutrients, and may promote lateritization

Table 2.5 The consequences of tropical deforestation

1 Reduced biological diversity

   (a) Species extinctions
   (b) Reduced capacity to breed improved crop varieties
   (c) Inability to make some plants economic crops
   (d) Threat to production of minor forest products

2 Changes in local and regional environments

   (a) More soil degradation
   (b) Changes in water flows from catchments
   (c) Changes in buffering of water flows by wetland forests
   (d) Increased sedimentatioin of rivers, reservoirs etc.
   (e) Possible changes in rainfall characteristics

3 Changes in global environments

   (a) Reduction in carbon stored in the terrestrial biote
   (b) Increase in carbon dioxide content of atmosphere
   (c) Changes in global temperature and rainfall patterns by greenhouse effects
   (d) Other changes in global climate due to changes in land surface processes

Source: from Grainger, 1992

Some traditional societies have developed means of exploiting the rain forest environment which tend to minimize the problems posed by soil fertility deterioration, soil erosion and vegetation degradation. Such a system is known as shifting agriculture (*swidden*). As Geertz (1963: 16) has remarked:

In ecological terms, the most distinctive positive characteristic of swidden agriculture . . . is that it is integrated into and, when genuinely adaptive, maintains the general structure of the pre-existing natural ecosystem into which it is projected, rather than creating and sustaining one organised along novel lines and displaying novel dynamics.

The tropical rain forest and the swidden plots have certain common characteristics. Both are closed cover systems, in part because in swidden some trees are left standing, in part because some tree crops (such as banana, papaya, areca, etc.) are planted, but also because food plants are not planted in an open field, crop-row manner, but helter-skelter in a tightly woven, dense botanical fabric. It is in Geertz's words (1963: 25) 'a miniaturized tropical forest'. Secondly, swidden agriculture normally involves a wide range of cultigens, thereby having a high diversity index like the rain forest itself. Thirdly, both swidden plots and the rain forest have high quantities of nutrients locked up in the biotic community (Douglas, 1969) compared to that in the soil. The primary concern of 'slash-and-burn' activities is not merely the clearing of the land, but rather the transfer of the rich store of nutrients locked up in the prolific vegetation of the rain forest to a botanical complex whose yield to people is a great deal larger. If the period of cultivation is not too long and the period of fallow is long enough, an equilibrated, non-deteriorating and reasonably productive farming regime can be sustained in spite of the rather impoverished soil base upon which it rests.

Unfortunately this system often breaks down, especially when population increase precludes the maintenance of an adequately long fallow period. When this happens, the rain forest cannot recuperate and is replaced by a more open vegetation assemblage which is often dominated by the notorious *Imperata* savanna grass which has turned so much of South-East Asia into a green desert. *Imperata cylindrica* is a tall grass which springs up from rhizomes. Because of its rhizomes it is fire-resistant, but because it is a tall grass it helps to spread fire (Gourou, 1961).

One particular type of tropical forest ecosystem coming under increasing pressure from various human activities is the mangrove forest characteristic of inter-tidal zones. These ecosystems constitute a reservoir, refuge, feeding ground and nursery for many useful and unusual plants and animals (Mercer and Hamilton, 1984). In particular, because they export decomposable plant debris into adjacent coastal waters, they provide an important energy source and nutrient input to many tropical estuaries. In addition they can serve as buffers against the erosion caused by tropical storms – a crucial consideration in low-lying areas like Bangladesh. In spite of these advantages, mangrove forests are being degraded and destroyed on a large scale in many parts of the world, either through exploitation of their wood resources or because of their conversion to single-use systems such as agriculture, aquaculture, salt-evaporation ponds or housing developments. To give two examples: mangrove areas in the Philippines converted to fish ponds have increased from less than 90 000 hectares in the early 1950s to over 244 000 hectares in the early 1980s; while in Indonesia logging operations are claiming 200 000 hectares of mangrove each year.

On a global basis Richards (1991: 164) has calculated that since 1700 about 19 per cent of the world's forests and woodlands have been removed. Over the same period the world's cropland area has increased by over four and a half times.

Deforestation is not an unstoppable or irreversible process. In the United States the forested area has increased substantially since the 1930s and 1940s. This 'rebirth of the forest' (Williams, 1988) has a variety of causes: new timber growth and planting was rendered possible because the old forest had been removed, forest fires have been suppressed and controlled, farmland has been abandoned and reverted to forest, and there has been a falling demand for lumber and lumber derived products. It is, however, often very difficult to disentangle the relative importance of grazing impacts, fire and fluctuating climates in causing changes in the structure of forested landscapes. In some cases, as for example the Ponderosa Pine forests of the American south-west, all three factors may have contributed to the way in which the parklike forest of the nine-teenth century has become significantly denser and younger today (Savage, 1991).

Having considered the importance of the three basic processes of fire, grazing and deforestation, we can now turn to a consideration of some of the major changes in vegetation types that have taken place over extensive areas as a result of such activities.

Secondary rain forest

When an area of rain forest that has been cleared for cultivation or timber exploitation is abandoned by humans, the forest begins to regenerate; but for an extended period of years the type of forest that occurs – secondary forest – is very different in character from the virgin forest it replaces. The features of such secondary forest, which is widespread in many tropical regions, have been sum-marized by Richards (1952) and Ellenberg (1979).

First, secondary forest is lower and consists of trees of smaller average dimensions than those of primary forest; but since it is comparatively rare that an area of primary forest is clear-felled or completely destroyed by fire, occasional trees much larger than average are usually found scattered through secondary forest. Secondly, very young secondary forest is often remarkably regular and uniform in structure, though the abundance of small climbers and young saplings gives it a dense and tangled appearance which is unlike that of primary forest and makes it laborious to penetrate. Thirdly, secondary forest tends to be much poorer in species than primary, and is sometimes, though by no means always, dominated by a single species, or a small number of species. Fourthly, the dominant trees of secondary forest are light-demanding and in-tolerant of shade, most of the trees possess efficient dispersal mechanisms (having seeds or fruits well adapted for transport by wind or animals), and most of them can grow very quickly. Some species are known to grow at rates of up to 12 m in 3 years, but they tend to be short-lived and to mature and reproduce early. One consequence of their rapid growth is that their wood often has a soft texture and low density.

The human role in the creation and maintenance of savanna

The savannas can be defined, following Hills (1965: 218–19), as:

a plant formation of tropical regions, comprising a virtually continuous ecologically dominant stratum of more or less xeromorphic plants, of which herbaceous plants, especially grasses and sedges, are frequently the principal, and occasionally the only, components, although woody plants often of the dimension of trees or palms generally occur and are present in varying densities.

Plate 2.6
The Serengeti area of Tanzania provides a fine example of the savanna vegetation that covers large areas of Africa

They are extremely widespread in low latitudes (Harris, 1980), covering about 18 million km^2 (an area about 2.6 million km^2 greater than that of the tropical rain forest). Their origin has been the subject of great contention in the literature of biogeography, though most savanna research workers agree that no matter what savanna origins may be, the agent which seems to maintain them is intentional or inadvertent burning (Scott, 1977). As with most major vegetation types a large number of interrelated factors are involved in causing savanna, and too many arguments about origins have neglected this fact (figure 2.6). Confusion has also arisen because of the failure to distinguish clearly between predisposing, causal, resulting and maintaining factors (Hills, 1965). It appears, for instance, that in the savanna regions around the periphery of the Amazon basin, the climate *predisposes* the vegetation towards the development of savanna rather than forest. The geomorphic evolution of the landscape may be a *causal* factor; increased laterite development a *resulting* factor; and fire, *a maintaining* factor.

Originally, however, savanna was envisaged as a predominantly natural vegetation type of climatic origin (see, for example, Schimper, 1903), and the climatologist Köppen used the term

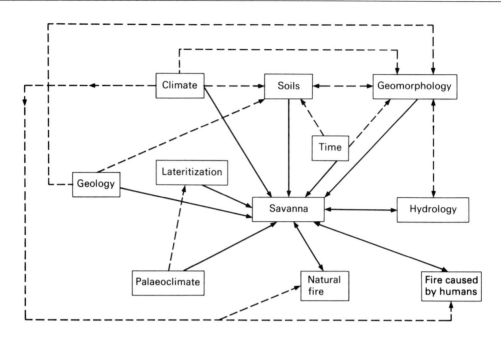

Figure 2.6
The interrelated factors
involved in the formation of
savanna vegetation

'savanna climate', implying that a specific type of climate is associated with all areas of savanna origin. According to supporters of this theory, savanna is better adapted than other plant formations to withstand the annual cycle of alternating soil moisture: rain forests could not resist the extended period of extreme drought, while dry forests could not compete successfully with perennial grasses during the equally lengthy period of large water surplus (Sarmiento and Monasterio, 1975).

Other workers have championed the importance of edaphic (soil) conditions, including poor drainage, soils which have a low water-retention capacity in the dry season, soils with a shallow profile due to the development of a lateritic crust, and soils with a low nutrient supply (either because they are developed on a poor parent rock such as quartzite, or because the soil has undergone an extended period of leaching on an old land surface). Associated with soil characteristics, ages of land surfaces and degree of drainage is the geomorphology of an area (as stressed, for example, by Cole, 1963). This may also be an important factor in savanna development.

Some other researchers – for example, Eden (1974) – find that savannas are the product of former drier conditions (such as late Pleistocene aridity) and that, in spite of a moistening climate, they have been maintained by fire. He points to the fact that the patches of savanna in southern Venezuela occur in forest areas of similar humidity and soil infertility, suggesting that neither soil, nor drainage, nor climate was the cause. Moreover, the present 'islands' of savanna are characterized by species which are also present elsewhere in tropical American savannas and whose disjunct distribution conforms to the hypothesis of previous widespread continuity of that formation.

The importance of fire in maintaining and originating some savannas is suggested by the fact that many savanna trees are fire-resistant. Controlled experiments in Africa (Hopkins, 1965) demonstrate that some tree species, such as *Burkea africana* and *Lophira lanceolata*, withstand repeated burning better than others. There are also many observations of the frequency with which, for example, African herdsmen and agriculturalists burn over much of tropical Africa and thereby maintain grassland. Certainly the climate of savanna areas is conducive to fire for, as Gillon (1983: 617) put it, 'Large scale grass fires are more likely to take place in areas having a climate moist enough to permit the production of a large amount of grass, but seasonally dry enough to allow the dried material to catch fire and burn easily.' On the other hand Morgan and Moss (1965) express some doubts about the role of fire in western Nigeria, and point out that fire is not itself necessarily an independent variable:

The evidence suggests that some notions of the extent and destructiveness of savanna fires are rather exaggerated. In particular the idea of an annual burn which affects a large proportion of the area in each year would seem to be false. It is more likely that some patches, peculiarly susceptible to fire, as a result of edaphic or biotic influences upon the character of the community itself, are repeatedly burned, whereas others are hardly, if ever, affected . . . It is also important to note that there is no evidence anywhere along the forest fringe . . . to suggest that fire sweeps into the forest, effecting notable destruction of forest trees.

Some savannas are undoubtedly natural, for pollen analysis in South America shows that savanna vegetation was present before the arrival of human civilization. None the less even natural savannas, when subjected to human pressures, change their characteristics. For example, the inability of grass cover to maintain itself over long periods in the presence of heavy stock grazing may be documented from many of the warm countries of the world (Johannessen, 1963). Heavy grazing tends to remove the fuel (grass) from much of the surface. The frequency of fires is therefore significantly reduced, and tree and bush invasion take place. As Johannessen (p. 111) wrote:

Without intense, almost annual fires, seedlings of trees and shrubs are able to invade the savannas where the grass sod has been opened by heavy grazing . . . the age and size of the trees on the savannas usually confirm the relative recency of the invasion.

In the case of the savannas of interior Honduras, he reports that they only had a scattering of trees when the Spaniards first encountered them, whereas now they have been invaded by an assortment of trees and tall shrubs.

Elsewhere in Latin America changes in the nature and density of population have also led to a change in the nature and distribution of savanna grassland. C. F. Bennett (1968: 101) noted that in

Panama the decimation of the Indian population saw the re-establishment of trees:

Plate 2.7 (opposite)
The pindan bush of
north-western Australia,
composed of *Eucalyptus,
Acacia* and grasses, is
frequently burned. The pattern
of the burning shows up
clearly on Landsat imagery

Large areas which today are covered by dense forest were in farms, grassland or low second growth in the early sixteenth century when the Spaniards arrived. At that time horses were ridden with ease through areas which today are most easily penetrated by river, so dense has the tree growth become since the Indians died away.

Such bush encroachment is a serious cause of rangeland deterioration. Overgrazing reduces the vigour of favourable perennial grasses, which tend to be replaced, as we have seen, by less reliable annuals and by woody vegetation. Annual grasses do not adequately hold or cover the soil, especially early in the growing season. Thus runoff increases and topsoil erosion occurs. Less water is then available in the topsoil to feed the grasses, so that the woody species, which depend on deeper water, become more competitive relative to the grasses which can only use water within their shallow root zones. The situation is exacerbated when there are few browsers in the herbivore population and, once established, woody vegetation competes effectively for light and nutrients. It is also extremely expensive to remove established dense scrub cover by mechanical means, though some success has been achieved by introducing animals that have a browsing or bulldozing effect, such as goats, giraffes and elephants. Experiments in Zimbabwe have shown that, if bush clearance is carried out, a threefold increase in the sustainable carrying capacity can take place in semi-arid areas (Child, 1985).

The spread of desert vegetation on desert margins

One of the most contentious and important environmental issues of recent years has been the debate on the question of the alleged expansion of deserts.

The term 'desertification' was first used but not formally defined by Aubréville (1949), and for some years the term 'desertization' was also employed, as, for example, by Rapp (1974: 3) who defined it as: 'The spread of desert-like conditions in arid or semi-arid areas, due to man's influence or to climatic change.'

An alternative expression 'land aridiziation' has also been used by the Soviet pedologist, Kovda (1980: 15): 'The phrase "land aridization" means a complex of diverse processes and trends that reduce the effective moisture content over large areas and decrease the biological productivity of the soils and plants of an ecosystem.'

There has been some variability in how 'desertification' itself is defined. Some definitions stress the importance of human causes (e.g. Dregne, 1986: 6–7):

Desertification is the impoverishment of terrestrial ecosystems under the impact of man. It is the process of deterioration in these ecosystems that can be measured by reduced productivity of desirable plants, undesirable alterations in the biomass and the diversity of the micro and macro fauna and flora, accelerated soil deterioration, and increased hazards for human occupancy.

Others admit the possible importance of climatic controls but give them a relatively inferior role (e.g. Sabadell et al., 1982: 7):

The sustained decline and/or destruction of the biological productivity of arid and semi arid lands caused by man made stresses, sometimes in conjunction with natural extreme events. Such stresses, if continued or unchecked, over the long term may lead to ecological degradation and ultimately to desert-like conditions.

Yet others, more sensibly, are even-handed or open-minded with respect to natural causes (e.g. Warren and Maizels, 1976: 1):

A simple and graphic meaning of the word 'desertification' is the development of desert like landscapes in areas which were once green. Its practical

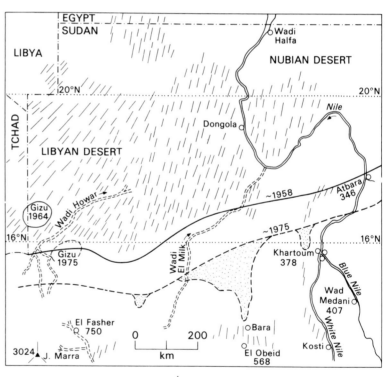

Figure 2.7
Desert encroachment in the northern Sudan 1958–75, as represented by the position of the boundary between sub-desert scrub and grassland and the desert (after Rapp et al., 1976, figure 8.5.3)

~1958	Approximate desert boundary 1958
~1975	Approximate desert boundary 1975
	Protruding area with many mobile dunes in 1975
	National boundary
	Wadi, dry stream bed
o Atbara 346	Town – altitude in metres
▲ 3024	Mountain – altitude in metres
///////	Direction of longitudinal dunes, recent and fossil

meaning . . . is a sustained decline in the yield of useful crops from a dry area accompanying certain kinds of environmental change, both natural and induced.

It is also by no means clear how extensive desertification is or how fast it is proceeding. Indeed, the lack of agreement on the former makes it impossible to determine the latter. As Grainger (1990: 145) has remarked in a well balanced review, 'Desertification will remain an ephemeral concept to many people until better estimates of its extent and rate of increase can be made on the basis of actual measurements.' He continues (p. 157): 'The subjective judgements of a few experts are insufficient evidence for such a major component of global environmental change . . . Monitoring desertification in the drylands is much more difficult than monitoring deforestation in the humid tropics, but it should not be beyond the ingenuity of scientists to devise appropriate instruments and procedures.'

There are relatively few reliable studies of the rate of supposed desert advance. Lamprey (1975) attempted to measure the shift of vegetation zones in the Sudan (see figure 2.7) and concluded that the Sahara had advanced by 90 to 100 km between 1958 and 1975, an average rate of about 5.5 km per year. However, on the basis of analysis of remotely sensed data and ground observation, Helldén (1984) found sparse evidence that this had in fact happened. One problem is that there may be very substantial fluctuations in

Plate 2.8
In the Sahel zone of West Africa and on other desert margins human activities such as wood collection, over-grazing and cultivation in drought-prone areas has led to the process of desertization

biomass production from year to year. This has been revealed by meteorological satellite observations of green biomass production levels on the south side of the Sahara (Dregne and Tucker, 1988).

The spatial character of desertification is also the subject of some controversy (Helldén, 1985). Contrary to popular rumour, the spread of desert-like conditions is not an advance over a broad front in the way that a wave overwhelms a beach. Rather, it is like a 'rash' which tends to be localized around settlements. It has been likened to Dhobi's itch – a ticklish problem in difficult places. Fundamentally, as Mabbutt (1985: 2) has explained, 'the extension of desert-like conditions tends to be achieved through a process of accretion from without, rather than through expansionary forces acting from within the deserts'. This distinction is important in that it influences perceptions of appropriate remedial or combative strategies, which are discussed in Goudie (1990).

Woodcutting is an extremely serious cause of vegetation decline around almost all towns and cities of the Sahelian and Sudanian zones of Africa. Likewise, the installation of modern boreholes has enabled rapid multiplication of livestock numbers and large-scale destruction of the vegetation in a radius of 15–30 km around boreholes (figure 2.8). Given this localization of degradation, amelioration schemes such as local tree-planting may be partially effective, but ideas of planting green belts as a 'cordon sanitaire' along the desert edge (whatever that is) would not halt deterioration of the situation beyond this Maginot line (Warren and Maizels, 1977: 222). The deserts are not invading from without; the land is deteriorating from within.

There has been considerable debate as to whether the vegetation change and environmental degradation associated with desertization is irreversible.

In many cases where ecological conditions are favourable because of the existence of such factors as deep sandy soils or beneficial hydrological characteristics, vegetation recovers once

Figure 2.8
Relation between spacing of wells and over-grazing:
(a) Original situation. End of dry season 1. Herd size limited by dry season pasture. Small population can live on pastoral economy
(b) After well-digging but no change in traditional herding. End of dry season 2. Large-scale erosion. Larger total subsistence herd and more people can live there until a situation when grazing is finished in a drought year; then the system collapses (after Rapp, 1974)

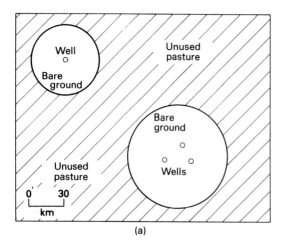

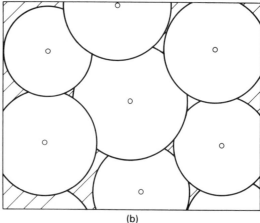

(a)   (b)

excess pressures are eliminated. There is evidence of this in arid zones throughout the world where temporary or permanent enclosures have been set up (Le Houérou, 1977). The speed of recovery will depend on how advanced deterioration is, the size of the area which is degraded, the nature of the soils and moisture resources, and the character of local vegetation. It needs to be remembered in this context that much desert vegetation is adapted to drought and to harsh conditions, and that it often has inbuilt adaptations which enable a rapid response to improved circumstances.

None the less, elsewhere experiments and observations of natural conditions tend to reveal that in certain specific circumstances recovery is slow and so limited that it may be appropriate to talk of 'irreversible desertization'. Le Houérou (1977: 419), for example, has pointed to such a case in North Africa:

in southern Tunisia, tracks made by tanks and wheeled vehicles of Allied and Axis armies are still apparent on the ground and in the devastated and unregenerated vegetation 35 years after the conclusion of the fighting. The perennial species have not re-established themselves in spite of several series of years with long-term, above-average rainfall in the 1950s, the late 1960s, and the early 1970s, although in this area grazing pressure is very low due to the absence of permanent water.

Plate 2.9
A combination of severe droughts with the grazing and cultivation of susceptible soils creates wind erosion and dust palls in Somalia, particularly when the soil is almost without vegetation

The causes of desertification are also highly controversial. The question has been asked whether this process is the result of temporary drought periods of high magnitude, whether it is due to long-term climatic change towards aridity (either as alleged post-Glacial progressive desiccation or as part of a 200-year cycle), whether it is caused by anthropogenic climatic change (see p. 312), or whether it is the result of human action degrading the biological environments in arid zones. There is little doubt that severe droughts do take place, and have taken place (Nicholson, 1978), and that their effects become worse as human and domestic animal populations increase. The devastating drought in the African Sahel from 1968 to 1984 caused greater ecological stress than the broadly comparable droughts of 1910–15 and 1944–8, largely because of the increasing anthropogenic pressures (plates 2.8 and 2.9).

The venerable idea that climate is deteriorating through the mechanism of post-Glacial progressive desiccation is now discredited (see Goudie, 1972a, for a critical analysis), though the idea that the Sahel zone is currently going through a 200-year cycle of drought has been proposed (Winstanley, 1973). However, numerous studies of available meteorological data (which in some cases date back as far as 130–150 years) do not allow any conclusions to be reached on the question of systematic long-term changes in rainfall, and the case for climatic deterioration – whether natural or aggravated by humans – is not proven. Indeed, in a judicious review, Rapp (1974: 29) wrote that after consideration of the evidence for the role of climatic change in desertization his conclusion was 'that the reported desertization northwards and southwards from the Sahara could not be explained by a general trend towards drier climate during this century'.

It is evident, therefore, that it is largely a combination of human activities (figure 2.9) with occasional series of dry years that leads to presently observed desertization. The process also seems to be fiercest not in desert interiors, but on the less arid marginal areas around them. It is in semi-arid areas – where biological productivity is much greater than in extremely arid zones, where precipitation is frequent and intense enough to cause rapid erosion of unprotected soils, and where humans are prone to mistake short-term economic gains under temporarily favourable climatic conditions for long-term stability – that the combination of circumstances particularly conducive to desert expansion can be found. It is in these marginal areas that dry farming and cattle rearing can be a success in good years, so that susceptible areas are ploughed and cattle numbers become greater than the vegetation can support in dry years. In this way, a depletion of vegetation occurs which sets in train such insidious processes as water erosion and deflation. The vegetation is removed by clearance for cultivation, by the cutting and uprooting of woody species for fuel, by over-grazing and by the burning of vegetation for pasture and charcoal.

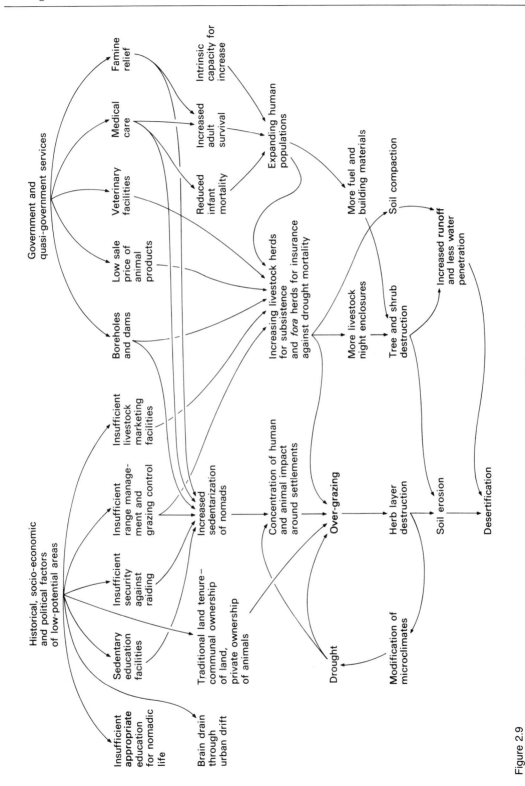

Figure 2.9
Some causal factors in desert encroachment in northern Kenya (after Lamprey, 1978, figure 2)

These tendencies towards bad land-use practices result in part from the restrictions imposed on many nomadic societies through the imposition of national boundaries across their traditional migration routes, or through various schemes implemented for political and social reasons to encourage their establishment in settled communities. Some of their traditional grazing lands have been taken over by cash-crop farmers. In Niger, for example, there was a sixfold increase in the acreage of peanuts grown between 1934 and 1968. The traditional ability to migrate enabled pastoral nomads and their cattle to emulate the natural migration of such wild animals as wildebeest and kob, and thereby to make flexible use of available resources according to season and according to yearly variations in rainfall. They could also move away from regions that had become exhausted after a long period of use. As soon as migrations are stopped and settlements imposed, such options are closed, and severe degradation occurs (Sinclair and Fryxell, 1985).

The suggestion has sometimes been made that not only are deserts expanding because of human activity, but that the deserts themselves are created by human activity. There are authors who have suggested, for example, that the Thar Desert of India is a post-Glacial and possibly post-medieval creation (see Allchin et al., 1977, for a critique of such views), while Ehrlich and Ehrlich (1970) have written: 'The vast Sahara desert itself is largely man-made, the result of over-grazing, faulty irrigation, defor-estation, perhaps combined with a shift in the course of a jet stream.' Nothing could be further from the truth. The Sahara, while it has fluctuated greatly in extent, is many millions of years old, pre-dates human life, and is the product of the nature of the general atmospheric circulation, occupying an area of dry descending air.

The maquis of the Mediterranean lands

Around much of the Mediterranean basin there is a plant formation called *maquis* (plate 2.10). This consists of a stand of xerophilous non-deciduous bushes and shrubs which are evergreen and thick, and whose trunks are normally obscured by low-level branches. It includes such plants as holly oak (*Quercus ilex*), kermes oak (*Quercus coccifera*), tree heath (*Erica arborea*), broom heath (*Erica scoparia*) and strawberry trees (*Arbutus unedo*).

Some of the maquis may represent a stage in the evolution towards true forest in places where the climax has not yet been reached; but in large areas it represents the degeneration of forest. Considerable concern has been expressed about the speed with which degeneration to, and degeneration beyond, maquis is taking place as a result of human influences (Tomaselli, 1977), of which cutting, grazing and fire are probably the most important and long continued. Charcoal burners, goats and frequent outbreaks of fires

among the resinous plants in the dry Mediterranean summer have all taken their toll.

On the other hand, in some areas, particularly marginal mountainous and semi-arid portions of the Mediterranean basin, agricultural uses of the land have declined in recent decades, as local people have sought easier and more remunerative employment. In such areas scrubland, sometimes termed 'post-cultural shrub formations', may start to invade areas of former cultivation (May, 1991), while maquis has developed to become true woodland.

There is a considerable evidence that maquis vegetation is in part adapted to, and in part a response to, fire. One effect of fire is to reduce the frequency of standard trees and to favour species which after burning send up a series of suckers from ground level. Both *Quercus ilex* and *Quercus coccifera* seem to respond in this way. Similarly, a number of species (for example, *Cistus albidus*, *Erica arborea* and *Pinus halepensis*) seem to be distinctly advantaged by fire, perhaps because it suppresses competition or perhaps because (as with the comparable chaparral of the south-west USA) a short burst of heat encourages germination (Wright and Wanstall, 1977).

Plate 2.10
The maquis of the Mediterranean lands illustrated here in Corsica is a vegetation type in which humans have played a major role. It represents the degeneration of the natural forest cover

The prairie problem

The mid-latitude grasslands of North America – the prairies – are another major vegetation type that can be used to examine the human impact, although, as in the case of savanna grasslands in the tropics, the human role is the subject of controversy.

It was once fairly widely believed that the prairies were essentially a climatically related phenomenon. Workers like J. E. Weaver (1954) argued that under the prevailing conditions of soil and climate the invasion and establishment of trees was significantly hindered by the presence of a dense sod. High evapotranspiration levels combined with low precipitation were thought to give a competitive advantage to herbaceous plants with shallow, densely ramifying root systems, capable of completing their life cycles rapidly.

An alternative view was, however, put forward by Stewart (1956: 128):

The fact that throughout the tall-grass prairie planted groves of many species have flourished and have reproduced seedlings during moist years and, furthermore, have survived the most severe and prolonged period of drought in the 1930s suggests that there is no climatic barrier to forests in the area.

Other arguments along the same lines have been advanced. Wells (1965), for example, has pointed out that in the Great Plains a number of woodland species, notably the Junipers, are remarkably drought-resistant, and that their present range extends into the Chihuahua Desert where they often grow in association with one of the most xerophytic shrubs of the American deserts, the creosote bush (*Larrea divaricata*). He remarks (p. 247): 'There is no range of climate in the vast grassland climate of the central plains of North America which can be described as too arid for all species of trees native to the region.' Moreover, confirming Stewart, he points out that numerous plantations and shelter-belts have indicated that trees can survive for at least 50 years in a 'grassland' climate. One of his most persuasive arguments is that, in the distribution of vegetation types in the plains, a particularly striking vegetational feature is the widespread but local occurrence of woodlands along escarpments and other abrupt breaks in topography remote from fluvial irrigation. A probable explanation for this is the fact that fire effects are greatest on flat, level surfaces, where there are high wind speeds and no interruptions to the course of fire. It has also been noted that where burning has been restricted there has been extension of woodland into grassland. The reasons why fire tends to promote the establishment of grassland have been summarized by Cooper (1961: 150–1).

In open country fire favours grass over shrubs. Grasses are better adapted to withstand fire than are woody plants. The growing point of dormant grasses from which issues the following year's growth, lies near or beneath the ground, protected from all but the severest heat. A grass fire removes only one year's growth, and usually much of this is dried and dead. The living tissue of shrubs, on the other hand, stands well above the ground, fully exposed to fire. When it is burned, the growth of several years is destroyed. Even though many shrubs sprout vigorously after burning, repeated loss of their top growth keeps them small. Perennial grasses, moreover, produce seeds in abundance one or two years after germination; most woody plants require several years to reach seed-bearing age. Fires that are frequent enough to inhibit seed production in woody plants usually restrict the shrubs to a relatively minor part of the grassland area.

Thus, as with savanna, anthropogenic fires may be a factor which maintains, and possibly forms, grasslands in the Great Plains. Though again, following the analogy with savanna, it is possible that some of the American prairies may have developed in a post-Glacial dry phase, and that with a later increase in rainfall re-establishment of forest cover was impeded by humans through their use of fire and by grazing animals. The grazing animals concerned were not necessarily domesticated, however, for Larson (1940) has suggested that some of the short-grass plains were maintained by wild bison. These, he believed, stocked the plains to capacity so that the introduction of domestic livestock, such as cattle, after the destruction of the wild game was merely a substitution so far as the effect of grazing on plants is concerned. There is indeed pollen analytical evidence that shows the presence of prairie in the western mid-west over 11 000 years ago, prior to the arrival of human settlers (Bernabo and Webb, 1977). Therefore some of it, at least, may be natural.

Comparable arguments have attended the origins of the great Pampa grassland of Argentina. The first Europeans who penetrated the landscape were much impressed by the treeless open country and it was always taken for granted, and indeed became dogma, that the grassland was a primary climax unit. However, this interpretation was successfully challenged, notably by Schmieder (1927a, b) who pointed out that planted trees thrived, that precipitation levels were quite adequate to maintain tree growth and that, in topographically favourable locations such as the steep gullies (*barrancas*) near Buenos Aires, there were numerous endemic representatives of the former forest cover (*monte*). Schmieder believed the Pampa grasslands were produced by a pre-Spanish aboriginal hunting and pastoral population, the density of which had been underestimated, but whose efficiency in the use of fire was proven.

Post-Glacial vegetational change in Britain and Europe

The classic interpretation of the vegetational changes of post-Glacial (Holocene) times – that is, over the past 11 000 or so years – has been in terms of climate. That changes in the vegetation of Britain and other parts of Western Europe took place was identified

Table 2.6 Blytt and Sernander's division of the post-Glacial

Blytt and Sernander period	Climate	Date of boundary (BP)
Sub-Atlantic	Cold and wet Oceanic	
		c.2500
Sub-Boreal	Warm and dry Continental	
	Elm decline	c.5000
Atlantic	Warm and wet Oceanic	
		c.7500
Boreal	Warmer than before and dry	
		c.9600
Pre-Boreal	Sub-Arctic	

by pollen analysts and palaeo-botanists, and these changes were used to construct a model of climatic change – the Blytt–Sernander model (table 2.6). There was thus something of a chicken-and-egg situation: vegetational evidence was used to reconstruct past climates, and past climates were used to explain vegetational change. More recently, however, the importance of humanly induced vegetation changes in the Holocene has been described thus by Behre (1986: vii):

Human impact has been the most important factor affecting vegetation change, at least in Europe, during the last 7000 years. With the onset of agriculture, at the so-called Neolithic revolution, the human role changed from that of a passive component to an active element which impinged directly on nature. This change had dramatic consequences for the natural environment and landscape development. Arable and pastoral farming, the actual settlements themselves and the consequent changes in the economy significantly altered the natural vegetation and created the cultural land-scape with its many different and varying aspects.

Recent palaeo-ecological research has indicated the possibility that certain features of the post-Glacial pollen record in Britain can be attributed to the action of Mesolithic and Neolithic peoples. It is, for example, possible that the expansion of the alder (*Alnus glutinosa*) was not so much a consequence of supposed wetness in the Atlantic period as a result of Mesolithic colonizers. Their removal of natural forest cover helped the spread of alder by reducing competition, as possibly did the burning of reed swamp. It may also have been assisted in its spread by the increased runoff of surface waters occasioned by deforestation and burning of catch-ment areas. Furthermore, the felling of alder itself promotes vegetative sprouting and cloning, which could result in its rapid spread in swamp forest areas (Moore, 1986).

In the Yorkshire Wolds of northern England pollen analysis suggests that Mesolithic peoples may have caused forest disturb-ance as early as 8900 BP. They may even have so suppressed forest growth that they permitted the relatively open landscapes of the early post-Glacial to persist as grasslands even when climatic conditions favoured forest growth (Bush, 1988).

One vegetational change which has occasioned particular interest is the fall in *Ulmus* pollen – the so-called elm decline – which appears in all pollen diagrams from north-west Europe, though not from North America. A very considerable number of radiocarbon dates have shown that this decline was approximately synchronous over wide areas and took place at about 5000 BP (Pennington, 1974). It is now recognized that various hypotheses can be advanced to explain this major event in vegetation history: the original climatic interpretation; progressive soil deterioration; the spread of disease; or the role of people.

The climatic interpretation has been criticized on various grounds (Rackham, 1980: 265):

A deterioration of climate is inadequate to explain so sudden, universal and specific a change. Had the climate become less favourable for elm, this would not have caused a general decline in elm and elm alone; it would have wiped out elm in areas where the climate had been marginal for it, but would not have affected elm at the middle of its climatic range unless the change was so great as to affect other species also. A climatic change universal in Europe ought to have some effect on North American elms.

Humans may have contributed to soil deterioration and affected elms thereby. Troels-Smith (1956), however, postulated that around 5000 years ago a new technique of keeping stalled domestic animals was introduced by Neolithic peoples, and that these animals were fed by repeated gathering of heavy branches from those trees known to be nutritious – elms. This, it was held, reduced enormously the pollen production of the elms. In Denmark it was found that the first appearance of the pollen of a weed, Ribwort plaintain (*Plantago lanceolata*), always coincided with the fall in elm pollen levels, confirming the association with human settlements.

Rackham, however, doubts whether humans alone could have achieved the sheer extent of change in such a short period, and postulates that epidemics of elm disease may have played a role, aided by the fact that the cause of the disease, a fungus called *Ceratocystis*, is particularly attracted to pollarded elms (Rackham, 1980: 266).

Experiments have also shown that Neolithic peoples, equipped with polished stone axes, could cut down mature trees and clear by burning a fair-sized patch of established forest within about a week. Such clearings were used for cereal cultivation. In Denmark a genuine chert Neolithic axe was fitted into an ashwood shaft (figure 2.10). Three men managed to clear about 600 m² of birch forest in four hours. Remarkably, more than 100 trees were felled with one axe-head, which had not been sharpened for about 4000 years (Cole, 1970: 38).

Heathland is a vegetation type that is characteristic of temperate, oceanic conditions on acidic substrates, and is composed of ericoid low shrubs, which form a closed canopy at heights which are usually less than 2 m. Trees and tall shrubs are absent or scattered.

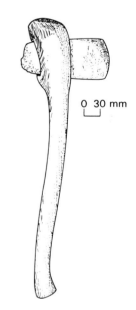

0 30 mm

Figure 2.10
A Neolithic chert axe-blade from Denmark, of the type which has been shown to be effective at cutting forest in experimental studies (after Cole, 1970, figure 25)

Some heathlands are natural: for example, communities at altitudes above the forest limit on mountains and those on exposed coasts. There are also well-documented examples of heath communities which appear naturally in the course of plant succession as, for example, where *Calluna vulgaris* (heather) colonizes *Ammophila arenaria* and *Carex arenaria* on coastal dunes.

However, at low and medium altitudes on the western fringes of Europe between Portugal and Scandinavia (figure 2.11) extensive areas of heathland occur. The origin of these lands is strongly disputed (Gimingham and de Smidt, 1983). Some areas were once thought to have developed where there were appropriate edaphic conditions (for example, well-drained loess or very sandy, poor soils), but pollen analysis showed that most heathlands occupy areas which were formerly tree-covered. This evidence alone, however, did not settle the question whether the change from forest to heath might have been caused by Holocene climatic change. However, the presence of human artefacts and buried charcoal, and the fact that the replacement of forest by heath has occurred at many different points in time between the Neolithic and the late nineteenth century, suggest that human actions established, and then maintained, most of the heathland areas. In particular, fire is an important management tool for heather in locations such as upland Britain, since the value of *Calluna* as a source of food for grazing animals increases if it is periodically burned.

The area covered by heathland in Western Europe reached a peak around 1860, but since then there has been a very rapid decline. For

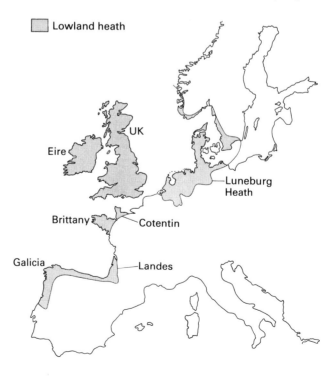

Figure 2.11
The lowland heath region of Western Europe (modified after Gimingham and de Smidt, 1983, figure 2)

example, by 1960 there had been a 60–70 per cent reduction in Sweden and Denmark (Gimingham, 1981). Reductions in Britain averaged 40 per cent between 1950 and 1984, and this was a continuation of a more long-term trend (figure 2.12). The reasons for this fall are many, and include unsatisfactory burning practices, peat removal, drainage, fertilization, replacement by improved grassland, conversion to forest, and sand and gravel abstraction. In England, the Dorset heathlands that were such a feature of Hardy's Wessex novels are now a fraction of their former extent.

Thus far we have considered the human impact on general assemblages of vegetation over broad zones. However, in turning to questions such as the range of individual plant species, the human role is no less significant.

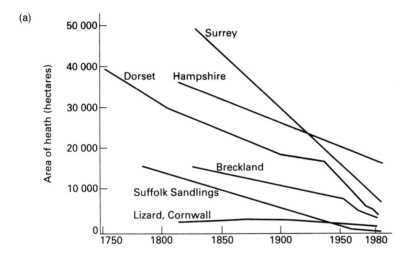

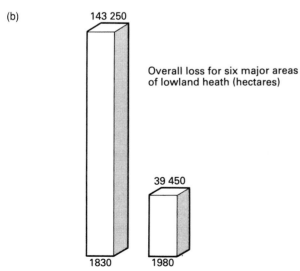

Figure 2.12
Losses of lowland heath in southern England (from Nature Conservancy Council, 1984)

Introduction, invasion and explosion

People are important agents in the spread of plants and other organisms (Bates, 1956). Some plants are introduced deliberately by humans to new areas; these include crops, ornamental and miscellaneous landscape modifiers (trees for reafforestation, cover plants for erosion control, etc.). Indeed, some plants, such as bananas and breadfruit, have become completely dependent on people for reproduction and dispersal, and in some cases they have lost the capacity for producing viable seeds and depend on human-controlled vegetation propagation. Most cultigens are not able to survive without human attention, partly because of this low capacity for self-propagation, but also because they cannot usually compete with the better-adapted native vegetation.

However, some domesticated plants have, when left to their own devices, shown that they are capable of at least ephemeral colonization, and a small number have successfully naturalized themselves in areas other than their supposed region of origin (Gade, 1976). Examples of such plants include several umbelliferous annual garden crops (fennel, parsnip and celery) which, though native to Mediterranean Europe, have colonized waste lands in California. The Irish potato, which is native to South America, grows unaided

Plate 2.11
Heaths, like those at Canford, in Dorset, southern England, have often been created by humans, but have become valued habitats. Many of them, however, are now under threat from housing expansion, reclamation, afforestation and other pressures

in the mountains of Lesotho. The peach (in New Zealand), the guava (in the Philippines), coffee (in Haiti) and the cocounut palm (on Indian Ocean island strands) are perennials that have established themselves as wild-growing populations, though the last-named is probably within the hearth region of its probable domestication. In Paraguay, orange trees (originating in South-East Asia and the East Indies) have demonstrated their ability to survive in direct competition with natural vegetation.

Plants that have been introduced deliberately because they have recognized virtues (Jarvis, 1979) can be usefully divided into an economic group (for example, crops, timber trees, etc.), and an ornamental or amenity one. In the British Isles, Jarvis believes that the great bulk of deliberate introductions before the sixteenth century had some sort of economic merit, but that only a handful of the species introduced thereafter were brought in because of their utility. Instead, plants were introduced increasingly out of curiosity or for decorative value.

A major role in such deliberate introductions was played by European botanic gardens and those in the colonial territories from the sixteenth century onwards. Many of the 'tropical' gardens (such as those in Calcutta, Mauritius and Singapore) were often more like staging posts or introduction centres than botanic gardens in the modern sense (Heywood, 1989).

Many plants, however, have been dispersed accidentally as a result of human activity: some by adhesion to moving objects, such as individuals themselves or their vehicles; some among crop seed; some among other plants (like fodder or packing materials); some among minerals (such as ballast or road metal); and some by the carriage of seeds for purposes other than planting (as with drug plants). As illustrated in figure 2.13, based on California, the establishment of alien species proceeds rapidly. In the Pampa of Argentina, Schmeider (1927b) estimates that the invasion of the country by European plants has taken place on such a large scale that at present only one-tenth of the plants growing wild in the Pampa are native.

The accidental dispersal of such plants and organisms can have serious ecological consequences. In Britain, for instance, many elm trees died in the 1970s because of the accidental introduction of the Dutch elm disease fungus which arrived on imported timber at certain ports, notably Avonmouth and the Thames Estuary ports (Sarre, 1978) – see plate 2.12.

There are also other examples of the dramatic impact of some introduced plant pathogens (von Broembsen, 1989). The American chestnut *Castanea dentata* was, following the introduction of the chestnut blight fungus *Cryphonectria parasitica* in ornamental nursery material from Asia late in the 1890s, almost eliminated throughout its natural range in less than 50 years. In western Australia the great jarrah forests have been invaded and decimated by a root fungus, *Phytophthora cinnamomi*. This was probably introduced on diseased nursery material from eastern Australia,

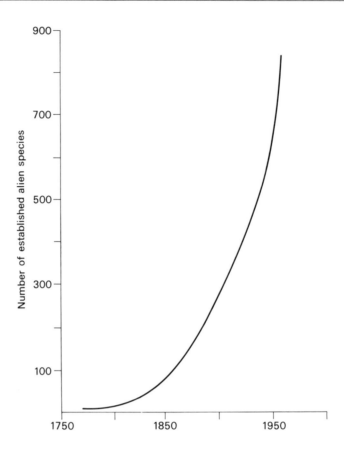

Figure 2.13
Estimation of the establishment
of alien plant species in
California, USA, since 1750
(modified after Frenkel, 1970,
figure 3)

Plate 2.12
Dutch Elm disease has
transformed the landscape of
much of lowland England in
the 1970s and 1980s.
Accidentally introduced fungus
caused the death of these
elms in Leicestershire.

and the spread of the disease within the forests was facilitated by roadbuilding, logging and mining activities that involved movement of soil or gravel containing the fungus. More than 3 000 000 ha of forest have been affected.

Ocean islands have often been particularly vulnerable. The simplicity of their ecosystems inevitably leads to diminished stability, and introduced species often find that the relative lack of competition enables them to broaden their ecological range more easily than on the continents. Moreover, because the natural species inhabiting remote islands have been selected primarily for their dispersal capacity, they have not necessarily been dominant or even highly successful in their original continental setting. Therefore, introduced species may prove more vigorous and effective (Holdgate and Wace, 1961). There may also be a lack of indigenous species to adapt to conditions such as bare ground caused by humans. Thus introduced weeds may catch on.

Table 2.7 illustrates clearly the extent to which the flora of selected islands now contain alien species, with the percentage varying between about one-quarter and two-thirds of the total number of species present.

Table 2.7 Alien plant species on oceanic islands

Island	Number of native species	Number of alien species	% of alien species in flora
New Zealand	1200	1700	58.6
Campbell Island	128	81	39.0
South Georgia	26	54	67.5
Kerguelen	29	33	53.2
Tristan da Cunha	70	97	58.6
Falklands	160	89	35.7
Tierra del Fuego	430	128	23.0

Source: from data in Moore, 1983

The introduction of new animals can have an adverse effect on plant species. A clear demonstration of this comes from the atoll of Laysan in the Hawaii group. Rabbits and hares were introduced in 1903 in the hope of establishing a meat cannery. The number of native species of plants at this time was 25; by 1923 it had fallen to 4. In that year all the rabbits and hares were systematically exterminated to prevent the island turning into a desert, but recovery has been slower than destruction. By 1930 there were nine species, and by 1961 sixteen species, on the island (Stoddart, 1968).

Pigs are another animal that has been introduced extensively to the islands of the Pacific and long-established feral populations are known on many islands. Like rabbits they have caused considerable damage, not least because of their non-fastidious eating habits and their propensity for rooting into the soil. This is a theme that is reviewed by Nunn (1991).

Plate 2.13
Feral animals (introduced
domestic stock that have gone
wild) have had a major impact
on large portions of Australia.
Feral buffalo, for example,
range widely in parts of
northern Australia

It has often been proposed that the introduction of exotic
terrestrial mammals has had a profound effect on the flora of New
Zealand. Among the reasons that have been put forward for this
belief are that the absence of native terrestrial mammalian herbivores permitted the evolution of a flora highly vulnerable to
damage from browsing and grazing, and that the populations of
wild animals (including deer and opossums) that were introduced
in the nineteenth century grew explosively because of the lack of
competitors and predators. It has, however, proved difficult to
determine the magnitude of the effects which the introduced
mammals had on the native forests (Veblen and Stewart, 1982).

There are many other examples of ecological explosions caused
by humans creating new habitats. Some of the most striking are
associated with the establishment of artificial lakes in place of
rivers. Riverine species which cannot cope with the changed
conditions tend to disappear, while others that can exploit the new
sources of food, and reproduce themselves under the new conditions, multiply rapidly in the absence of competition (Lowe-
McConnell, 1975). Vegetation on land flooded as the lake waters
rise decomposes to provide a rich supply of nutrients which allow
explosive outgrowth of organisms as the new lake fills. In particular, floating plants may form dense mats of vegetation, which in
turn support large populations of invertebrate animals, may cause
fish deaths by deoxygenating the water, and can create a serious
nuisance for turbines, navigators and fishermen. On Lake Kariba in
Central Africa there were dramatic growths in the communities of
the South American water fern (*Salvinia molesta*) (plate 2.14),
bladder-wort (*Utricularia*) and the African water lettuce (*Pistia
stratiotes*); on the Nile behind the Jebel Aulia Dam there was a huge
increase in the number of water hyacinths (*Eichhornia crassipes*);

Plate 2.14
Water weed, *Salvinia molesta*,
which has developed since the
construction of the Kariba Dam
across the Zambezi river in
central Africa

and in the Tennessee Valley lakes there was a massive outbreak of the Eurasian water-millfoil (*Myriophyllum*).

Roads have been of major importance in the spread of plants. As Frenkel (1970) has pointed out in a valuable survey of this aspect of anthropogenic biogeography:

By providing a route for the bearers of plant propagules – man, animal and vehicle – and by furnishing, along their margins, a highly specialized habitat for plant establishment, roads may facilitate the entry of plants into a new area. In this manner roads supply a cohesive directional component, cutting across physical barriers, linking suitable habitat to suitable habitat.

Roadsides tend to possess a distinctive flora in comparison with the natural vegetation of an area. As Frenkel has again written:

Roadsides are characterized by numerous ecologic modifications includ-ing: treading, soil compaction, confined drainage, increased runoff, removal of organic matter and sometimes additions of litter or waste material of frequently high nitrogen content (including urine and faeces), mowing or crushing of tall vegetation, herbicide application, removal of woody vegetation but occasionally the addition of wood chips or straw, substrate maintained in an ecologically open condition by blading, inten-sified frost action, rill and sheetwash erosion, snow deposition (together with accumulated dirt, gravel, salt and cinder associated with winter maintenance), soil and rock additions related to slumping and rock-falls, and altered microclimatic conditions associated with pavement and right-of-way structures. Furthermore, road rights-of-way may be used for driving stock in which case unselective, hurried but often close grazing may constitute an additional modification. Where highway landscaping or stability of cuts and fills is a concern, exotic or native plants may be planted and nurtured. (1970: 1)

The speed with which plants can invade roadsides is impressive. A study by Helliwell (1974) demonstrated that the M1 motorway in

England, less than twelve years after its construction, had on its cuttings and embankments not only the thirty species which had been deliberately sown or planted, but more than 350 species that had not been introduced there.

Railways have also played their role in plant dispersal. The classic example of this is provided by the Oxford ragwort, *Senecio squalidus*, a species native to Sicily and southern Italy. It spread from the Oxford Botanical Garden (where it had been established since at least 1690) and colonized the walls of Oxford. Much of its dispersal to the rest of Britain (Usher, 1973) was achieved by the Great Western Railway, in the vortices of whose trains and in whose cargoes of ballast and iron ore the plumed fruits were carried. The distribution of the plant was very much associated with railway lines, railway towns and waste ground.

Indeed, by clearing forest, cultivating, depositing rubbish and many other activities, humans have opened up a whole series of environments which are favourable to colonization by a particular group of plants. Such plants are generally thought of as weeds. In fact it has often been said that the history of weeds is the history of human society (though the converse might equally be true), and that such plants follow people like flies follow a ripe banana or a gourd of unpasteurized beer (see Harlan, 1975b).

One weed which has been causing especially severe problems in upland Britain in the 1980s is bracken (*Pteridium aquilinum*), although the problems it poses through rapid encroachment are of wider geographical significance. Indeed, Taylor (1985: 53) maintains that it 'may justifiably be dubbed as the most successful international weed of the twentieth century', for, 'It is found, and is mostly expanding, in all of the continents.' This tolerant, aggressive opportunist follows characteristically in the wake of evacuated settlement, deforestation or reduced grazing pressure, and estimated encroachment rates in the upland parts of the UK average 1 per cent per annum. This encroachment results from reduced use of bracken as a resource (e.g. for roofing) and from changes in grazing practices in marginal areas. As bracken is hostile to many other plants and animals, and generates toxins, including some carcinogens, this is a serious issue.

Air pollution and its effects on plants

Air pollutants exist in gaseous or particulate forms. The gaseous pollutants may be separated into primary and secondary forms. The primary pollutants, such as sulphur dioxide, most oxides of nitrogen, and carbon monoxide, are those directly emitted into the air from, for example, industrial sources. Secondary air pollutants, like ozone and products of photochemical reactions, are formed as a consequence of subsequent chemical processes in the atmosphere, involving the primary pollutants and other agents such as sunlight. Particulate pollutants consist of very small solid- or liquid-suspended droplets (for example, dust, smoke, and aerosolic salts), and contain a wide range of insoluble components (for example, quartz) and of soluble components (for example, various common cations, together with chloride, sulphate and nitrate).

Some of the air pollutants humans have released into the atmosphere have had detrimental impacts on plants: sulphur dioxide, for example, is toxic to them. This was shown by Cohen and Rushton (in Barrass, 1974) who grew plants in containers of similar soil in different parts of Leeds, England, in 1913. They found a close relationship between the amount of sulphate in the air and in the plants, and in the yield obtained (figure 2.14). In the more polluted areas of the city the leaves were blackened by soot and there was a smaller leaf area. The whole theme of urban vegetation has been studied in the North American context by Schmid (1975).

Lichens are also sensitive to air pollution and have been found to be rare in central areas of cities such as Bonn, Helsinki, Stockholm, Paris and London. There appears to be a zonation of lichen types around big cities as shown for north-east England (figure 2.15a) and Belfast (figure 2.15b). Moreover, when a healthy lichen is transplanted from the country to a polluted atmosphere, the algal component gradually deteriorates and then the whole plant dies (see Gilbert, 1970). Overall it has been calculated (Rose, 1970) that more than one-third of England and Wales, extending in a belt from the London area to Birmingham, broadening out to include the industrial Midlands and most of Lancashire and West Yorkshire, and reaching up to Tyneside, has lost nearly all its epiphytic lichen flora, largely because of sulphur dioxide pollution.

Hawksworth (1990: 50), indeed, believes 'the evidence that sulphur dioxide is the major pollutant responsible for impoverishment of lichen communities over wide areas of Europe is now overwhelming', but he also points out that sulphur dioxide is not necessarily always the cause of impoverishment. Other factors, such as fluoride or photochemical smog, acting independently or synergistically, can also be significant in particular locations.

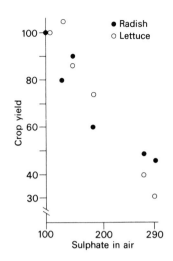

Figure 2.14
The effect of air quality on plant growth in Leeds, England, in 1913. Values for yield and for air sulphate in a low pollution area taken as 100 and other values are scaled in proportion (data of Cohen and Rushton in Barrass, 1974, p. 187)

Figure 2.15
(a) Air pollution in north-east England and its impact upon growth areas for lichens (after Gilbert in Barrass, 1974, figure 73)
(b) The increase of lichen cover on trees outside the city of Belfast, Northern Ireland (after Fenton in Mellanby, 1967, figure 3)

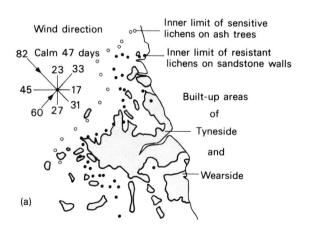

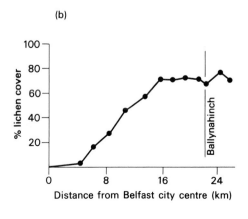

Local concentrations of industrial fumes also kill vegetation. In the case of the smelters of the Sudbury mining district of Canada, 2 million tonnes of noxious gases annually affect a 1900 km^2 area and White pine now only exists on about 7–8 per cent of the productive land. Likewise in the lower Swansea Valley, Wales, the fumes from a century of coal-burning resulted in almost complete destruction of the vegetation, with concomitant soil erosion. The area became a virtual desert. In Norway a number of the larger Norwegian aluminium smelters built immediately after the Second World War were sited in deep, narrow, steep-sided valleys at the heads of fjords. The relief has not proved conducive to the rapid dispersal of fumes, particularly of fluoride. In one valley a smelter with a production of 1 110 000 tonnes per year causes the death of pines (*Pinus sylveststris*) for over 13 km in each direction up and down the valley. To about 6 km from the source all pines are dead. Birch, however, seems to be able to withstand these conditions, and to grow vigorously right up to the factory fence (Gilbert, 1975).

Photochemical smog is also known to have adverse effects on plants both within cities and on their outskirts. In California, Ponderosa pines in the San Bernadino mountains as much as 129 km to the east of Los Angeles have been extensively damaged by smog. In summer months in Britain ozone concentrations produced by photochemical reactions have reached around 17 pphm (parts per hundred million) compared with a maximum of 4 pphm associated with clean air; while in Los Angeles ozone concentrations may reach 70 pphm (Marx, 1975). Fumigation experiments in the USA show that plant injury can occur at levels only marginally above the natural maximum and well within the summertime levels now known to be present in Britain and the USA (table 2.8). Ozone appears to reduce photosynthesis and to inhibit flowering and germination (W. H. Smith, 1974). It also seems to predispose conifers to bark-beetle infestation and to microbial pathogens.

Vegetation will also be adversely affected by excessive quantities of suspended particulate matter in the atmosphere. The particles,

Table 2.8 Examples of low levels of ozone producing plant injury in the course of fumigation experiments in the USA

Ozone concentration (parts per hundred million)	Exposure time	Species	Type of injury
10	4 hours	*Solanum tuberosum L.*	Leaf damage
5	3 days	*Phaseolus vulgaris L.*	Reduction in leaf growth and early senescence
5	5 weeks (40 hours per week)	*Raphanus sativus L.*	Fresh weight reduction
5	18 weeks	*Pinus Elliotti; pinus taeda L.*	Reduction in photosynthesis
5	10 days (12 hours per day)	*Pinus strobus*	Needle spotting
6.5	4 hours	*Pinus strobus*	Needle-tip burn

Source: data from various sources in Bell and Cox, 1975

by covering leaves and plugging plant stomata, reduce both the absorption of carbon dioxide from the atmosphere and the intensity of sunlight reaching the interior of the leaf. Both tendencies may suppress the growth of some plants. This and other consequences of air pollution are well reviewed by Elsom (1987).

The adverse effects of pollution on plants are not restricted to air pollution, for water and soil pollution can also be serious. Excessive amounts of heavy metals may prove toxic to them (Hughes et al., 1980) and, as a consequence, distinctive patterns of plant species may occur in areas contaminated with the waste from copper, lead, zinc and nickel mines (Cole and Smith, 1984). Heavy metals in soils may also be toxic to microbes and especially to fungi which may in turn change the environment by reducing rates of leaf-litter decomposition (W. H. Smith, 1974). In many areas it has proved extremely difficult to undertake effective re-establishment of plant communities on mining spoil tips, though toxicity is only one of the problems. Kent (1982) has listed some of the other obstacles that have been encountered in reclaiming colliery spoil:

1 Erosion of spoil causes instability that hinders plant growth
2 Black shale surfaces may create undesirably high surface temperatures
3 Spontaneous combustion
4 Concentrations of clay rich sediments can produce areas of waterlogging or compaction
5 Areas of coarse sediment with an open structure can create conditions of extreme soil drought
6 Rock spoil may lack key nutrients (particularly phosphorus)
7 The weathering of pyrites may create toxicity, acidity and release other toxic elements (for example, aluminium).

Other types of industrial effluents may smother and poison some species. Salt marshes, mangrove swamps and other kinds of wetlands are particularly sensitive to oil spills, for they tend to be anaerobic environments in which the plants must ventilate their root systems through pores or openings that are prone to coating and clogging (Lugo et al., 1981). The situation is especially serious if the system is not subjected to flushing by, for example, frequent tidal inundation. There are many case studies of the consequences of oil spills. For example, at the Fawley Oil Refinery on Southampton Water in England, Dicks (1977) has shown how the salt marsh vegetation has been transformed by the pollution created by films of oil, and how, over extensive areas, *Spartina anglica* has been killed off.

Forest decline

Forest decline (plate 2.15), now often called *Waldsterben* or *Waldschäden* (the German words for 'forest death' and 'forest decline') is an environmental issue that attained considerable

Plate 2.15
Forest decline is a serious
threat to various types of
mid-latitude forest, and these
trees in the Hartzgebirge
Forest, Germany display some
of the major symptoms.
However the causes of
dieback are still a matter of
considerable debate

prominence in the 1980s. The common symptoms of this phenomenon (modified from *World Resources*, 1986, table 12.1) are:

1 *Growth-decreasing Symptoms*
   Discoloration and loss of needles and leaves
   Loss of feeder-root biomass (especially in conifers)
   Decreased annual increment (width of growth rings)
   Premature ageing of older needles in conifers
   Increased susceptibility to secondary root and foliar path-
   ogens

Death of herbaceous vegetation beneath affected trees
Prodigious production of lichens on affected trees
Death of affected trees.

2 *Abnormal Growth Symptoms*
Active shedding of needles and leaves while still green, with
no indication of disease
Shedding of whole green shoots, especially in spruce
Altered branching habit
Altered morphology of leaves.

3 *Water-stress Symptoms*
Altered water balance
Increased incidence of wet wood disease.

Germany is probably the most seriously affected country in Europe
and in 1985 55 per cent of the forest stands in West Germany were
reported to be damaged. The decline is, however, widespread in
much of Europe (see table 2.9) and the process is now undermining
the health of North America's high elevation eastern coniferous
forests (*World Resources*, 1986, chapter 12). In Germany it was the
white fir, *Abies alba*, which was afflicted initially, but since then the
symptoms have spread to at least ten other species in Europe,
including Norway spruce (*Picea abies*), Scotch pine (*Pinus sylvestris*), European larch (*Larix decidua*), and seven broad-leaved
species.

Many hypotheses have been put forward to explain this dieback
(Wellburn, 1988): poor forest management practices, ageing of
stands, climatic change, severe climatic events (such as the severe
droughts in Britain during 1976), nutrient deficiency, viruses,
fungal pathogens and pest infestation. However, particular attention is being paid to the role of pollution, either by gaseous
pollutants (sulphur dioxide, nitrous oxide or ozone), acid deposition on leaves and needles, soil acidification and associated
aluminium toxicity problems and excess leaching of nutrients (for
example, magnesium), over-fertilization by deposited nitrogen,
and trace metal or synthetic organic compound (e.g. pesticide,
herbicide) accumulation as a result of atmospheric deposition.

Table 2.9 Reported percentage of different tree species affected by 'Waldschäden' in Western European countries
in 1984

Tree species	Country				
	West Germany	Eastern France	Switzerland	Austria	Italy (Southern Tyrol)
Norway spruce	51	16	11	29	16
Silver fir	87	26	13	28	35
Scots pine	59	17	18	30	6
Beech	50	3	8	–	–
Oak	43	4	9	–	–
Others	31	6	9	–	–

Source: Wellburn, 1988, table 7.3

The arguments for and against each of these possible factors have been expertly reviewed by Innes (1987), who believes that in all probability most cases of forest decline are the result of the cumulative effects of a number of stresses. He draws a distinction between predisposing, inciting and contributing stresses (p. 25):

Predisposing stresses are those that operate over long time scales, such as climatic change and changes in soil properties. They place the tree under permanent stress and may weaken its ability to resist other forms of stress. Inciting stresses are those such as drought, frost and short-term pollution episodes, that operate over short time scales. A fully healthy tree would probably have been able to cope with these, but the presence of predisposing stresses interferes with the tree's mechanisms of natural recovery. Contributing stresses appear in weakened plants and are frequently classed as secondary factors. They include attack by some insect pests and root fungi. It is probable that all three types of stress are involved in the decline of trees.

As with many environmental problems, interpretation of forest decline is bedevilled by a paucity of long-term data and detailed surveys. Given that forest condition oscillates from year to year in response to variability in climatic stress (e.g. drought, frost, wind throw) it is dangerous to infer long-term trends from short-term data (Innes and Boswell, 1990).

There may also be differences in causation in different areas. Thus while widespread forest death in eastern Europe may result from high concentrations of sulphur dioxide combined with extreme winter stress, this is a much less likely explanation in Britain, where sulphur dioxide concentrations have shown a marked decrease in recent years. Indeed, in Britain Innes and Boswell (1990: 46) suggest that the direct effects of gaseous pollutants appear to be very limited.

It is also important to recognize that some stresses may be particularly significant for a particular tree species. Thus, in 1987 a survey of ash trees in Great Britain showed extensive die-back over large areas of the country. Almost one-fifth of all ash trees sampled showed evidence of this phenomenon. Hull and Gibbs (1991) indicated that there was an association between dieback and the way the land is managed around the tree, with particularly high levels of damage being evident in trees adjacent to arable land. Uncontrolled stubble burning, the effects of drifting herbicides, and the consequences of excessive nitrate fertilizer applications to adjacent fields were seen as possible mechanisms. However, the prime cause of dieback was seen to be root disturbance and soil compaction by large agricultural machinery. Ash has shallow roots and if these are damaged repeatedly the tree's uptake of water and nutrients might be seriously reduced, while broken root surfaces would be prone to infection by pathogenic fungi.

Innes (1992: 51) also suggests that there has been some modification in views about the seriousness of the problem since the mid-1980s:

The extent and magnitude of the forest decline is much less than initially believed. The use of crown density as an index of tree health has resulted in very inflated figures for forest 'damage' which cannot now be justified . . . If early surveys are discounted on the basis of inconsistent methodo-logy . . . then there is very little evidence for a large-scale decline of tree health in Europe.

. . . the term 'forest decline' is rather misleading in that there are relatively few cases where entire forest ecosystems are declining. Forest ecosystems are dynamic and may change through natural processes.

He also suggests that the decline of certain species has been associated with climatic stress for as long as records have been maintained.

Miscellaneous causes of plant decline

Some of the main causes of plant decline have already been referred to: deforestation, grazing, fire and pollution. However, there are many records of species being affected by other forms of human interference. For example, casual flower-picking has resulted in local elimination of previously common species, and has been held responsible for decreases in species such as the primrose (*Primula vulgaris*) on a national scale in England. In addition, serious naturalists or plant collectors using their botanical knowledge to seek out rare, local and unusual species can cause the eradication of rare plants in an area.

More significantly, agricultural 'improvements' mean that many types of habitat are disappearing or that the range of such habitats is diminishing (see figure 3.8). Plants associated with distinctive habitats suffer a comparable reduction in their range. This is certainly the case for two British plants (figure 2.16a and b), the corncockle (*Agrostemma githago*) and the pasque flower (*Pulsatilla vulgaris*). The former was characteristic of unmodified cereal fields, while the latter was characteristic of traditional chalk downland and limestone pastures. These calcareous grasslands were semi-natural swards formed by centuries of grazing by sheep and rabbits, and covered large areas underlain by Cretaceous chalk and Jurassic limestone until the Napoleonic Wars. Since then (figure 2.17) they have been subjected to increasing amounts of cultivation, reclama-tion and reseeding, and the loss of such grasslands between 1934 and 1972 has been estimated at around 75 per cent (Nature Conservancy Council, 1984). Other plants have suffered a reduc-tion in range because of drainage activities (see figure 2.18).

The introduction of pests, either deliberately or accidentally, can also lead to a decrease in the range and numbers of a particular species. Reference has already been made to the decline in the fortunes of the elm in Britain because of the unintentional establish-ment of Dutch Elm disease. There are, however, many cases where 'pests' have been introduced deliberately to check the explosive

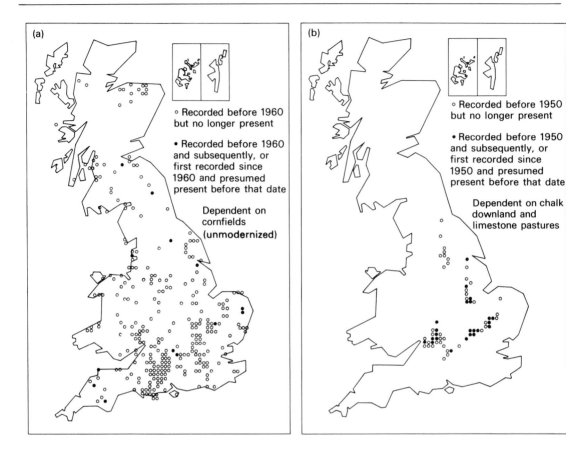

Figure 2.16
Reduction in the range of species related to habitat loss:
(a) corncockle (*Agrostemma githago*)
(b) pasque flower (*Pulsatilla vulgaris*) (after Nature Conservancy Council, 1977, figures 8 and 9)

invasion of a particular plant. One of the most spectacular examples of this involves the history of the Prickly Pear (*Opuntia*) imported into Australia from the Americas (plate 2.16). It was introduced some time before 1839 (Dodd, 1959), and spread dramatically. By 1900, it covered 4 million hectares, and by 1925 more than 24 million. Of the latter figure approximately one-half was occupied by dense growth (1200–2000 tonnes ha^{-1}) and other more useful plants were excluded. To combat this menace one of *Opuntia*'s natural enemies, a South American moth, *Cactoblastus*, was introduced to remarkable effect. 'By the year 1940 not less than 95 per cent of the former 50 million acres (20 million ha) of prickly pear in Queensland had been wiped out' (Dodd, 1959: 575).

A further good example comes from Australia. By 1952 the aquatic fern, *Salvinia molesta*, which originated in south-eastern Brazil, appeared in Queensland and spread explosively as a result of clonal growth, accompanied by fragmentation and dispersal. Significant pests and parasites appear to have been absent. Under optimal conditions *Salvinia* has a doubling time for biomass production of only 2.5 days. In June 1980 possible control agents from the *Salvinia*'s native range in Brazil – the black, long-snouted

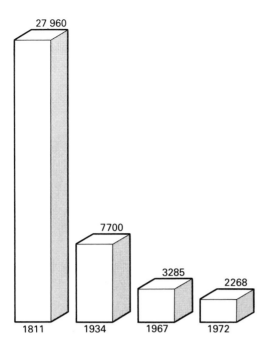

Figure 2.17
The loss of lowland chalk
grasslands in Dorset, southern
England, as depicted by the
area remaining (in hectares)
(from Nature Conservancy
Council, 1984, figure 12.2.1.2)

weevil (*Cyrtobagous* sp.) – were released on to Lake Moon Darra
(which carried an infestation of 50 000 tonnes fresh weight of
Salvinia, covering an area of 400 hectares). By August 1981 there
was estimated to be less than 1 tonne of the weed left on the lake.

Finally, the growth of leisure activities is placing greater pressure
on increasingly fragile communities, notably in tundra and high-
altitude areas. These areas tend to recover slowly from disturbance,
and both the trampling of human feet and the actions of vehicles
can be severe (see, for example, Bayfield, 1979).

The change in genetic and species diversity

The application of modern science, technology and industry to
agriculture has led to some spectacular progress in recent decades
through such developments as the use of fertilizers and the selective
breeding of plants and animals. The latter has caused some
concern, for in the process of evolution domesticated plants have
become strikingly different from their wild progenitors. Plant
species that have been cultivated for a very long time and are widely
distributed demonstrate this particularly clearly. Crop evolution
through the millennia has been shaped by complex interactions
reflecting the pressures of both artificial and natural selection.
Alternate isolation of stocks followed by migration and seed
exchanges brought distinctive stocks into new environments and
permitted new hybridizations and the recombination of characteristics. Great genetic diversity resulted.

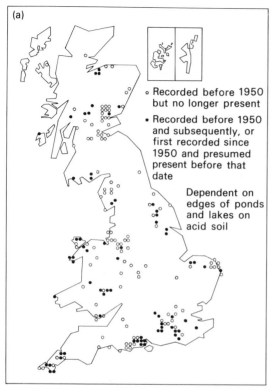

(a) Dependent on edges of ponds and lakes on acid soil

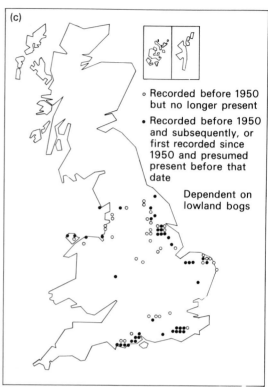

(c) Dependent on lowland bogs

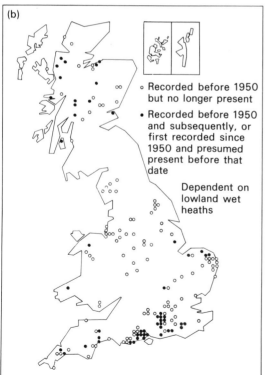

(b) Dependent on lowland wet heaths

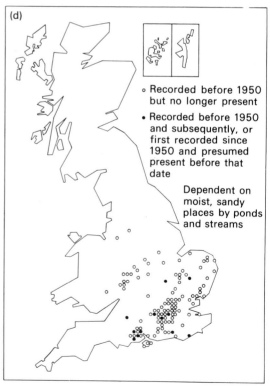

(d) Dependent on moist, sandy places by ponds and streams

Plate 2.16
The Prickly Pear *(Opuntia)* in Queensland is a plant which was introduced from the Americas and spread explosively. Recently this has been controlled by the introduction of moths and beetles

There are fears, however, that since the Second World War the situation has begun to change (Harlan, 1975a). Modern plant-breeding programmes have been established in many parts of the developing world in the midst of genetically rich centres of diversity. Some of these programmes, associated with the so-called Green Revolution, have been successful, and new, uniform high-yielding varieties have begun to replace the wide range of old, local strains that have evolved over the millennia. This may lead to a serious decline in the genetic resources which could potentially serve as reservoirs of variability. Ehrlich et al. (1977: 344) have warned:

Aside from nuclear war, there is probably no more serious environmental threat than the continued decay of the genetic variability of crops. Once the process has passed a certain point, humanity will have permanently lost the coevolutionary race with crop pests and diseases and will no longer be able to adapt crops to climatic change.

New, high-yielding crop varieties need continuous development if they are to avoid the effects of crop pests, and Ehrlich and Ehrlich (1982: 65) have summarized the situation thus:

The life of a new cultivated wheat variety in the American Northwest is about five years. The rusts (fungi) adapt to the strain, and a new resistant one must be developed. That development is done through artificial selection: the plant breeder carefully combines genetic types that show promise of giving resistance.

Figure 2.18
Reduction in the range of species related to habitat loss associated with drainage activities:
(a) pillwort (*Pilularia globulifera*)
(b) marsh clubmoss (*Lycopodiella inundata*)
(c) marsh gentian (*Gentiana pneumonanthe*)
(d) small fleabane (*Pulicaria vulgaris*)
(after Nature Conservancy Council, 1977, figures 4–7)

The impact of human activities on species diversity, while clearly negative on a global scale (as evidenced by extinction rates) is not so cut-and-dried on the local scale. Under certain conditions chronic stress caused by humans can lead to extremely high numbers of coexisting species within small areas. By contrast, site enrichment or fertilization can result in a decline of species density (Peet et al., 1983).

Certain low productivity grasslands which have been grazed for long periods have high species densities in Japan, the UK and the

Netherlands. The same applies to Mediterranean scrub vegetation in Israel, and to savanna in Sri Lanka and North Carolina (USA). Studies have confirmed that species densities may increase in areas subject to chronic mowing, burning, domestic grazing, rabbit grazing or trampling. It is likely that in such ecosystems humans encourage a high diversity of plant growth by acting as a 'keystone predator', a species which prevents competitive exclusion by a few dominant species.

Grassland enrichment experiments employing fertilizers have suggested that in many cases high growth rates result in the competitive exclusion of many plants. Thus the tremendous increase in fertilizer use in agriculture has had what in retrospect is the predictable result – a widespread decrease in the species diversity of grasslands.

One other area in which major developments may take place in the coming years, which will have implications for plant life, is the field of genetic engineering. This involves the manipulation of DNA, the basic chromosomal unit that exists in all cells and contains genetic information that is passed on to subsequent generations. Recombinant DNA technology (also known as *in vitro* genetic manipulation and gene cloning) enables the insertion of foreign DNA (containing the genetic information necessary to confer a particular target characteristic) into a vector. Fears have been expressed that this technology could produce pathogens that might interact detrimentally with naturally occurring species.

THREE

Human Influence on Animals

Introduction

The range of impacts that humans have had on animals, though large, can be grouped conveniently into five main categories: domestication, dispersal, extinction, expansion and contraction. As with plants, people have helped to disperse animals deliberately, though many have also been dispersed accidentally.

The number of animals that accompany people without their leave is enormous, especially if we include the clouds of microorganisms that infest their land, food, clothes, shelter, domestic animals and their own bodies.

Humans have also domesticated many animals (a project that greatly intrigued Count Buffon), to the extent that, as with many plants, those animals depend on humans for their survival and, in some cases, for their reproduction. The extinction of animals by human predators has been extensive over the past 20 000 years, and in spite of recent interest in conservation continues at a high rate. In addition the presence of humans has led to the contraction in the distribution and welfare of many animals (because of factors like pollution), though in other cases human alteration of the environment and modification of competition has favoured the expansion of some species, both numerically and spatially. Humans now have the highest biomass of any animal species, some 200 million tonnes (Myers, 1979: ix).

Domestication of animals

We have already referred briefly to one of the great themes in the study of human influence on nature: domestication. This has been one of the most profound ways in which humans have affected animals, for during the ten or eleven millennia that have passed since this process was initiated the animals that human societies have selected as useful to them have undergone major changes. The consequences are so substantial that the differences between breeds of animals of the same species often exceed those between different species under natural conditions. A cursory and superficial comparison of the tremendous range of shapes and sizes of modern dog breeds (as, for example, between a wolfhound and a chihuahua) is sufficient to establish the extent of alteration brought about by domestication, and the speed at which domestication has accelerated the process of evolution. In particular humans have changed and enhanced the characteristics for which they originally chose to domesticate animals. For example, the wild ancestors of cattle gave no more than a few hundred millilitres of milk; today the best milk cow can yield up to 15 000 litres of milk during its lactation period. Likewise, sheep have changed enormously (Ryder, 1966). Wild sheep have short tails, while modern domestic sheep have long tails which may have arisen during human selection of a fat tail. Wild sheep also have an overall brown colour, whereas domestic sheep tend to be mainly white. Moreover, the woolly undercoats of wild sheep have developed at the expense of bristly outer coats. With the ancestors of domestic sheep, wool (which served as protection for the skin and as insulation) consisted mainly of thick rough hairs and a small amount of down; the total weight of wool grown per year probably never reached 1 kg. The wool of present-day fine-fleeced sheep consists of uniform, thin down fibres and the total yearly weight may now reach 20 kg. Wild sheep also undergo a complete spring moult, while domestic sheep rarely shed wool.

Indeed, one of the most important consequences or manifestations of the domestication of animals consists of a sharp change in the seasonal biology. Whereas wild ancestors of domesticated beasts are often characterized by relatively strict seasonal reproduction and moulting rhythms, most domesticated species can reproduce at almost any season of the year and tend not to moult to a seasonal pattern.

Dispersal and invasions of animals

Most textbooks on zoogeography devote much time to dividing the world into regions with distinctive animal life – the great 'Faunal Realms' of Wallace and subsequent workers. This pattern of wildlife distribution evolved slowly over geological time. The most striking dividing line between such realms is probably that between Australasia and Asia – Wallace's Line. Australasia developed its distinctive fauna – its relative absence of placental mammals, its well-developed marsupials, and the egg-laying monotremes (echidna and the platypus) – because of its isolation from the Asian land mass.

Modern societies, by moving wildlife from place to place, consciously or otherwise, are breaking down these classic distinct Faunal Realms. Highly adaptable and dispersive forms are spreading, perhaps at the expense of more specialized organisms.

Humans have introduced a new order of magnitude into distances over which dispersal takes place, and through the transport by design or accident of seeds or other propagules, through the disturbance of native plant and animal communities and of their habitat, and by the creation of new habitats and niches, the invasion and colonization by adventive species is facilitated.

Di Castri (1989) has identified three main stages in the process of biological invasions stimulated by human actions (table 3.1). In the first stage, covering several millennia up to about AD 1500, human historical events favoured invasions and migrations primarily *within* the Old World. The second stage commenced about AD 1500, with the exploration, discovery and colonization of new territories, and the initiation of 'the globalization of exchanges'. It shows the occurrence of flows of invaders from, to and within the Old World. The third stage, which only covers the last 100 to 150 years, sees an even more extensive 'multifocal globalization' and an increasing rate of exchanges. The Europocentric focus has diminished.

Deliberate introductions of new animals to new areas has been carried out for many reasons (Roots, 1976): for food, for sport, for revenue, for sentiment, for control of other pests, and for aesthetic purposes. Such deliberate actions probably account, for instance, for the widespread distribution of trout (see figure 3.1). There have been many accidental introductions, especially since the development of ocean-going vessels. The rate is increasing for, whereas in the eighteenth century there were few ocean-going vessels of more than 300 tonnes, today there are thousands. Because of this, in the words of Elton (1958: 31), 'we are seeing one of the great historical convulsions in the world's fauna and flora'. Indeed, many animals are introduced with vegetable products, for 'just as trade followed the flag, so animals have followed the plants'.

One of the most striking examples of the accidental introduction of animals given by Elton is the arrival of some chafer beetles, *Popillia japonica* (the Japanese beetle), in New Jersey in a consignment of plants from Japan. From an initial population of about one dozen beetles in 1916, the centre of population grew rapidly outwards to cover many thousands of square kilometres in only a few decades (figure 3.2).

A more recent example of the spread of an introduced insect in the Americas is provided by the Africanized honey bee. A number of these were brought to Brazil from South Africa in 1957 as an experiment and some escaped. Since then (figure 3.3) they have moved northwards to Central America and Texas, spreading at a rate of 300–500 km per year, and competing with established populations of European honey bees.

Some animals arrive accidentally with other beasts that are imported deliberately. In northern Australia, for instance, water buffalo were introduced (McKnight, 1971) and brought their own bloodsucking fly, a species which bred in cattle dung and transmitted an organism sometimes fatal to cattle. Australia's native

Table 3.1 Human-history driving forces in the Old World as related to biological invasions

Before AD 1500	After AD 1500	From last century in a worldwide perspective
Forest clearing	Exploration, discovery and early colonization by Europeans of other territories and continents	Improvement of transportation systems (roads, railways, internal navigation canals)
Primeval agriculture	Establishment of new market economies and crossroads places (e.g. Amsterdam, London) favouring the 'globalization' of trade exchanges	Large engineering works for irrigation and hydropower
Sheep and cattle raising	Large 'colonies' under the rule of Europeans, often entailing introduction of European-like agriculture and increasing inter alia intertropical exchanges	Opening of inter-oceanic canals (e.g. Suez, Panama, Volga-Don)
Migrations and nomadism	Revolution of food customs in wealthy Europe (e.g. increased use of tea, coffee, chocolate, rice, sugar, potatoes, maize, beef and lamb)	Aircraft transportation
Inshore coastal traffic	Increased demand in Europe of products such as cotton, tobacco, wool, etc.	World wars and displacement of human populations
Settlements of islands (e.g. Corsica)	Negro slavery: Indian and Chinese migrations	'Decolonization': international aid to newly independent countries following 'western' patterns
Intensification of agriculture by ploughing	Missionary establishments	Emergence of multinational companies
Offshore traffic and trade	Occupation by Russians of northern and part of central Asia, up to Siberia	Tropical deforestation and resettlement schemes
Coastal 'colonies' (e.g. Phoenician and Greek colonies)	Intentional introduction into the Old World of exotic species through activities of acclimatization societies, botanical gardens and zoos, and for agricultural, forestry, fishery or ornamental purposes	Afforestation of arid lands with exotic species
Building up of large empires (e.g. Persian, Roman, Arab, Mogul) with considerable expansion of communication and transportation systems	Large-scale emigration from the Old World due to persecution during religious conflicts, civil and independence wars, and to increased demography, unemployment and famine.	Environment impacts decreasing ecosystems resilience
Long-ranging wars and military expansion		Increased urbanization and creation of ruderal habitats
Invasions of German and Asian people, mainly from east to west		International interdependence of markets
Long-distance shipping trade		Release of genetically engineered organisms
Establishment of 'market economies' (e.g. Venice) covering the 'known world' up to the Far East		

Source: Di Castri, 1989, table 1.2

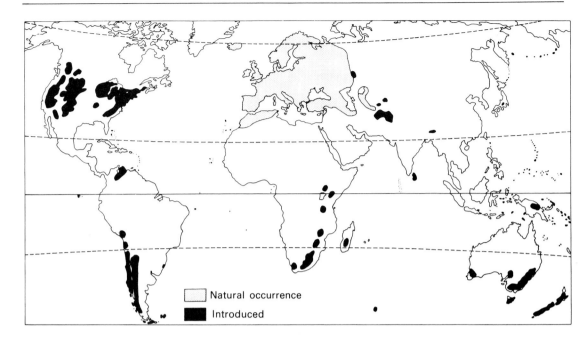

Natural occurrence

Introduced

dung beetles, accustomed only to the small sheep-like pellets of the grazing marsupials, could not tackle the large dung pats of the buffalo. Thus untouched pats abounded and the flies were able to breed undisturbed. Eventually African dung beetles were introduced to compete with the flies (Roots, 1976).

A study conducted in California (Moyle, 1976) gives an indication of the causes and significance of fish introductions in the state, for no fewer than 50 of the 133 fish species are introduced, and the rate of introduction appears to be increasing (figure 3.4). The prime reason for introductions, both authorized and unauthorized, has been sport fishing, but introductions have also occurred through the provision of forage for game fish, the escape of live bait, the release of pets, and the provision of fish for the biological control of insects.

While domesticated plants have, in most cases, been unable to survive without human help, the same is not so true of domesticated animals. There are a great many examples of cattle, horses (see, for example, McKnight, 1959), donkeys and goats which have effectively adapted to new environments and have virtually become wild. Frequently they have both ousted native animals and, particularly in the case of goats on ocean island, caused desertization. Sometimes, however, introduced animals have spread so thoroughly and rapidly, and have led to such a change in the environment to which they were introduced, that they have sown the seeds of their own demise. Reindeer, for example, were brought from Lapland to Alaska in 1891–1902 to provide a new resource for the Eskimos, and the herds increased and spread to over half a million animals. By the 1950s, however, there was less than a twentieth of

Figure 3.1
The original area of distribution of the brown trout and areas where it has been artificially naturalized (after MacCrimmon in Illies, 1974, figure 3.2)

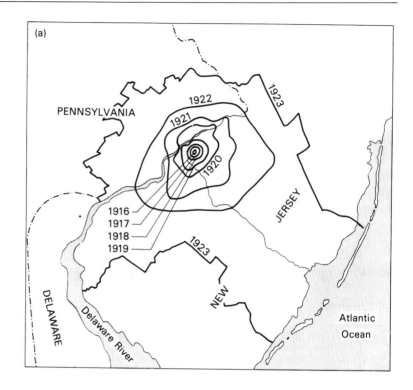

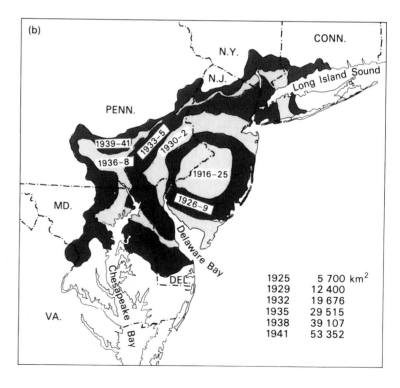

Figure 3.2
The spread of the Japanese
beetle, *Popillia japonica*, in the
eastern USA:
(a) from its point of
    introduction in New Jersey,
    1916–23 (after Elton, 1958,
    figure 14, and Smith and
    Hadley, 1926)
(b) from its point of
    introduction to elsewhere,
    1916–41 (after Elton, 1958,
    figure 15, and US Bureau
    of Entomology, 1941)

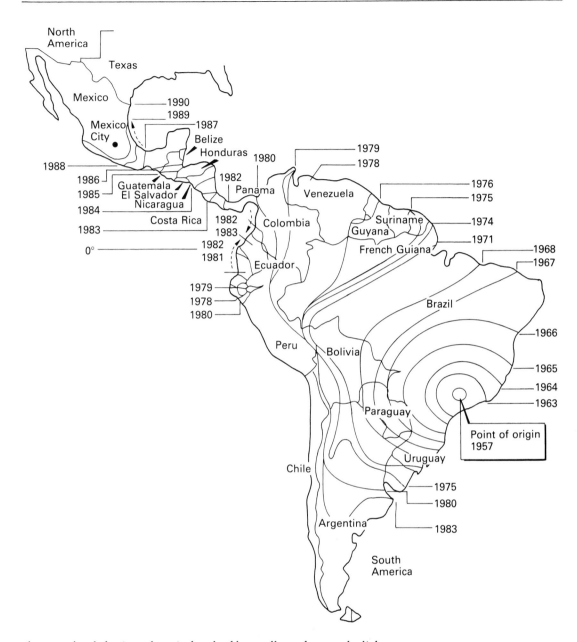

Figure 3.3
The spread of the Africanized honey bee in the Americas between 1957 (when it was introduced to Brazil) and 1990 (modified after Texas Agricultural Experiment Station, in *Christian Science Monitor*, September 1991)

that number left, since the reindeer had been allowed to eat the lichen supplies that are essential to winter survival; as lichen grows very slowly their food supply was drastically reduced (Elton, 1958: 129).

The accidental dispersal of animals can be facilitated by means other than transport on ships or introduction with plants. This applies particularly to aquatic life which can be spread through human alteration of waterways by methods such as canalization. Wooster (1969) gives the example of the way in which the construction of the Suez Canal has enabled the exchange of animals

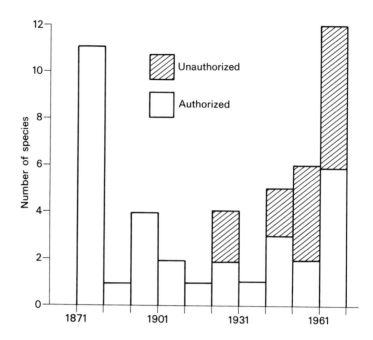

Figure 3.4
The introduction of fish species
into California (adapted from
Moyle, 1976). 1871, when the
graph begins, was the period
of the railroad arrival, the US
Fish Commission and the
California Fish Commission

between the Red Sea and the eastern Mediterranean. Initially, the
high salinity of the Great Bitter Lake acted to prevent movement,
but the infusion of progressively fresher waters (figure 3.5) through
the Suez Canal has meant that this barrier has gradually become
less effective. Some thirty-nine Red Sea fish immigrants have now
been identified in the Mediterranean, and they are especially
important in the Levant basin where they comprise about 12 per
cent of the fish population. Menacing jellyfish (*Rhopilema nomadi-
ca*) have invaded Levantine beaches (Spanier and Galil, 1991). This
type of movement has recently been termed 'Lessepsian migration'.
The construction of the Welland Canal, linking the Atlantic with
the Great Lakes, has permitted similar movements with more
disastrous consequences. Much of the native fish fauna has been
displaced by alewife (*Alosa pseudoharengus*) through competition

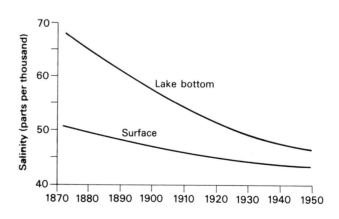

Figure 3.5
Decrease in the salinity of the
Great Bitter Lake, Egypt,
resulting from the infusion of
fresher water by way of the
Suez Canal (after Wooster,
1969)

for food, and by the sea-lamprey (*Petromyzon marinum*) as a predator, so that once common Atlantic salmon (*Salmo salar*), lake trout (*Salvelinus namaycush*) and lake herring (*Leucichtys artedi*) have been nearly exterminated (Aron and Smith, 1971).

Can one make any generalizations about the circumstances that enable successful invasion by exotic vertebrates? Brown (1989) suggests that there may be 'five rules of biological invasions':

Rule 1
'Isolated environments with a low diversity of native species tend to be differentially susceptible to invasion.'

Rule 2
'Species that are successful invaders tend to be native to continents and to extensive, nonisolated habitats within continents.'

Rule 3
'Successful invasion is enhanced by similarity in the physical environment between the source and target areas.'

Rule 4
'Invading exotics tend to be more successful when native species do not occupy similar niches.'

Rule 5
'Species that inhabit disturbed environments and those with a history of close association with humans tend to be successful in invading man-modified habitats.'

Miscellaneous causes of animal contractions and decline

The extreme effect of human interference with animals is extinction, but before that point is reached humans may cause major contractions in both animal numbers and animal distribution. This decline may be brought about partly by intentional killings for subsistence and commercial purposes, but much wildlife decline occurs indirectly (Doughty, 1974), for example, through pollution (Waldichuk, 1979; Holdgate, 1979).

Particular concern has been expressed about the role of certain pesticides in creating undesirable and unexpected changes. The classic case of this concerns DDT and related substances. These were introduced on a worldwide basis after the Second World War and proved highly effective in the control of insects such as malarial mosquitoes. However, evidence accumulated that DDT was persistent, capable of wide dispersal, and reached high levels of concentration in certain animals at the top trophic levels (see figure 5.26). The tendency for DDT to become more concentrated as one moves up the food chain is illustrated further by the data in table 3.2c. DDT levels in river and estuary waters may be low, but the zooplankton and shrimps contain higher levels, the fish that feed on

Table 3.2 Examples of biological magnification

(a) Enrichment factors for the trace element compositions of shellfish compared with the marine environment

	Enrichment factor		
Element	Scallops	Oysters	Mussels
Ag	23 002	18 700	330
Cd	260 000	318 000	100 000
Cr	200 000	60 000	320 000
Cu	3 000	13 700	3 000
Fe	291 500	68 200	196 000
Mn	55 000	4 000	13 500
Mo	90	30	60
Ni	12 000	4 000	14 000
Pb	5 300	3 300	4 000
V	4 500	1 500	2 500
Zn	28 000	110 300	9 100

Source: Merlini, 1971, in King, 1975, table 8.8, p. 303

(b) Mean methyl mercury concentrations in organisms from a contaminated salt marsh

Organism	Parts per million
Sediments	<0.001
Spartina	<0.001–0.002
Echinoderms	0.01
Annelids	0.13
Bivalves	0.15–0.26
Gastropods	0.25
Crustaceans	0.28
Fish muscle	1.04
Fish liver	1.57
Mammal muscle	2.2
Bird muscle	3.0
Mammal liver	4.3
Bird liver	8.2

Source: Gardner et al., 1978, in Bryan, 1979

(c) The concentration of DDT in the food chain

Source	Parts per million
River water	0.000003
Estuary water	0.00005
Zooplankton	0.04
Shrimps	0.16
Insects–*Diptera*	0.30
Minnows	0.50
Fundulus	1.24
Needlefish	2.00
Tern	2.80–5.17
Cormorant	26.40
Immature gull	75.50

Source: King, 1975, table 8.7, p. 301

them higher levels still, while fish-eating birds have the highest
levels of all.

DDT has a major effect on sea-birds. Their egg shells become
thinned to the extent that reproduction fails in fish-eating birds.
Similar correlations between eggshell thinning and DDT residue
concentrations have been demonstrated for various raptorial land-
birds. The bald eagle (*Haliaeetus leucocephalus*), peregrine falcon
(figure 3.6, plate 3.1) (*Falco peregrinus*), and osprey (*Pandion
haliaetus*) all showed decreases in eggshell thickness and population
decline from 1947 to 1967, a decline which correlated with DDT
usage and subsequent reproductive and metabolic effects upon the
birds (Johnston, 1974).

Another example serves to make the same point about the
'magnification' effects associated with pesticides (Southwick, 1976:
45–6). At Clear Lake in California, periodic appearances of large
numbers of gnats caused problems for tourists and residents. An
insecticide, DDD, was applied at a low concentration (up to
0.05 ppm) and killed 99 per cent of the larvae. Following the
application of DDD, Western grebes (*Alchmorphorus occidentalis*)
were found dead, and tissue analysis showed concentrations of
DDD in them of 1600 ppm, representing a concentration of 32 000
times the application rate. DDD had accumulated in the insect-
eating fish at levels of 40–2500 ppm and the grebes feeding on the
fish received lethal doses of the pesticide. There are also records
elsewhere of trout reproduction ceasing when DDT levels build up
(Chesters and Konrad, 1971).

However, appreciation of the undesirable side-effects of DDD
and DDT brought about by ecological concentration has led to
severe controls on their use. For example, DDT reached a peak in
terms of utilization in the USA in 1959 (35×10^6 kg) and by 1971
was down to 8.1×10^6 kg. The monitoring of birds in Florida over
the same period indicated a parallel decline in the concentration of

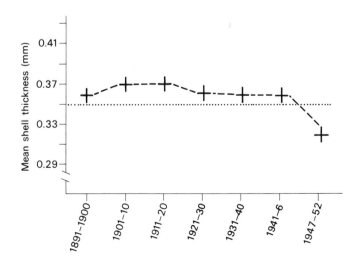

Figure 3.6
Changing eggshell thickness
versus time, associated with
DDT use since the Second
World War (modified after
Hickey and Anderson, 1968)

Plate 3.1
Carnivores, including peregrine falcons, are prone to the effects of biological magnification, whereby pesticides may become concentrated in their bodies. One consequence of this for perergrines has been a reduction in eggshell thickness, which in turn has reduced the number of successful hatchings

DDT and its metabolites (DDD and DDE) in their fat deposits (Johnston, 1974).

Other substances are also capable of concentration. For example, heavy metals and methyl mercury may build up in marine organisms, and filter feeders like shellfish have a strong tendency to concentrate the metals from very dilute solutions, as is clear in table 3.2a and b.

It is possible, however, that the importance of biological accumulation and magnification has been overstated in some textbooks. G. W. Bryan (1979) reviews the situation and notes that, although the absorption of pollutants from foods is often the most important route for bioaccumulation and transfer along food chains certainly occurs, this does not automatically mean that predators at high trophic levels will always contain the highest levels. He writes (p. 497):

although, for a number of contaminants, concentrations in individual predators sometimes exceed those of their prey, when the situation overall is considered only the more persistent organochlorine pesticides, such as DDT and its metabolites and methyl mercury, show appreciable signs of being biologically magnified as a result of food-chain transfer.

Attempts in Israel to increase agricultural productivity by the chemical control of rodents and jackals have had certain unintended impacts on other beasts. Thallium-sulphate-coated wheat was used for rodent control but also afflicted birds of prey which lived on the rodents. All the birds found dead were discovered to contain large amounts of thallium. By 1955–6, breeding as well as wintering populations of those species which fed mainly on rodents had been almost entirely eliminated. The only bird of prey to succeed in keeping its number at about the same level was the short-toed eagle, *Circaetus gallicus*, which feeds entirely on reptiles. Attempts were also made to exterminate jackals (*Canis aureus*) by means of poisoned baits, mainly strychnine. The undesirable results of this included the poisoning (and serious decline) of many Griffon vultures (*Gyps fulvus*), hooded crows (*Corvus corone*), crested cuckoos (*Clamator glandarius*) and the mongoose (*Herpestes auropunctatus*). This in turn meant that there was an enormous increase in the number of hares (*Lepus europaeus*) and Palestine vipers (*Vipera xanthina palestinae*) – the latter thriving because of the reduction in the mongoose population (Mendelssohn, 1973).

Oil pollution is a serious problem for marine and coastal fauna and flora (plate 3.2), although some of it derives from natural seeps (Landes, 1973; Blumer, 1972). Sea-birds are especially vulnerable since oil clogs their feathers; while the ingestion of oil when birds attempt to preen themselves leads to enteritis and other complaints. Local bird populations may be seriously diminished though probably no extinction has resulted (Bourne, 1970). Fortunately, though oil is toxic it becomes less so with time, and the oil spilt from the *Torrey Canyon* in March 1967 was almost biologically inert when it was stranded on the Cornish beaches (Cowell, 1976). There are 'natural' oil-degrading organisms in nature. Much of the damage caused to marine life as a result of this particular disaster was caused not by the oil itself but by the use of 2.5 million gallons of detergents to disperse it (J. E. Smith, 1968).

It is possible that oil spills pose a particular risk in cold areas, for biodegradation processes achieved by microbes appear to be slow. This is probably because of a combination of low temperatures and limited availability of nitrogen, phosphorus, and oxygen. The last of these is a constraint because compared to temperate ecosystems, arctic tundra and coastal ecosystems are relatively stagnant – ice dampens reaeration due to wave action in marine ecosystems, while standing water in tundra soils limits inputs of oxygen to them. Detailed reviews are provided in Engelhardt (1985).

There is far less information available on the effects of oil spills in freshwater environments, though they undoubtedly occur. Freshwater bodies have certain characteristics which, compared with marine environments, tend to modify the effects of oil spills. The most important of these are the smaller and shallower dimensions

of most rivers, streams and lakes, which means that dilution and spreading effects are not as vigorous and significant in reducing surface slicks as would be the case at sea. In addition, the confining dimensions of ponds and lakes are likely to cause organisms to be subjected to prolonged exposure to dissolved and dispersed hydrocarbons. Information on this problem is summarized in Vandermeulen and Hrudey (1987).

Other industrial pollutants have a clear impact on aquatic systems but only a few examples can be given here (from Southwick, 1976: 19–22). In Biscayne Bay, Florida, industrial and municipal wastes from Miami are thought to be responsible for a high prevalence of dermal tumours in several species of marine fish, while in Bellingham Bay, north of Seattle, sulphite wastes from pulp

Plate 3.2 (opposite)
Oil spills are generally perceived as a major cause of pollution on coastlines. In January 1993 the tanker *Braer* ran aground on the Shetland Isles (Scotland) and large quantities of oil were liberated, causing some distress and mortality to sea birds

Plate 3.3
Pollution adversely affects fish and can make them subject to severe fungal infections

mills lead to the abnormal growth of oyster larvae. The lower
Mississippi River, from which New Orleans obtains its drinking
water, contains (even after water treatment) at least thirty-six
chemical compounds of industrial origin known to be harmful to
laboratory animals. Fish kills can sometimes eliminate massive
populations. In San Diego Harbour in 1962 an estimated 37.8
million fish were killed by pollution. Such kills are often produced
by industrial accidents.

One particular type of aquatic ecosystem where pollution is an
increasingly serious problem is the coral reef (see Kuhlmann, 1988).
Coral reefs (plate 3.4) are important because they are among the
most diverse, productive and beautiful communities in the world.
They also provide coastal protection, opportunities for recreation,
and potential sources of substances like drugs. Accelerated sedimentation resulting from poor land management, together with
dredging, is probably responsible for more damage to reef communities than all the other forms of human insult combined, for
suspended sediments restrict the light penetration necessary for
coral growth. Also, the soft, shifting sediments may not favour
colonization by reef organisms. Sewage is the second worst form of
stress to which coral reefs are exposed, for oxygen-consuming
substances in sewage result in reduced levels of oxygen in the water

Plate 3.4
Coral Reefs, like the Maldive
Islands shown on this Shuttle
photograph of part of the
Indian Ocean, are diverse,
productive and beautiful
habitats that are coming under
increasing human pressure

of lagoons. The detrimental effects of oxygen starvation are compounded by the fact that sewage may cause nutrient enrichment to stimulate algal growth, which in turn can overwhelm coral. Poor land management can also cause salinity levels to be reduced below the level of tolerance of reef communities as a consequence of accelerated runoff of freshwater from catchments draining into lagoons. One of the reasons why all these stresses may be especially serious for reefs, is that corals live and grow for several decades or more, so that it can take a long time for them to recover from damage.

As yet the effect of pollution from nuclear industries on plants and animals seems to be limited (although its health effects on humans are proven). Mellanby (1967: 63) has written:

I do not myself think that, except in the vicinity of nuclear test explosions, fall-out has caused measurable damage so far to animals and plants . . . up to the present man-made radiation can seldom have had important ecological effects.

Since the late nineteenth century it has been clear that industrial air pollution has had an adverse effect on domestic animals, but there is less information about its influence on wildlife. None the less, there is some evidence that it does affect wild animals (Newman, 1979). Arsenic emissions from silver foundries are known to have killed deer and wild rabbits in Germany; sulphur emissions from a pulp mill in Canada are known to have killed many song-birds; industrial fluorosis has been found in deer living in the USA and Canada; asbestosis has been found in free-living baboons and rodents in the vicinity of asbestos mines in South Africa; and oxidants from air pollution are recognized causes of blindness in bighorn sheep in the San Bernadino Mountains near Los Angeles in California.

In Britain considerable concern has arisen over the effects of lead poisoning on wildlife, particularly on wildfowl. Poisoning occurs in swans and mallards, for example, when they feed and seek grit, since they ingest spent shotgun pellets or discarded anglers' weights in the process. Such pellets are eroded in the bird's gizzard and the absorbed lead causes a variety of adverse physiological effects that can result in death: damage to the nervous system, muscular paralysis, anaemia, and liver and kidney damage (Mudge, 1983).

One could list many other indirect causes of wildlife decline. Vehicle speed, noise and mobility upset remote and sensitive wildlife populations, and this problem has intensified with the rapid development in the use of off-road recreation vehicles. In California an estimated 1.2 million motorcyles and 500 000 dune buggies are regularly in use in the fragile desert ecosystems (Busack and Bury, 1974). The effects of roads are becoming ever more pervasive as a result of increasing traffic levels, vehicle speed and road width. A fine-meshed road network is a highly effective cause of habitat isolation, acting as a series of barriers to movement, particularly of small, cover-loving animals. As Oxley et al. (1974)

have put it, 'a four-lane divided highway is as effective a barrier to the dispersal of small forest mammals as a body of fresh water twice as wide'. This barrier effect results from a variety of causes (Mader, 1984): roads interrupt microclimatic conditions; they are a broad band of emissions and disturbance; they are zones of instability due to cutting and spraying, etc.; they provide little cover against predators; and they subject animals to death and injury by moving wheels. Leisure activities in general may create problems for some species (Speight, 1973). A survey of the breeding status of the little tern (*Sterna albifrons*) in Britain gave a number of instances of breeding failure by the species, apparently caused by the presence of fishermen and bathers on nesting beaches. The presence of even a few people inhibited the birds from returning to their nests. Likewise, species building floating nests on inland waters, such as great crested grebes (*Podiceps cristatus*), are very prone to disturbance by water-skiers and power-boats.

New types of construction can create problems. As Doughty (1974) has remarked '"wirescapes", tall buildings, and towers can become to birds what dams, locks, canals and irrigation ditches are to the passage of fish'. The construction of canals can cause changes in aquatic communities.

Likewise turbines associated with hydroelectric schemes can cause numerous fish deaths, either directly through ingesting them, or through gas-bubble disease. This resembles the 'bends' in divers and is produced if a fish takes in water supersaturated with gases. The excess gas may come out of solution as bubbles which can lodge in various parts of the fish's body causing injury or death. Supersaturation of water with gas can be produced in turbines or when a spillway plunges into a deep basin (Baxter, 1977).

Agricultural change can also result in animal decline. One of the most important habitat changes in Britain is the removal of many of the hedgerows that form the patchwork quilt so characteristic of large tracts of the country (Sturrock and Cathie, 1980) (plate 3.5). Their removal (figure 3.7) has taken place for a variety of reasons: to create larger fields, so that larger machinery can be employed and so that it spends less time turning; because as farms are amalgamated boundary hedges may become redundant; because as farms become more specialized (for example, just arable) hedges may not be needed to control stock or to provide areas for lambing and calving; and because drainage improvements may be more efficiently executed if hedges are absent. In 1962 there were almost 1 million km of hedge in Britain, probably covering the order of 180 000 hectares, an area greater than that of the National Nature Reserves, but the figure before Enclosure would not have been this high (see table 3.3). Bird numbers and diversity are likely to be severely reduced if the removal of hedgerows continues, for most British birds are essentially of woodland type, and since woods only cover about 5 per cent of the land surface of Britain they very much depend on hedgerows (Moore et al., 1967).

Other types of habitat have also been disappearing at a quick rate with the agricultural changes of recent years (Nature Conservancy

Plate 3.5
An air photograph of part of
Suffolk in eastern England
shows the traces of hedgerows
that have been removed in the
interests of modern farm
technology. Hedgerows
provide cover for many birds
and animals.

Council, 1977) (see figure 3.8). For example, the botanical diversity
of old grasslands is often reduced by replacing the lands with grass
leys or by treating them with selective herbicides and fertilizers; this
can mean that the habitat does not contain some of the basic
requirements essential for many species. One can illustrate this by
reference to the larva of the common blue butterfly (*Polyommatus
icarus*) which feeds upon bird-foot trefoil (*Lotus corniculatus*). This

County (selected areas in each county only)

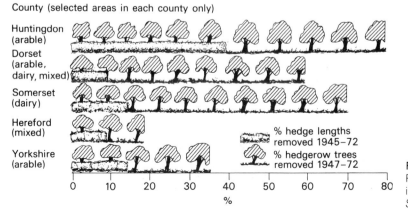

Figure 3.7
Recent changes in hedgerows
in parts of England (from *New
Society*, June 1979, p. 650)

Table 3.3 Changes in length of hedgerows in three parishes in
Huntingdonshire, England over 600 years

Date	Length of hedge (km)	Date	Length of hedge (km)
Prior to 1364	32	1780	93
1364	40	1850	122
1500	45	1946	114
1550	52	1963	74
1680	74	1965	32

Source: from Moore et al., 1967

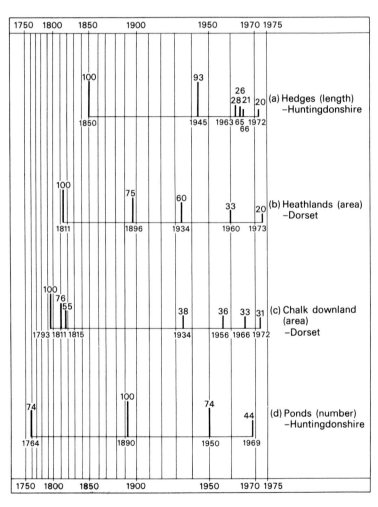

Figure 3.8
Examples of habitat loss in England since the mid-eighteenth century (after Nature Conservancy Council, 1977, figure 3). All values are expressed as percentages of the largest value recorded in each case

plant disappears when pasture is ploughed and converted into a ley, or when it is treated with a selective herbicide. Once the plant has gone the butterfly vanishes too because it is not adapted to feeding on the plants grown in leys of improved pasture. Figure 3.9

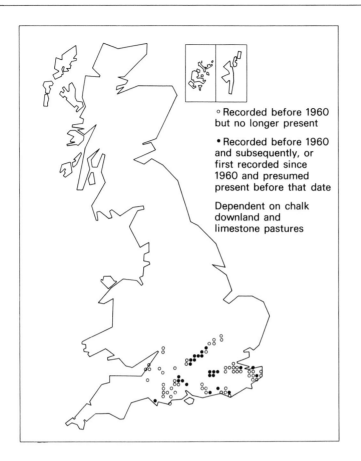

Figure 3.9
Reduction in the range of the silver-spotted skipper butterfly (*Hesperia comma*) as a result of the reduction in the availability of the chalk downland and limestone pastures upon which it depends (after Nature Conservancy Council, 1977, figure 3.8)

illustrates the way in which the distribution of the silver-spotted skipper butterfly (*Hesperia comma*) has contracted with the diminished availability of chalk downland and limestone pastures upon which it depends. Likewise, the large blue butterfly (*Maculinea arion*) has decreased in Britain. Its larvae live solely on the wild thyme (*Thymus druccei*), a plant which thrives on close-cropped grassland. Since the decimation of the rabbit by myxomatosis, conditions for the thyme have been less favourable so that both thyme and the large blue butterfly have declined. Another group of animals which has suffered from agriculture is the species implicated in disease cycles involving farm stock. Recently in Britain it has been thought necessary to eliminate some badgers (*Meles meles*) from a number of farms in the west of England in an attempt to remove the threat of them transmitting tuberculosis to cattle.

Overall, as table 3.4 shows, the number of species occurring in England on unmodernized farms compared to modernized farms is relatively great.

The replacement of natural oak-dominated woodlands in Britain by conifer plantations, another major land-use change of recent decades, also has its implications for wildlife. It has been estimated

Table 3.4 Approximate numbers of species occurring in equivalent habitats in unmodernized and modern farms

Group of animals	Unmodernized farms		Modern farms	
	Habitat	No. of species	Habitat	No. of species
	(a) Hedges* and semi-natural grass verges		(a) Wire fences with (i) sown grass (ii) semi-natural grass verges	
Mammals		20		(i) 5 (ii) 6
Birds		37		(i) 6 (ii) 9
Lepidoptera (butterflies only)		17		(i) 0 (ii) 8
	(b) Permanent pasture † (untreated)		(b) Grass leys	
Lepidoptera (butterflies only)		20		0
	(c) Permanent ponds and ditches		(c) Temporary ditches and piped water	
Mammals (aquatic)		2		0
Amphibia		5		2
Fish		9		0
Odonata (dragonflies)		11		0
Mollusca (gastropods only)		25		3

* Includes hedgerow trees
† Includes grasslands on chalk
Source: Nature Conservancy Council, 1977, table 2

that where this change has taken place the number of species of birds found has been approximately halved. This is illustrated by some data for Wales presented in table 3.5. Likewise the replacement of upland sheep walks with conifer plantations has led to a sharp decline in raven numbers in southern Scotland and northern England. The raven (*Corvus corax*) obtains much of its carrion from open sheep country (Marquiss et al., 1978). Other birds that have suffered from planting moorland areas with forest plantations are miscellaneous types of wader, golden eagles, peregrine falcons and buzzards (Grove, 1983). A succinct summary of the present position of species changes brought about in Britain by these miscellaneous alterations to habitat is provided by the Nature Conservancy Council (1984), and its main findings are reproduced in figure 3.10.

Agricultural intensification on British farms also appears to have had a deleterious effect on species of bird, many of which have shown a downward trend in numbers over the last two decades (table 3.6). This is because of habitat changes – lack of fallows, less mixed farming, new crops, modern farm management, biocide use, hedgerow removal, etc. (Fuller et al., 1991).

On a global basis the loss of wetland habitats (marshes, bogs, swamps, fens, mires, etc.) is a cause of considerable concern (Maltby, 1986; Williams, 1990). In all, wetlands cover about 6 per cent of the earth's surface (not far short of the total under tropical rain forest), and so they are far from being trivial, even though they

Table 3.5 Estimated numbers of breeding songbirds in two broad-leaved and two coniferous woods in Wales

	Broad-leaved woods		Coniferous woods	
	(1) (275 trees/ha, relict)	(2) (375 trees/ha, relict)	(3) (450 trees/ha, planted)	(4) (275 trees/ha, planted)
Chaffinch	7	4	4	7
Wren	13	15	34	9
Robin	7	4	3	3
Dunnock	–	–	2	–
Goldcrest	2	4	9	31
Coal Tit	6	3	7	4
Blue Tit	5	2	–	–
Great Tit	6	2	–	–
Marsh Tit	–	–	–	–
Nuthatch	3	–	–	–
Tree Creeper	1	1	–	–
Blackbird	3	–	–	–
Song Thrush	–	–	–	–
Mistle Thrush	–	–	–	3
Wood Warbler	3	3	–	–
Willow Warbler	4	10	–	–
Blackcap	–	–	–	–
Chiffchaff	–	–	–	–
Pied Flycatcher	6	7	–	–
Redstart	2	2	–	–
TOTAL	68	57	59	57
No. of species	14	12	6	6
Index of diversity	2.47	2.21	1.31	1.37

Source: Edington and Edington, 1977, table 2.1, p. 8

tend to occur in relatively small patches. However, they also account for about one-quarter of the earth's total net primary productivity, have a very diverse fauna and flora, and provide crucial wintering, breeding and refuge areas for wildlife. According to some sources, the world may have lost half of its wetlands since 1900, and the USA alone has lost 54 per cent of its original wetland area, primarily because of agricultural developments. There are, however, other threats, including drainage, dredging, filling, peat removal, pollution, and channelization.

In the former USSR, the ambitious ploughing-up of the so-called Virgin Lands of Central Kazakhstan in the late 1950s and early 1960s led to the replacement of herbaceous steppe and grass steppe by cultivated fields. As a result, the animal inhabitants of the steppe became increasingly restricted to smaller areas of suitable habitat.

Losses of the heritage of nature

Habitat
Lowland herb-rich hay meadows: 95% now lacking significant wild-life interest and only 3% left undamaged by agricultural intensification.

Lowland grasslands of sheep walks. On chalk and Jurassic limestone: 80% loss, largely by conversion to arable or improved grassland (mainly since 1940), but some scrubbed over through lack of grazing.

Lowland heaths on acidic soils: 40% loss, largely by conversion to arable or improved grassland, afforestation and building; some scrubbed over through lack of grazing.

Limestone pavements in northern England: 45% damaged or destroyed, largely by removal of weathered surfaces for sale as rockery stone, and only 3% left completely undamaged.

Ancient lowland woods composed of native, broad-leaved trees: 30–50% loss, by conversion to conifer plantation or grubbing out to provide more farmland.

Lowland fens, valley and basin mires: 50% loss or significant damage through drainage operations, reclamation for agriculture and chemical enrichment of drainage water.

Lowland raised mires: 60% loss or significant damage through afforestation, peat-winning, reclamation for agriculture or repeated burning.

Upland grasslands, heaths and blanket bogs: 30% loss or significant damage through coniferous afforestation, hill land improvement and reclamation, burning and over-grazing.

Species
The large blue butterfly became extinct in 1979, but ten more species are vulnerable or even seriously endangered, and out of a total British list of 55 resident breeding species of butterfly another 13 have declined and contracted in range substantially since 1960.

Three or four of our 43 species of dragonflies have become extinct since 1953, six are vulnerable or endangered, and five have decreased substantially.

Four of our 12 reptiles and amphibians are endangered.

At least 36 breeding species of bird have shown appreciable long-term decline during the last 35 years as a result of habitat loss or deterioration, 30 in the lowlands and six in the uplands.

The otter has become rare or has disappeared in many parts of England and Wales.

Bats in general have decreased and several of our 15 species, notably the greater horseshoe and mouse-eared bats are at risk of extinction. Others such as Bechstein's, Leisler's and the barbastelle are rare, and even the pipistrelle is no longer common. Problems are food supply, destruction of breeding and hibernation roosts and pollution.

There have been local increases and extensions of range in some vertebrate populations (e.g. wild cat and pine marten), but the overall balance of change is on the debit side.

Figure 3.10
Summary of current habitat and species changes in Britain (from Nature Conservancy Council, 1984, Summary of objectives and strategy, p. 7)

The population of the fur-bearing marmot or bobac (*Marmot bobac*) steadily declined from about 3 million in the middle 1950s to 318 000 in 1969. The animals are further threatened because the rise in the human population following the extension of agriculture has increased trapping activity to obtain pelts (Zimina et al., 1972).

Soil erosion, by increasing stream turbidity, may adversely affect fish habitats (Ritchie, 1972). The reduction of light penetration inhibits photosynthesis, which in turn leads to a decline in food and a decline in carrying capacity. Decomposition of the organic matter, which is frequently deposited with sediment, uses dissolved oxygen, thereby reducing the oxygen supply around the fish; sediment restricts the emergence of fry from eggs; and turbidity

Table 3.6 Decline in farmland birds in Britain between 1970 and 1988 as shown by the Common Birds Census

Species	% decline
Corn bunting	69
Grey partridge	67
Tree sparrow	67
Lapwing	59
Bullfinch	58
Song thrush	54
Turtle dove	48
Linnet	36
Skylark	33
Spotted flycatcher	31
Blackbird	28

Source: from data in Fuller et al., 1991

reduces the ability of fish to find food (though conversely it may also allow young fish to escape predators).

Turbidity has also been increased by mining waste (plate 3.6), by construction (Barton, 1977), and by dredging. In the case of some of the Cornish rivers in china-clay mining areas, river turbidities reach 5000 mg l^{-1} and trout are not present in streams so affected.

Indeed, the non-toxic turbidity tolerances of river fish are much less than that figure. Alabaster (1972), reviewing data from a wide range of sources, believes that in the absence of other pollution, fisheries are unlikely to be harmed at chemically inert suspended sediment concentrations of less than 25 mg l^{-1}, that there should be good or moderate fisheries at 25–80 mg l^{-1}, that good fisheries are unlikely at 80–400 mg l^{-1} and that at best only poor fisheries would exist at more than 400 mg l^{-1}.

The effect of mining on animals is referred to in the world's first mining textbook, *De Res Metallica* by Georgius Agricola (1556), which contains an excellent description of the destruction caused by mining in Germany (Down and Stocks, 1977: 7):

the strongest argument of the detractors is that the fields are devastated by mining operations . . . The woods and groves are cut down, for there is need of an endless amount of wood for timber, machines and the smelting of metals. And when the woods and groves are felled, then are exterminated the beasts and birds, very many of which furnish a pleasant and agreeable food for man. Further, when the areas are washed, the water which has been used poisons the brooks and streams and either destroys the fish or drives them away.

It needs to be stressed, however, that the changes of habitat brought about by urbanization, industry (Davis, 1976) and mining need not be detrimental. Indeed Ratcliffe (1974: 368) has described the present situation:

I think it is true to say that while mineral exploitation has caused some damage to nature conservation interest in Britain, this has been mostly on a

Plate 3.6
Gold mining in Brazil has caused serious deforestation in certain parts of the Amazonian rainforest, and this in turn has increased the sediment loads of Amazonian streams. Furthermore, the mining process itself delivers large quantities of sediment into stream channels

local and minor scale, and on the whole the gains have outweighed the losses, especially in the creation of interesting new habitats.

He does, however, express some doubts about the future:

The anxiety about future mineral extraction stems from the scale and method of operation. Many mineral workings in the past were mainly subterranean, and surface disturbance was limited largely to waste tipping. Future workings are likely to use strip or opencast mining techniques increasingly, and the demand to rehabilitate new workings is often disadvantageous from the wildlife angle.

It might also be thought that human use of fire could be directly detrimental to animals, though the evidence is not conclusive. Many forest animals appear to be able to adapt to fire, and fire also tends to maintain habitat diversity. Thus, after a general review of the available ecological literature Bendell (1974) found that fires did not seem to produce as much change in birds or small mammals as one might have expected. Some of his data are presented in table 3.7. The amount of change is fairly small, though some increase in species diversities and animal densities is evident. Indeed Vogl (1977) has pointed to the diversity of fauna associated with fire-affected ecosystems (p. 281):

Birds and mammals usually do not panic or show fear in the presence of fire, and are even attracted to fires and smoking or burned landscapes . . .

Table 3.7 Environmental effects of burning

(a) Change in number of species of breeding birds and small mammals after fire

	Foraging zone	Before fire	After fire	Gained (%)	Lost (%)
BIRDS	Grassland and shrub	48	62	38	8
	Tree trunk	25	26	20	16
	Tree	63	58	10	17
	TOTALS	136	146	21	14
SMALL MAMMALS	Grassland and shrub	42	45	17	10
	Forest	16	14	13	25
	TOTALS	58	59	16	14

(b) Change in density and trend of population of breeding birds and small mammals after fire

		Density (%)			Trend (%)		
	Foraging zone	Increase	Decrease	No change	Increase	Decrease	No change
BIRDS	Grassland and shrub	50	9	41	24	10	66
	Tree trunk	28	16	56	4	8	88
	Tree	24	19	57	6	6	88
	TOTALS	35	15	50	12	8	80
SMALL MAMMALS	Grassland and shrub	24	13	63	20	5	75
	Forest	23	42	35	0	11	89
	TOTALS	23	25	52	14	1	80

The greatest arrays of higher animal species, and the largest numbers per unit area, are associated with fire-dependent, fire-maintained, and fire-initiated ecosystems.

Purposeful hunting, particularly when it involves modern firearms, means that the distribution of many animal species has become smaller very rapidly. This clearly illustrated by a study of the North American bison (figure 3.11), which on the eve of European colonization still had a population of some 60 million in spite of the presence of a small Indian population. By 1850 only a few dozen examples of bison still survived, though conscious protective and conservation measures have saved it from extinction.

Animals native to remote ocean islands have been especially vulnerable since many have no flight instinct. They also provided a convenient source of provisions for seafarers. Thus the last example of the dodo (*Raphus cucullatus*) was slaughtered on Mauritius in 1681 and the last Steller's sea-cow (*Hydrodamalis gigas*) was killed

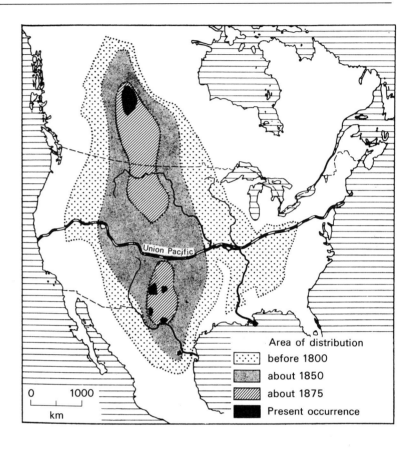

Figure 3.11
The former and present
distribution of the bison in
North America (after Ziswiler in
Illies, 1974, figure 3.1)

on Bering Island in 1768 (Illies, 1974: 96). The land-birds, which
are such a feature of coral atoll ecosystems, rapidly become extinct
when cats, dogs and rats are introduced. The endemic flightless rail
of Wake Island, *Rallus wakensis*, became extinct during the
prolonged siege of the island in the Second World War, and the
flightless rail of Laysan, *Porzanula palmeri*, is also now extinct.
Sea-birds are more numerous on the atolls, but many colonies of
terns, noddies and boobies have been drastically reduced by
vegetation clearance and by introduced rats and cats in recent times
(Stoddart, 1968). Other atoll birds, including the albatross, have
been culled because of the threat they pose to aircraft using military
airfields on the islands. Some birds, including the sulids, have
declined because they are large, vulnerable and edible (Feare,
1978).

The fish resources of the oceans are still exploited by hunting and
gathering techniques, and about 70 million tonnes of fish are
caught in the world each year. There are countless examples of fish
decline brought about by reckless over-exploitation. Overall (see
Ehrlich et al., 1977: 358) it has been estimated, for instance, that
whale stocks have been reduced by at least half through uncontrolled hunting during the past half-century. Populations of the

Plate 3.7
The Moa is an extinct land-bird from New Zealand. This reconstructed version was placed in the Dunedin Public Gardens

large varieties, those that have been most heavily exploited, have been cut to a tiny fraction of their former sizes, and the hunting, especially by the Russians and the Japanese, still goes on. As Fisher et al. (1969) have put it, 'It may not be long before the blue whale joins the dinosaurs in the museum of oblivion.'

The introduction of a new animal species can cause the decline of another, whether by predation, competition or by hybridization. This last mechanism has been found to be a major problem with fish in California, where many species have been introduced by humans. Hybridization results when fish are transferred from one basin to a neighbouring basin, since closely related species of the type likely to hybridize usually exist in adjacent basins. One species can eliminate another closely related to it through genetic swamping (Moyle, 1976).

Human influence on the expansion of animal populations

Although most attention tends to be directed towards the decline in animal numbers and distribution brought about by human agency,

Plate 3.8
The dodo. Top: the skin and skull remains. Bottom: its reconstructed form. The last example disappeared three hundred years ago in Mauritius

there are many circumstances where alteration of the environment and modification of competition has favoured the expansion of some species. Such expansion is not always welcome or expected, as Marsh (1864: 34) appreciated.

Insects increase whenever the birds which feed upon them disappear. Hence in the wanton destruction of the robin and other insectivorous birds, the *bipes implumis*, the featherless biped, man, is not only exchanging the vocal orchestra which greets the rising sun for the drowsy beetle's evening drone, and depriving his groves and his fields of their fairest ornament, but he is waging a treacherous warfare on his natural allies.

Human actions, however, are not invariably detrimental, and even great cities may have effects on animal life which can be considered desirable or tolerable. This has been shown in the studies of bird populations in several urban areas. For example, Nuorteva (in Jacobs, 1975) examined the bird fauna in the city of Helsinki (Finland), in agricultural areas near rural houses, and in uninhabited forests (see table 3.8). The city supported by far the highest biomass and the highest number of birds, but exhibited the lowest number of species and the lowest diversity. In the artificially created rural areas the number of species, and hence diversity, were much

higher than in the uninhabited forest, and so was biomass. Altogether, human civilization appeared to have brought about a very significant increase of diversity in the whole area: there were thirty-seven species in city and rural areas that were not found in the forest. Similarly, after a detailed study of suburban neighbourhoods in west-central California, Vale and Vale (1976) found that in suburban areas the number of bird species and the number of individuals increased with time. Moreover, when compared to the pre-suburban habitats adjacent to the suburbs, the residential areas were found to support a larger number of both species and individuals. Horticultural activities appear to provide more luxuriant and more diverse habitats than do pre-suburban environments.

Plate 3.9
Whale butchering, whaling station Hvalfijordur bay, Iceland. If reckless over-exploitation continues some species will soon become extinct

Table 3.8 Dependence of bird biomass and diversity on urbanization

	City (Helsinki)	Near rural houses	Uninhabited forest
Biomass (kg km^{-2})	213	30	22
No. of birds km^{-2}	1089	371	297
Number of species	21	80	54
Diversity	1.13	3.40	3.19

Source: data from Nuorteva, 1971, modified after Jacobs, 1975, table 2

Some beasts other than the examples of birds given here are also favoured by urban expansion (Schmid, 1974). Animals that can tolerate disturbance, are adaptable, utilize patches of open or woodland-edge habitat, creep about inside buildings, tap people's food supply surreptitiously, avoid recognizable competition with humans, or attract human appreciation and esteem, may increase in the urban milieu. For these sorts of reasons the north-east megalopolis of the United States hosts thriving populations of squirrels, rabbits, racoons, skunks and opossums, while some African cities are now frequently blessed with the scavenging attention of hyenas.

Jacobs (1975) provides another apposite example of how humans can increase species diversity inadvertently. The saline Lake Nakuru in Kenya before 1961 was not a particularly diverse ecosystem. There were essentially one or two species of algae, one copopod, one rotifer, corrixids, notonectids and some 500 000 flamingos belonging virtually to one species. In 1962, however, a fish (*Tilapia grahami*) was introduced in order to check mosquitoes. In the event it established itself as a major consumer of algae, and its numbers increased greatly. As a consequence, some thirty species of fish-eating birds (pelicans, anhingas, cormorants, herons, egrets, grebes, terns and fish-eagles) have colonized the area to make Lake Nakuru a much more diverse system. None the less, such fish introductions are not without their potential ecological costs. In Lake Atitlan in Central America the introduction of the game fish *Micropterus salmoides* (largemouth bass) and *Pomoxis nigromaculatus* (black crappie), both voracious eaters, led to the diminution of local fish and crab populations. Similarly, the introduction to the Gatun Lake in the Panama Canal Zone in 1967 of the cichlid fish, *Cichla ocellaris* (a native of the Amazon River), led to the elimination of six of the eight previously common fish species within five years, and the tertiary consumer populations such as the birds, formerly dependent on the small fishes for food, appeared less frequently (Zaret and Paine, 1973).

Human economic activities may lead to a rise in the number of examples of a particular habitat which can lead to an expansion in the distribution of certain species, though often humans have prompted a contraction in the range of a particular species because of the removal or modification of its preferred habitat. In Britain the range of the little ringed plover (*Charadrius dubius*), a species virtually dependent on anthropogenic habitats, principally wet gravel and sandpits, has greatly expanded as mineral extraction has been accelerated in recent decades (see figure 3.12).

Very often human activities do not lead to species diversity, but to important increases in numbers of individuals, by creating new and favourable environments. Two especially serious examples of this from the human point of view are the explosions in the prevalence of both mosquitoes and bilharzia snails as a result of the extension of irrigation.

One of the most remarkable examples of the consequences of creating new environments is provided by the European rabbit

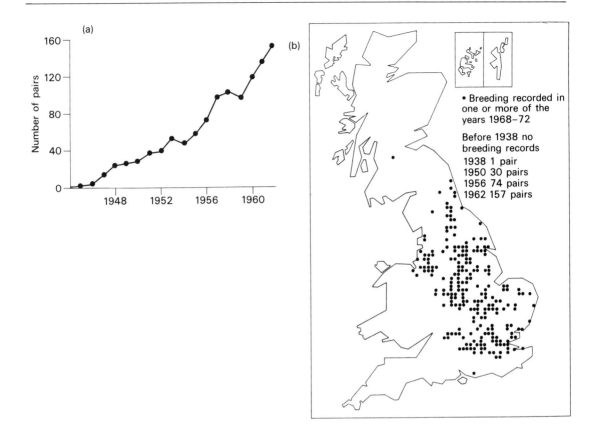

Figure 3.12
The changing range of the little ringed plover *Charadrius dubius*:
(a) the increase in the number of little ringed plovers summering in Britain (after Murton, 1971, figure 6)
(b) the increase in the range related to habitat change, especially as a result of the increasing number of gravel pits (after Nature Conservancy Council, 1977, figure 11)

(*Oryctolagus cuniculus*). Introduced into Britain in early medieval times, and originally an inhabitant of the western Mediterranean lands, it was kept for food and fur in carefully tended warrens. Agricultural improvements, especially to grassland, together with the increasing decline in the numbers of predators such as hawks and foxes, brought about by game-guarding landlords (Sheail, 1971), enabled the rabbit to become one of the most numerous of mammals in the British countryside. By the early 1950s there were 60–100 million rabbits in Britain. Frequently, as many contemporary reports demonstrated, it grazed the land so close that in areas of light soils, like the Breckland of East Anglia or in coastal dune areas, wind erosion became a serious problem. Similarly, the rabbit flourished in Australia, especially after the introduction of the merino sheep which created favourable pasture-lands. Erosion in susceptible lands like the Mallee was severe. Both in England and Australia an effective strategy developed to control the rabbit was the introduction of a South American virus, *Myxoma*.

Some of the most familiar British birds have benefited from agricultural expansion. Formerly, when Britain was an extensively wooded country, the starling (*Sturnus vulgaris*) was a rare bird. The lapwing (*Vanellus vanellus*) is yet another component of the grassland fauna of Central Europe which has benefited from

agriculture and the creation of open country with relatively sparse
vegetation (Murton, 1971). There is also a large class of beasts
which profit so much from the environmental conditions wrought
by humans that they become very closely linked to them. These
animals are often referred to as synanthropes. Pigeons and spar-
rows now form permanent and numerous populations in almost all
the large cities of the world; human food supplies are the food
supplies for many synanthropic rodents (rats and mice); a once shy
forest-bird, the blackbird (*Turdus merula*), has in the course of a
few generations become a regular and bold inhabitant of many
gardens; and the squirrel (*Sciurus vulgaris*) in many places now
occurs more frequently in parks than in forests (Illies, 1974: 101).
The English house sparrow (*Passer domesticus*), a familiar bird in
towns and cities, currently occupies approximately one-quarter of
the earth's surface, and over the past one hundred years it has
doubled the area that it inhabits as settlers, immigrants and others
have carried it from one continent to another (figure 3.13). It is
found around settlements both in Amazonia and the Arctic Circle
(Doughty, 1978). The marked increase in many species of gulls in

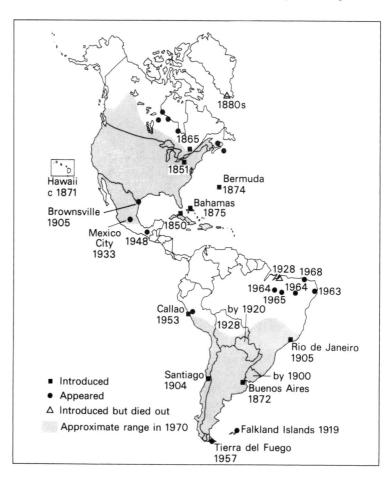

Figure 3.13
The spread of the English
house sparrow in the New
World (after Doughty, 1978,
p. 14)

temperate regions over recent decades is largely attributable to their growing utilization of food scraps on refuse tips. This is, however, something of a mixed blessing, for most of the gull species which feed on urban rubbish dumps breed in coastal areas, and there is now good evidence to show that their greatly increased numbers are threatening other less common species, especially the terns (*Sterna* spp.). It may seem incongruous that the niches for scavenging birds which exploit rubbish tips should have been filled by sea-birds (Murton, 1971). But the gulls have proved ideal replacements for the kites which humans had removed, as they have the same ability to watch out for likely food sources from aloft and then to hover and plunge when they spy a suitable victim.

Most of the examples given so far to illustrate how human actions can lead to expansion in the numbers and distributions of certain animal species, have been used to make the point that such expansion frequently occurs as an incidental consequence of human activities. There are, of course, many ways in which people have intentionally and effectively promoted the expansion of particular species (table 3.9 presents data for some introduced mammals in Britain). This may sometimes be done deliberately to reduce the numbers of a species which has expanded as an unwanted consequence of human actions. Perhaps the best known exemplification of this is biological control using introduced predators and parasites. Thus an Australian insect, the cottony-cushion scale, *Icerya*

Table 3.9 Introduced mammals in Britain

Species	Date of introduction of present stock	Reason for introduction
House mouse *Mus musculus*	Neolithic?	Accidental
Wild goat *Capra hircus*	Neolithic?	Food
Fallow deer *Dama dama*	Roman or earlier	Food, sport
Domestic cat *Felis catus*	Early Middle Ages	Ornament, pest control
Rabbit *Oryctolagus cuniculus*	Mid to late twelfth century	Food, sport
Black rat *Rattus rattus*	Thirteenth century?	Accidental
Brown rat *Rattus norvegicus*	Early eighteenth century (1728–9)	Accidental
Sika deer *Cervus nippon*	1860	Ornament
Grey squirrel *Sciurus carolinensis*	1876	Ornament
Indian muntjak *Muntiacus muntjak*	1890	Ornament
Chinese muntjak *Muntiacus reevesi*	1900	Ornament
Chinese water deer *Hydropetes inermis*	1900	Ornament
Edible dormouse *Glis glis*	1902	Ornament
Musk rat *Ondatra zibethica*	1929 (–1937)	Fur
Coypu *Myocastor coypus*	1929	Fur
Mink *Mustela vison*	1929	Fur
Bennett's wallaby *Macropus rufogriseus bennetti*	1939 or 1940	Ornament
Reindeer *Rangifer tarandus*	1952	Herding, ornament
Himalayan (Hodgson's) porcupine *Hystrix hodgsoni*	1969	Ornament
Crested porcupine *Hystrix cristata*	1972	Ornament
Mongolian gerbil *Meriones unguiculatus*	1973	Ornament

Source: from Jarvis, 1979, table 1, p. 188

purchasi, was found in California in 1868. By the mid-1880s it was effectively destroying the citrus industry, but its ravages were quickly controlled by deliberately importing a parasitic fly and a predatory beetle (the Australian ladybird) from Australia. Because of the uncertain ecological effects of synthetic pesticides, such biological control has its attractions but the importation of natural enemies is not without its own risks. For example, the introduction into Jamaica of mongooses to control rats led to the undesirable decimation of many native birds and small land mammals.

One further calculated method of increasing the numbers of wild animals, of conserving them, and of gaining an economic return from them, is game cropping. In some circumstances, because they are better adapted to, and utilize more components of, the environment, wild ungulates provide an alternative means of land use to domestic stock. The exploitation of the saiga antelope, *Saiga tatarica* (plate 3.10) in the CIS is the most successful story of this kind (Edington and Edington, 1977). Hunting (much overdone because of the imagined medical properties of its horns) had led to its near demise, but this was banned in the 1920s and a system of controlled cropping was instituted. Under this regime (which produces appreciable quantities of meat and leather) total numbers have risen from about 1000 to over 2 million, and in the process the herds have reoccupied most of their original range.

Animal extinctions

As the size of human populations has increased and technology has developed, humans have been responsible for the extinction of many species of both birds and mammals. There appears to be a close correlation, for example, between the curve of population growth since the mid-seventeenth century and the curve of the number of species that have become extinct (figure 3.14).

Some workers, notably Martin (1967, 1974), believe, however, that the human role in animal extinctions goes back to the Stone Age and to the Late Pleistocene. They believe that extinction closely follows the chronology of the spread of prehistoric civilization and the development of big-game hunting. They would also maintain that there are no known continents or islands in which accelerated extinction definitely pre-dates the arrival of human settlements. An alternative interpretation, however, is that the Late Pleistocene extinctions of big mammals were caused by the rapid and substantial changes of climate at the termination of the last Glacial (Martin and Wright, 1967; Martin and Klein, 1984). Martin (1982) argues that the global pattern of extinctions of the large land mammals follows the footsteps of Palaeolithic settlements. He suggests that Africa and parts of Southern Asia were affected first, with substantial losses at the end of the Acheulean, around 200 000 years ago. Europe and northern Asia were affected between 20 000 and 10 000 years ago, while North and South America were stripped of large herbivores between 12 000 and 10 000 years ago. Extinctions continued into the Holocene on oceanic islands and in the Galápagos Islands, for example, virtually all extinctions took place after the first human contact in AD 1535 (Steadman et al., 1991). Likewise the complete deforestation of Easter Island in the Pacific between 1200 and 800 years BP was an ecological disaster that led to the demise of much of the native flora and fauna and also

Figure 3.14
Geometric increase in the human population (a) parallelled by increasing numbers of extinctions in birds and mammals (b) suggesting that as human numbers increase more and more living species become exterminated (after Ziswiler in National Academy of Sciences, 1972, figure 3.2)

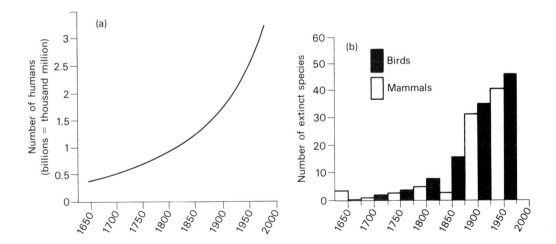

Plate 3.11
The inhabitants of Easter
Island who carved and erected
these famous monuments may,
through deforestation, have
sown the seeds of their own
demise

precipitated the decline of the megalithic civilization that had
erected the famous statues on the islands (Flenley et al., 1991).

The dates of the major episodes of Pleistocene extinction are
crucial to the discussion (table 3.10). Martin (1967: 111–15)
writes:

Radiocarbon dates, pollen profiles associated with extinct animal remains,
and new stratigraphic and archaeological evidence show that, depending
on the region involved, late-Pleistocene extinction occurred either after,
during or somewhat before worldwide climatic cooling of the last maxi-

Table 3.10 Dates of major episodes of generic extinction in the Late
Pleistocene

Area	Date (years BP)
North America	11 000
South America	10 000
West Indies	Mid-post Glacial
Australia	13 000
New Zealand	900
Madagascar	800
Northern Eurasia	13 000–11 000
Africa and South-East Asia (?)	50 000–40 000

Source: data in Martin, 1967: 111

mum . . . of glaciation . . . Outside continental Africa and South East Asia, massive extinction is unknown before the earliest known arrival of prehistoric man. In the case of Africa, massive extinction coincides with the final development of Acheulean hunting cultures which are widespread throughout the continent . . . The thought that prehistoric hunters ten to fifteen thousand years ago (and in Africa over forty thousand years ago) exterminated far more large animals than has modern man with modern weapons and advanced technology is certainly provocative and perhaps even deeply disturbing.

The nature of the human impact on animal extinctions can conveniently be classified into three types (Marshall, 1984): the 'blitzkrieg effect', which involves rapid deployment of human populations with big-game hunting technology so that there is very rapid demise of animal populations; the 'innovation effect', whereby long-established human population groups adopt new hunting technologies and erase fauna that have already been stressed by climatic changes; and the 'attrition effect', whereby extinction takes place relatively slowly after a long history of human activity because of loss of habitat and competition for resources.

The arguments that have been used in favour of an anthropogenic interpretation can be summarized as follows. First, massive extinction in North America seems to coincide in time with the arrival of humans in sufficient quantity and with sufficient technological skill in making suitable artefacts (the bifacial Clovis blades) to be able to kill large numbers of animals (Krantz, 1970). Second, in Europe the efficiency of Upper Palaeolithic hunters is attested by such sites as Solutré in France, where a late-Perigordian level is estimated to contain the remains of over 100 000 horses. Third, many beasts unfamiliar with people are remarkably tame and stupid in their presence, and it would have taken them a considerable time to learn to flee or seek concealment at the sight or scent of human life.

Plate 3.12
Prehistoric hunters may have been very effective at killing large numbers of mammals in relatively short periods of time. Upper Palaeolithic hunters from Solutré in France used superb 'laurel-leaf' points for this purpose

Fourth, humans, in addition to hunting animals to death, may also have competed with them for particular food or water supplies. Fifth, Coe (1981, 1982) has suggested that the supposed preferential extinction of the larger mammals could also lend support to the role of human actions. He argues that while large body size has certain definite advantages, especially in terms of avoiding predation and being able to cover vast areas of savanna in search of food, large body size also means that these herbivores are required to feed almost continuously to sustain a large body mass. Furthermore, as the size and generation time of a mammal increases so the rate at which they turn over their biomass decreases (figure 3.15). Hence a population of Thompson's gazelle will turn over up to 70 per cent of their biomass each year, a wildebeeste over 27 per cent, but a rhino only 10 per cent and the elephant just 9 per cent. The significance of this is that, since large mammals can only turn over a small percentage of their population biomass each year, the rate of slaughter that such a population can sustain in the face of even a primitive hunter is very low indeed.

Certain objections have been levelled against the climatic change model and these tend to support the anthropogenic model. It has been suggested, for instance, that changes in climatic zones are generally sufficiently gradual for beasts to be able to follow the shifting vegetation and climatic zones of their choice. Similar environments are available in North America today as were present, in different locations and in different proportions, during Late Pleistocene times. Second, it can be argued that the climatic

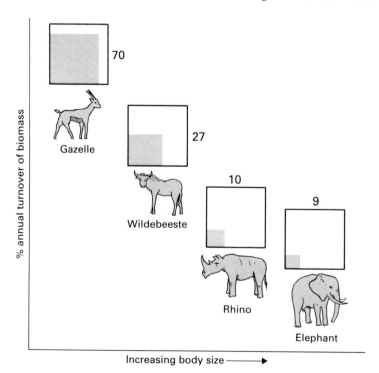

Figure 3.15
Decreasing rate of population biomass turnover with increasing body size in warm- and cold-blooded vertebrates (from Coe, 1982)

changes associated with the multiple glaciations, interglacials, pluvials and interpluvials do not seem to have caused the same striking degree of elimination as those in the Late Pleistocene. A third difficulty with the climatic-cause theory is that animals like the mammoth occupied a broad range of habitats from arctic to tropical latitudes, so that it is unlikely all would perish as a result of a climatic change (Martin, 1982).

This is not to say, however, that the climatic hypothesis is without foundation or support. The migration of animals in response to rapid climatic change could be halted by geographical barriers such as high mountain ranges or seas. The relatively rich state of the African big mammalian fauna is due, according to this point of view, to the fact that the African biota is not, or was not, greatly restricted by an insuperable geographical barrier. Another way in which climatic change could cause extinction is through its influence on disease transmission. It has been suggested that during glacials animals would be split into discrete groups cut off by ice sheets but that, as the ice melted (before 11 000 BP in many areas), contacts between groups would once again be made enabling diseases to which immunity might have been lost during isolation to spread rapidly. Large mammals, because of their low reproduction rates, would recover their numbers only slowly, and it was large mammals (according to Martin) that were the main sufferers in the Late Pleistocene extinctions.

The detailed dating of the European megafauna's demise lends further credence to the climatic model (Reed, 1970). The Eurasiatic boreal mammals, such as mammoth, woolly rhinoceros, musk ox and steppe bison, were associated with and adapted to the cold steppe which was the dominant environment in northern Europe during the glacial phases of the last Glaciation. Each of these forms, especially the mammoth and the steppe bison, had been hunted by humans for tens of thousands of years, yet managed to survive through the last Glacial. They appear to have disappeared, according to Reed, within the space of a few hundred years when warm conditions associated with the Allerød interstadial led to the restriction and near disappearance of their habitat.

Grayson (1977) has added to the doubts expressed about the anthropogenic overkill hypothesis. He suggests that the overkill theory, because it states that the end of the North American Pleistocene was marked by extraordinary high rates of extinction of mammalian genera, requires terminal Pleistocene mammalian generic extinctions to have been relatively greater than generic extinctions within other classes of vertebrates at this time. When he examined the extinction of birds he found that an almost exactly comparable proportion became extinct at the end of the Pleistocene as one finds for the megafauna. Moreover, as the radiocarbon dates for early societies in countries like Australia are pushed back, it becomes increasingly clear that humans and several species of megafauna were living together for quite long periods, thereby undermining the notion of rapid overkill (Gillespie et al., 1978).

Further arguments can be marshalled against the view that humans as predators played a critical role in the Late Pleistocene extinctions in North America (Butzer, 1972: 509–10). There are, for example, relatively few Palaeo-Indian sites over an immense area, and the majority of these have a very limited cultural inventory. In addition there is no clear evidence that Palaeo-Indian subsistence was necessarily based, in the main part, on big-game hunting; if it was, only two genera were hunted intensively: mammoth and bison. A final point that militates against the argument that humans were primarily responsible for the waves of extinction is the survival of many big-game species well into the nineteenth century, despite a much larger and more efficient Indian population.

Thus the human role in the great Late Pleistocene extinction is still a matter of debate. The problem is complicated because certain major cultural changes in human societies may have occurred in response to climatic change. The cultural changes may have assisted in the extinction process, and increasing numbers of technologically competent humans may have delivered in the final *coup de grâce* to isolated remnants already doomed by rapid post-Glacial environmental changes (Guilday, 1967). In this context it is worth noting that Haynes (1991) has suggested that at around 11 000 years ago conditions were dry in the interior of the USA and that Clovis hunters may as a result have found large game animals easier prey when concentrated around waterholes and under stress.

Actual extinction may have occurred concurrently with a dwarfing in the size of animals, and this too is a subject of controversy. On the one hand, there are those who believe that the dwarfing could result from a reduction in food availability brought about by climatic deterioration. On the other, it can be maintained that this phenomenon derives from the fact that small animals, being more adept at hiding and being a less attractive target for a hunter, are more likely to survive human predators, so that a genetic selection towards reduced body size takes place (Marshall, 1984).

Although the debate about the cause of this extinction spasm has now persisted for a long time, there are still great uncertainties. This is particularly true with regard to the chronologies of extinction and of human colonization. This is particularly so in the context of North America, as noted by Grayson (1988: 118):

While most of us dealing with the extinctions issue have assumed that all of the extinctions occurred between 12,000 and 10,000 years ago, and were, by that assumption, drawn into thinking that the extinction event had occurred rapidly, there is surprisingly little hard evidence to support that assumption. Much effort needs to be placed into the construction of a detailed extinction chronology. Did the extinctions occur in a brief – say 500 year – period of time, with no major decreases in population numbers before then, or were they scattered across the millennia, with major population increases leading slowly to the final losses? Were mammoths so greatly diminished in numbers by 12,000 years ago that Clovis peoples could have made all the difference 500 years later?

What a difference to our proposed explanation if, as the current chronology perhaps suggests and surely does not deny, the extinctions were well under way prior to Clovis times.

The only way I can conceive of showing overkill to be incorrect is by showing that significant losses had already occurred by the time people had arrived south of the glacial ice. Even then, overkill would be falsified only for those taxa already extinct by the time of human occupation. Unfortunately, the statement that 35 genera of mammals became extinct in North America at about 11,000 years ago has about the same status as the statement that Clovis represents the first human occupation of the New World south of the glacial ice. We simply do not know if either is true, and there are reasons to suspect that both are false.

It is probable that the Late Pleistocene extinctions themselves had major ecological consequences. As Birks (1986: 49) has put it:

The ecological effects of rapid extinction of over 75% of the New World's large herbivores ... must have been profound, for example on seed dispersal, browsing, grazing, trampling and tree regeneration ... large grazers and browsers such as bison, mammoth and woolly rhinoceros may have been important in delaying or even inhibiting tree growth.

When considering modern-day extinctions the role of human agency is much less controversial. None the less, some modern extinctions of species are natural. Extinction is a biological reality: it is part of the process of evolution. In any period, including the present, there are naturally doomed species, which are bound to disappear, either through over-specialization or an incapacity to adapt themselves to climatic change and the competition of others, or because of natural cataclysms such as earthquakes, eruptions and floods. Fisher et al. (1969) believe that probably one-quarter of the species of birds and mammals that have become extinct since 1600 may have died out naturally. In spite of this, however, the rate of animal extinctions brought about by humans in the last 400 years is of a very high order when compared to the norm of geological time. In 1600 there were approximately 4226 living species of mammals: since then 36 (0.85 per cent) have become extinct; and at least 120 of them (2.84 per cent) are presently believed to be in some (or great) danger of extinction. A similar picture applies to birds. In 1600 there were about 8684 living species; since then 94 (1.09 per cent) have become extinct; and at least 187 of them (2.16 per cent) are presently, or have very lately been, in danger of extinction (Fisher et al., 1969).

Some beasts appear to be more prone to extinction than others, and a distinction is now often drawn between r-selected species and K-selected species. These are two ends of a spectrum. The former have a high rate of increase, short gestation periods, quick maturation, and the advantage that they have the ability either to react quickly to new environmental opportunities or to make use of transient habitats (such as seasonal ponds). The life duration of individuals tends to be short, populations tend to be unstable, and the species may over-exploit their environment to their eventual

Plate 3.13
In recent years many
populations of African
elephants have suffered
extreme reductions in their
numbers as a result of
poaching for ivory. The
situation is less grave in those
countries of southern and
central Africa where
conservation policies have
been rigorously enforced (e.g.
South Africa, Zimbabwe and
Botswana), but because of
their large territorial
requirements, long gestation
periods, and small litter sizes,
elephants have many qualities
that make them especially
susceptible to extinction

detriment. Many pests come into this category. The other end of the
spectrum, the *K*-selection species, are those which tend to be
endangered or to become extinct. Their prime characteristics are
that they are better adapted to physical changes in their environment (such as seasonal fluctuations in temperature and moisture)
and live in a relatively stable environment (Miller and Botkin,
1974). These species tend to have much greater longevity, longer
generation times, fewer offspring, but a higher probability of
survival of young and adults. They have traded a high rate of
increase and the ability to exploit transient environments for the
ability to maintain more stable populations with low rates of
increase, but correspondingly low rates of mortality and a closer
adjustment to the long-term capacity of the environment to support
their population.

Table 3.11, which is modified after Ehrenfield (1972), attempts
to bring together some of the characteristics of species which affect
their survival in the human world.

Possibly one of the most fundamental ways in which humans are
causing extinction is by reducing the *area* of natural habitat
available to a species. Even wildlife reserves tend to be small
'islands' in an inhospitable sea of artificially modified vegetation or
urban sprawl. We know from many of the classic studies in true
island biogeography that the number of species living at a particular location is related to area (see figure 3.16 for some examples);
islands support fewer species than do similar areas of mainland,
and small islands have fewer species than do large ones. Thus it
may well follow that if humans destroy the greater part of a vast
belt of natural forest, leaving just a small reserve, initially it will be
'supersaturated' with species, containing more than is appropriate
to its area when at equilibrium (Gorman, 1979). Since the population sizes of the species living in the forest will now be greatly

Table 3.11 Characteristics of species affecting survival

Endangered	Safe
Large size	Small size
Predator	Grazer, scavenger, insectivore
Narrow habitat tolerance	Wide habitat tolerance
Valuable fur, oil, hide, etc.	Not a source of natural products
Restricted distribution	Broad distribution
Lives largely in international waters	
Migrates across international boundaries	Lives largely in one country
Reproduction in one or two vast aggregates	Reproduction by solitary pairs or in many small aggregates
Long gestation period	Short gestation period
Small litters	Big litters and quick maturation
Behavioural idiosyncracies that are non-adaptive today	Adaptive
Intolerance of the presence of humans	Tolerance of humans
Dangerous to humans, livestock, etc.	Perceived as harmless

reduced, the extinction rate will increase and the number of species will decline towards equilibrium. For this reason it may be a sound principle to make reserves as large as possible; a larger reserve will support more species at equilibrium by allowing the existence of larger populations with lower extinction rates. Several small reserves will plainly be better than no reserves at all, but they will tend to hold fewer species at equilibrium than will a single reserve of the same area (figure 3.17). If it is necessary to have several small reserves, they should be placed as close to each other as possible so that each may act as a source area for the others. In this way their equilibrial number of species will be raised due to increased immigration rates.

There are situations when small reserves may have advantages over a single large reserve: they will be less prone to total decimation by some natural catastrophe such as fire; they may allow the the preservation of a range of rare and scattered habitats; and they may allow the survival of a group of competitors one of which would exclude the others from a single reserve.

Reduction in area leads to reduction in numbers, and this in turn can lead to genetic impoverishment through inbreeding (Frankel, 1984). The effect on reproductive performance appears to be particularly marked. Inbreeding degeneration is, however, not the only effect of small population size for, in the longer term, the depletion of genetic variance is more serious since it reduces the capacity for adaptive change. Space is therefore an important consideration, especially for those animals that require large expanses of territory. For example, the population density of the wolf is about one adult per 20 km^2, and it has been calculated that for a viable population to exist, one might need 600 individuals

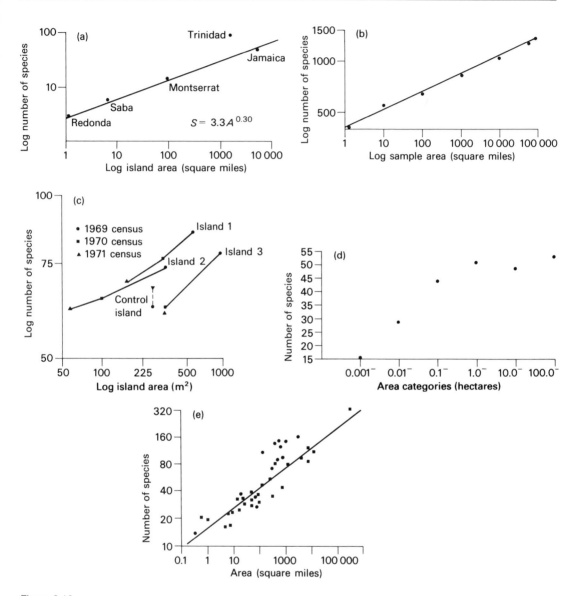

Figure 3.16
Some relationships between the size of 'islands' and numbers of species (after Gorman, 1979, figures 3.1, 3.2, 8.1 and 8.2):
(a) The number of amphibian and reptile species living on West Indian islands of various sizes. Trinidad, joined to South America 10 000 years ago, lies well above the area–species curve of other islands.
(b) The area–species curve for the number of species of flowering plants found in a sample area of England
(c) The effect on the number of arthropod species of reducing the size of mangrove islands. Islands 1 and 2 were reduced after both the 1969 and the 1970 census. Island 3 was reduced only after the 1969 Census. The control island was not reduced, the changes in species number being attributable to random fluctuation
(d) The number of species of birds living in British woods of various size categories
(e) The number of species of land-birds living in the lowland rain forest on small islands of New Guinea, plotted as a function of island area. The squares represent islands which have not had a land connection to New Guinea and whose avifaunas are at equilibrium. The regression line is fitted through these points. The circles represent former land-bridge islands connected to New Guinea some 10 000 years ago. Note that the large ones have more species than one would expect for their size

ranging over an area of 12 000 km^2. The significance of this is apparent when one realizes that most nature reserves are small: 93 per cent of the world's national parks and reserves have an area less than 5000 km^2, and 78 per cent less than 1000 km^2.

Equally, range loss, the shrinking of the geographical area in which a given species is found, often marks the start of a downward spiral towards extinction. Such a contraction in range may result from habitat loss or from such processes as hunting and capture. Particular concern has been expressed in this context about the pressures on primates, notably in South-East Asia. Of 44 species, 33 have lost at least half their natural range in the region. In two cases, those of the Javan leaf monkey and the Javan and grey gibbons, the loss of range is no less than 96 per cent. Recent figures produced by the International Union for the Conservation of Nature and the United Nations Environment Programme for wildlife habitat loss show the severity of the problem (table 3.12). In the Indomalayan countries 68 per cent of the original wildlife habitat has been lost, and the comparable figure for tropical Africa is 65 per cent. In these regions, only Brunei and Zambia have lost less than 30 per cent of their original habitat, while at the other end of the spectrum, Bangladesh, the most densely populated large country in the world, has suffered a loss of 94 per cent.

The rate of future extinctions is the cause of very considerable concern. As the Global 2000 report (Council on Environmental Quality, 1982) has suggested (pp. 327–9),

If present trends continue – as they certainly will in many areas – hundreds of thousands of species can be expected to be lost by the year 2000 . . . the extinctions projected for the coming decades will be largely human-generated and on a scale that renders natural extinction trivial by comparison. Efforts to meet human needs and rising expectations are likely to lead to the extinction of between one-fifth and one-seventh of all species (of plants and animals) over the next two decades.

The great bulk of these extinctions will take place because of the removal of areas of humid tropical moist forest, and the largest number of extinctions can be expected in the insect order.

According to Myers (1979: 31) 'during the last quarter of this century we shall witness an extinction spasm accounting for 1 million species'. This is a considerable proportion of the estimated number of species living in the world today, for which Myers gives a figure of between 3 and 9 million. He has calculated that from AD 1600 to 1900 humans were accounting for the demise of one species every four years, that from 1900 onwards the rate increased to an average of around one per year, that at present the rate is about one per day, and that within a decade we could be losing one every hour. By the end of the century, our planet could lose anywhere between from 20 to 50 per cent of its species (Lugo, 1988). The need to maintain *biodiversity* has become one of the crucial issues with which we must contend.

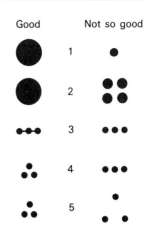

Figure 3.17
A set of general design rules for nature reserves based on theories of island biogeography. The designs on the left are preferable to those on the right because they should enjoy lower rates of species extinction:
(1) A large reserve will hold larger populations with lower probabilities of going extinct than will a small reserve
(2) A single reserve is preferable to a series of smaller reserves of equal total area, since these will support only small populations with relatively high probabilities of going extinct
(3–5) If reserves must be fragmented, then they should be connected by corridors of similar vegetation or be placed equidistant and as near to each other as possible. In this way immigration rates between the fragments will be increased, thereby maximizing the chances of extinct populations being replaced from elsewhere in the reserve complex (after Gorman, 1979, figure 8.4)

Table 3.12 Wildlife habitat loss

(a) In Indomalayan nations

	Original wildlife habitat (km²)	Amount remaining (km²)	Habitat loss (%)
Bangladesh	1 142 777	68 567	94
Bhutan	34 500	22 770	34
Brunei	5764	4381	24
Burma	774 817	225 981	71
China	423 066	164 996	61
Hong Kong	1066	32	97
India	3 017 009	615 095	80
Indonesia	1 446 433	746 861	49
Japan	320	138	57
Kampuchea	180 879	43 411	76
Laos	236 746	68 656	71
Malaysia	356 254	210 190	41
Nepal	117 075	53 855	54
Pakistan	165 900	39 816	76
Philippines	308 211	64 724	79
Sri Lanka	64 700	10 999	83
Taiwan	36 961	10 719	71
Thailand	507 267	130 039	74
Vietnam	332 116	66 423	80
TOTAL	8 169 860	2 487 683	68

(b) In Afrotropical nations, 1986

	Original wildlife habitat (km²)	Amount remaining (km²)	Habitat loss (%)
Angola	1 246 700	760 847	39
Benin	115 800	46 320	60
Botswana	585 400	257 576	56
Burkina Faso	273 800	54 760	80
Burundi	25 700	3598	86
Cameroon	469 400	192 454	59
Central African Rep.	623 000	274 120	56
Chad	720 800	172 992	76
Congo	342 000	172 420	49
Côte d'Ivoire	318 000	66 780	79
Djibouti	21 800	11 118	49
Equatorial Guinea	26 000	12 740	51
Ethiopia	1 101 003	30 300	70
Gabon	267 000	173 550	35
Gambia	11 300	1243	89
Ghana	230 000	46 000	80
Guinea	245 900	73 770	70
Guinea Bissau	36 100	7942	78
Kenya	569 500	296 140	48
Lesotho	30 400	9728	68
Liberia	111 400	14 482	87
Madagascar	595 211	148 803	75
Malawi	94 100	40 463	57
Mali	754 100	158 361	79
Mauritania	388 600	73 834	81
Mozambique	783 203	36 776	57
Namibia	823 200	444 528	46
Niger	566 000	127 880	77
Nigeria	919 800	229 950	75
Rwanda	25 100	3263	87
Senegal	196 200	35 316	82
Sierra Leone	71 700	10 755	85
Somalia	637 700	376 243	41
South Africa	1 236 500	531 695	57
Sudan	1 703 000	510 900	70
Swaziland	17 400	7656	56
Tanzania	886 200	505 134	43
Togo	56 000	19 040	66
Uganda	193 700	42 614	78
Zaire	2 335 900	1 051 155	55
Zambia	752 600	534 346	29
Zimbabwe	390 200	171 688	56
TOTAL	20 797 441	8 340 920	65

Source: IUCN/UNEP data in *World Resources 1988–9*, tables 6.4 and 6.5

FOUR

The Human Impact on the Soil

Introduction

Humans live close to and depend on the soil. It is one of the thinnest and most vulnerable human resources and is one upon which, both deliberately and inadvertently, humans have had a very major impact. Moreover, such an impact can occur with great rapidity in response to land-use change, new technologies or waves of colonization (see Russell and Isbell, 1986, for a review in the context of Australia).

Natural soil is the product of a whole range of factors and the classic expression of this is that of Jenny (1941):

$$s = f(cl, o, r, p, t \ldots)$$

where s denotes any soil property, cl is the regional climate, o the biota, r the topography, p the parent material, t the time (or period of soil formation), and the dots represent additional, unspecified factors. In reality soils are the product of highly complex interactions of many interdependent variables, and the soils themselves are not merely a passive and dependent factor in the environment. None the less, following Jenny's subdivision of the classic factors of soil formation, one can see more clearly the effects humans have had on soil be it detrimental or beneficial. These can be summarized as follows (adapted from the work of Bidwell and Hole, 1965):

1 *Parent Material*
   Beneficial: adding mineral fertilizers; accumulating shells and bones; accumulating ash locally; removing excess amounts of substances such as salts.
   Detrimental: removing through harvest more plants and animal nutrients than are replaced; adding materials in amounts toxic to plants or animals; altering soil constituents in a way to depress plant growth.
2 *Topography*
   Beneficial: checking erosion through surface roughening,

land forming and structure building; raising land level by accumulation of material; land levelling.
Detrimental: causing subsidence by drainage of wetlands and by mining; accelerating erosion; excavating.

3 *Climate*
Beneficial: adding water by irrigation; rainmaking by seeding clouds; removing water by drainage; diverting winds, etc.
Detrimental: subjecting soil to excessive insolation, to extended frost action, to wind, etc.

4 *Organisms*
Beneficial: introducing and controlling populations of plants and animals; adding organic matter including 'nightsoil'; loosening soil by ploughing to admit more oxygen; fallowing; removing pathogenic organisms, e.g. by controlled burning.
Detrimental: removing plants and animals; reducing organic content of soil through burning, ploughing, overgrazing, harvesting, etc.; adding or fostering pathogenic organisms; adding radioactive substances.

5 *Time*
Beneficial: rejuvenating the soil by adding fresh parent material or through exposure of local parent material by soil erosion; reclaiming land from under water.
Detrimental: degrading the soil by accelerated removal of nutrients from soil and vegetation cover; burying soil under solid fill or water.

Space precludes, however, that we can follow all these aspects of anthropogenic soil modification or, to use the terminology of Yaalon and Yaron (1966), of *metapedogensis*.

We will therefore concentrate on certain highly important changes which humans have brought about, especially chemical changes (such as salinization and lateritization), various structural changes (such as compaction), some hydrological changes (including the effects of drainage and the factors leading to peat-bog development), and, perhaps most important of all, soil erosion.

Salinity: natural sources

Many semi-arid and arid areas are naturally salty. By definition they are areas of substantial water deficit where evapotranspiration exceeds precipitation. Thus, whereas in humid areas there is sufficient water to percolate through the soil and to leach soluble materials from the soil and the rocks into the rivers and hence into the sea, in deserts this is not the case. Salts therefore tend to accumulate. This tendency is exacerbated by the fact that many desert areas are characterized by closed drainage basins (endoreic drainage) which act as terminal evaporative sumps for rivers.

Table 4.1 Soluble salt content of rivers in semi-arid areas

River	Total dissolved solids (parts per million)
Indus	250–300
Colorado	795
White Nile	174
Tigris–Euphrates	200–400
Niger	<60–80

The amount of natural salinity varies according to numerous factors, one of which is the source of salts. Some of the salts are brought in to the deserts by rivers (table 4.1), though it needs to be pointed out that most of these rivers do not have particularly high salt contents (generally less than 300 ppm). A second source of salts is the atmosphere – a source which in the past has often been accorded insufficient importance. Rainfall, coastal fogs and dust storms all transport significant quantities of soluble salts. This is shown by the data in table 4.4 for some forested catchments in south-western Australia, where up to 130 kg ha^{-1} yr^{-1} of chlorides are introduced in precipitation. Further soluble salts may be derived from the weathering and solution of bedrock. In the Middle East, for example, notably in Iran, there are extensive salt domes and evaporite beds within the bedrock which create locally high ground-water and surface-water salinity levels. In other areas, such as the Rift Valley of southern Ethiopia and northern Kenya, volcanic rocks may provide a large source of sodium carbonate (trona) to ground waters, while elsewhere the rocks in which ground water occurs may contain salt because they are themselves ancient desert sediments. Even in the absence of such localized sources of highly saline ground water it needs to be remembered that over a period of time most rocks will provide soluble products to ground water, and in a closed hydrological system such salts will eventually accumulate to significant levels.

A further source of salinity may be marine transgressions. At times of higher sea-levels, it has sometimes been proposed (see, for example, Godbole, 1972) that salts would have been laid down by the sea. Likewise in coastal areas, salts in ground-water aquifers may be contaminated by contact with sea water.

Human agency and increased salinity

It has already been stated that salinity is a normal characteristic of many desert areas. However, humans have increased the extent and degree of salinity in numerous different ways.

The extension of irrigation (see table 4.2) and the use of a wide range of different techniques (see table 4.3) for water abstraction and application, can lead to a build-up of salt levels in the soil

Table 4.2 Irrigated area in major regions (in thousand hectares)

Region	1961–5	1968	1978	% change 1968–78
World	149 474	162 634	200 913	23.5
Africa	5870	6772	7831	15.6
North and Central America	18 606	20 366	23 543	15.6
South America	4862	5403	6663	23.3
Asia	100 363	108 184	130 950	21.0
Europe	8957	10 266	13 670	33.2
Oceania	1198	1443	1656	14.8
USSR	9618	10 200	16 600	62.7

Source: Food and Agricultural Organization of the United Nations, 1980

through the mechanism of raising ground water so that it is near enough to the ground surface for capillary rise and subsequent evaporative concentration to take place. In the case of the semi-arid northern plains of Victoria in Australia, for instance, the water-table has been rising at around 1.5 m y^{-1} so that now in many areas it is almost within 1 m of the surface (figure 4.1). When ground water comes within 3 m of the surface in clay soils, but less for silty and sandy soils, capillary forces bring moisture to the surface where evaporation takes place (Currey, 1977). A survey of the problem in south-eastern Australia is provided by Grieve (1987).

Second, many irrigation schemes require the addition of large quantities of water over the soil surface. This is especially true for rice cultivation. Such surface water is readily evaporated so that salinity levels build up.

Third, the construction of large dams and barrages to control water flow and to give a head of water creates large reservoirs from which further evaporation can take place.

Fourth, notably in areas of soils with high permeability, water seeps laterally and downwards from irrigation canals so that further evaporation takes place (figure 4.2). Many distribution channels in a gravity scheme are located on the elevated areas of a flood plain or riverine plain to make maximum use of gravity. The elevated landforms selected are natural levées, river-bordering dunes and terraces, all of which are composed of silt and sand which may be particularly prone to seepage loss.

In coastal areas salinity problems are created by sea-water incursion brought about by over-pumping. This can be explained as follows. Fresh water has a lower density than salt water, such that a column of sea water can support a column of fresh water approximately 2.5 per cent higher than itself (or a ratio of about 40:41). So where a body of fresh water has accumulated in a reservoir rock which is also open to penetration from the sea, it does not simply lie flat on top of the salt water but forms a lens, whose thickness is approximately forty-one times the elevation of the piezometric

Table 4.3 Irrigation systems and methods

(a) IRRIGATION SYSTEMS
   i *Sources of water supply*
      1 Rivers
      2 Springs, artesian wells
      3 Rain water
      4 Ground water

   ii *Methods of raising water*
      1 Manual methods
         (a) Draw well, e.g. *shaduf* (Arabic), *cigonal* (Spanish)
      2 Worked by animals
         (a) Inclined plane, e.g. *gird* (Arabic)
         (b) Geared wheel, e.g. *noria* (Spanish/Arabic), *nora* (Portuguese), *sakien* (Arabic)
      3 Automatic devices
         (a) Impounding the flow of a river by a dam or weir
         (b) Kanat
         (c) Water wheels, e.g. *noria* (Spanish/Arabic), *naûra* (Arabic)
         (d) Windmill pumps
      4 Motor pumps
         (a) Diesel pumps
         (b) Electric pumps
         (c) Steam pumps

   iii *Water storage*
      1 Reservoirs
      2 Ponds
      3 Cisterns
      4 Tanks
      5 Water towers

   iv *Water distribution (supply and drainage)*
      1 Irrigation canals
         (a) Main canals
         (b) Secondary canals
         (c) Distributory canals
         (d) Furrows
      2 Drainage canals
         (a) Main drainage
         (b) Secondary drainage
         (c) Collector drains
      3 Levelling or terracing

(b) IRRIGATION METHODS
   i *Surface irrigation*
      1 Dam or barrage
      2 Dam/barrage and channel irrigation
      3 Channel irrigation
      4 Spraying

   ii *Underground irrigation*

Source: modified after Manshard, 1974, figure 5.3, p. 68

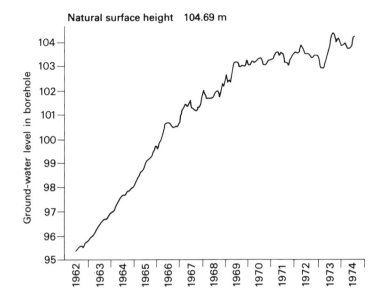

Natural surface height   104.69 m

Figure 4.1
Piezometer record to show
ground-water level changes in
the Murray Valley, Victoria,
Australia, from 1962 to 1974 as
a result of the extension of
irrigation (after Currey, 1977,
figure 2.13)

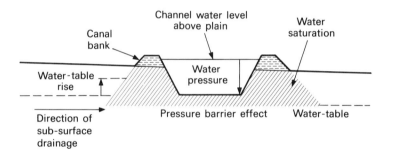

Figure 4.2
The effect of an irrigation
channel in causing a local rise
in the water-table, through the
development of a pressure
barrier which influences the
movement of sub-surface
drainage (after Currey, 1977,
figure 2.12)

surface above sea-level. This is called the Ghyben–Herzberg principle (see figure 4.3). The corollary of this rule is that if the
hydrostatic pressure of the fresh water falls as a result of over-
pumping in a well, then the underlying salt water will rise by forty
units for every unit by which the fresh-water table is lowered.

This problem presents itself on the coast of Israel, in California
(Banks and Richter, 1953), on the island of Bahrain, and in some of
the small coastal dune aquifers of the United Arab Emirates. A
comparable situation arises in the case of the Nile Delta where, as a
result of the construction of the Aswan Dam, ground-water levels
have dropped downstream, leading to the intrusion of coastal salt
water which has salinated the soil above, thus rendering it less
suitable for cultivation. This is one of the prices which have to be
paid for the undoubted gains brought by rationalizing the flow of
the Nile so that little of the water is wasted by draining out to sea.

The chemical changes in ground-water quality brought about by
sea-water incursion may also be considerable. An example from the

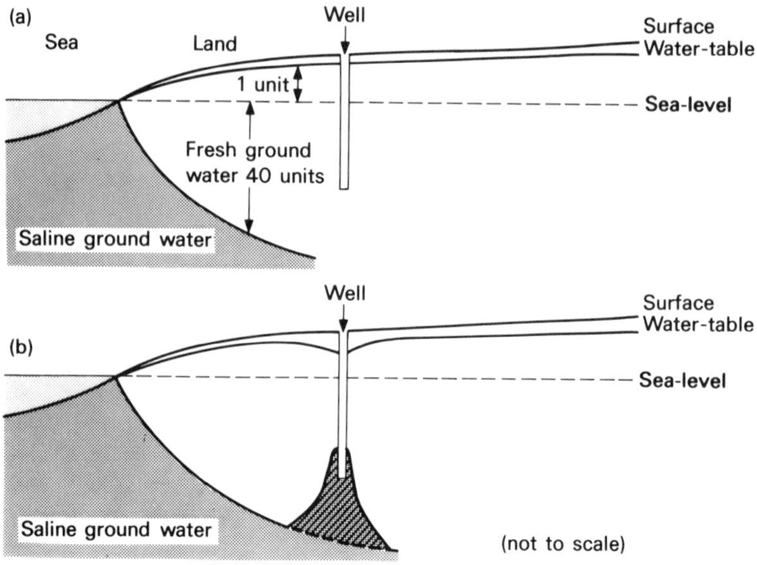

Figure 4.3
(a) The Ghyben–Herzberg relationship between fresh and saline ground water
(b) The effect of excessive pumping from the well. The diagonal hatching represents the increasing incursion of saline water (after Goudie and Wilkinson, 1977, figure 63)

Salinas valley of California is shown in figure 4.4. The most notable feature, apart from the general overall increase in the total dissolved solids content by approximately tenfold, is the change in the ratio of chloride to bicarbonate. As the coast is approached, as one might expect, the quantity of chloride relative to bicarbonate rises from a ratio of 0.5 to around 200.

The problem of sea-water incursion as a result of over-pumping of aquifers is not, however, restricted to arid and semi-arid areas. It was noted as early as the middle of the last century in London and Liverpool, England, and there are now many records of this phenomenon in Germany, the Netherlands, Japan and the eastern seaboard of the USA (Todd, 1959).

Increases in soil salinity are not restricted to irrigated areas. In certain parts of the world salinization has resulted from vegetation clearance (Peck, 1978). The removal of native forest vegetation allows a greater penetration of rainfall into deeper soil layers which causes ground-water levels to rise creating seepage of sometimes saline water in low-lying areas. Through this mechanism an estimated 2×10^5 hectares of land in southern Australia, which at the start of European settlement supported good crops of pasture, is now suitable only for halophytic species. Similar problems exist also in North America, notably in Manitoba, Alberta, Montana and North Dakota.

The clearance of the native evergreen forest (predominantly *Eucalyptus* forest) in south-western Australia has led both to an increase in recharge rates of ground water (table 4.4), and to an increase in the salinity of the streams.

The speed and extent of ground-water rise following such clearance of forest is shown in figure 4.5. Until late 1976, both the Wights and the Salmon catchments were forested, but then Wights

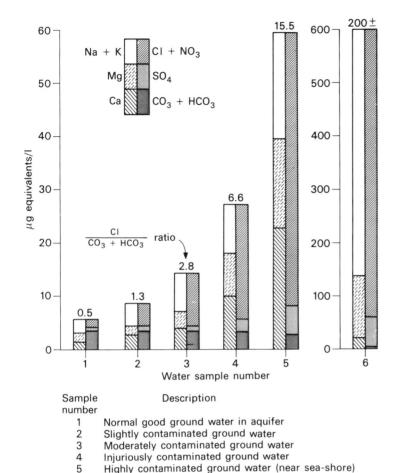

Figure 4.4
Evidence of salt-water intrusion exemplified by chemical analyses of a line of well-waters from the Salinas valley, California, extending from the centre of the aquifer to the coast.
Chloride–bicarbonate ratios are shown above each quality diagram (after Simpson in Todd, 1959, figure 12.14)

Sample number	Description
1	Normal good ground water in aquifer
2	Slightly contaminated ground water
3	Moderately contaminated ground water
4	Injuriously contaminated ground water
5	Highly contaminated ground water (near sea-shore)
6	Sea water

was cleared. Whereas prior to late 1976 both catchments showed a similar pattern of ground-water fluctuation, after that date there was a marked divergence, amounting to 5.7 m (Peck, 1983). Comparable studies have been undertaken in the State of Victoria (Williamson, 1983).

Replanting of the eucalyptus cover has been found to cause ground-water levels to fall and ground-water salinity to decline. The recovery can be swift. In Western Australia Bari and Schofield (1992) found that after a cleared area was reforested the level of the water-table fell by over 5 m in just nine years. At the same time ground-water salinity declined by 11 per cent.

The spread of salinity

Anthropogenic inceases in salinity are not new. Jacobsen and Adams (1958) have shown that they were a problem in Mesopotamian agriculture after about 2400 BC. Individual fields, which in

Table 4.4 Hydrological and chemical data for south-western Australia, illustrating the effects of deforestation on ground-water recharge and river-water chemistry

Catchment (type or name)	Chloride input in precipitation $(kg\ ha^{-1}\ yr^{-1})$	Chloride loss in stream flow $(kg\ ha^{-1}\ yr^{-1})$	Recharge before clearing $(mm\ yr^{-1})$	Recharge with farming $(mm\ yr^{-1})$
(a) Forested				
Julimar	53	78	–	–
Seldom seen	120	160	–	–
More seldom seen	120	140	–	–
Waterfall gully	110	180	–	–
North Dandalup	130	180	–	–
Davies	130	140	–	–
Yarragil	97	130	–	–
Harris	84	130	–	–
(b) Farmland				
Brockman	80	340	8	73
Wooroloo	78	420	4	61
Dale	24	460	0.8	24
Hotham	48	370	2	26
Williams	31	650	1	37
Collie East	50	740	2	60
Brunswick	110	350	70	500

Source: from data in Peck, 1975

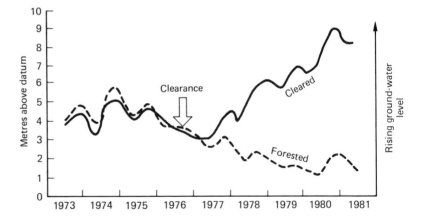

Figure 4.5
Comparison of hydrographs recorded from the boreholes in Wights (——) and Salmon (- - -) catchments in Western Australia. Both catchments were forested until late in 1976 when Wights was cleared (modified after Peck, 1983, figure 1)

2400 BC were registered as salt-free, can be seen in the records of ancient temple surveyors to have developed conditions of sporadic salinity by 2100 BC. Further evidence is provided by crop choice, for the onset of salinization strongly favours the adoption of crops which are most salt-tolerant. Counts of grain impressions in excavated pottery from sites in southern Iraq dated at about 3500 BC suggest that at that time the proportions of wheat and barley were nearly equal. A little more than 1000 years later the less salt-tolerant wheat accounted for less than 20 per cent of the crop, while by about 2100 BC it accounted for less than 2 per cent of the crop. By 1700 BC the cultivation of wheat had been abandoned completely in the southern part of the alluvial plain.

These changes in crop choice were accompanied by serious declines in yield which can also probably be attributed to salinity. At 2400 BC the yield was 2537 litres ha^{-1}, by 2100 BC it was 1460, and by 1700 BC it was down to 897. It seems likely that this played an important part in the break-up of Sumerian civilization.

Another area where salinization has posed a severe problem is the Indus valley of Pakistan, where Sir John Marshall, the archaeologist, described the salt as 'a Satanic mockery of snow'. Snelgrove (1967) reports that of the 25 million hectares of arable land in the country, 4.6 million are now mostly waterlogged or poorly drained, 1.9 million are predominantly severely saline, 4.5 million have saline patches, and some 0.4 million are being lost yearly through the spread of these conditions. The problem has largely followed the rise of ground-water levels by $0.15–0.60$ m yr^{-1} brought about by the building of major irrigation systems in areas of somewhat sandy, silty and permeable soils during the last hundred or so years.

In summary, it has been estimated that the percentage of salt-affected and waterlogged soils amounts to 50 per cent of the irrigated area in Iraq, 23 per cent of all Pakistan, 50 per cent in the Euphrates valley of Syria, 30 per cent in Egypt and over 15 per cent in Iran (Worthington, 1977: 30).

On a global basis, the calculations of Rozanov et al. (1991) make grim reading. They estimate (p. 120): 'From 1700 to 1984, the global area of irrigated land increased from 50,000 to 2,200,000 km^2, while at the same time some 500,000 km^2 were abandoned as a result of secondary salinization.'

Consequences of salinity

One consequence of the evaporative concentration of salts, and the pumping of saline waters back into rivers and irrigation canals from tubewells and other sources, is that river waters leading from

Plate 4.1
The extension of irrigation in the Indus valley of Pakistan by means of large canals has caused widespread salination of the soils. Waterlogging is also prevalent. The white efflorescence of salt in the fields has been termed 'a Satanic mockery of snow'

Table 4.5 Changes in water hardness brought about by irrigation in the USA

Location	Total hardness as mg 1^{-1} of $CaCo_3$		
	above irrigation area	below irrigation area	Increase
Rio Grande, Texas	111	631	× 5.7
Yakima, Washington	33	134	× 4.1
Sunnyside, Washington	40	299	× 7.5
Arkansas River	212	890	× 4.2
Sutter Basin, California	72	480	× 6.7
		MEAN	× 5.6

Source: Hotes and Pearson, 1977, p. 141

irrigation areas show higher levels of dissolved salts. These, particularly when they contain nitrates, can make the water undesirable for human consumption. Table 4.5 illustrates the changes in total hardness levels that have been found for river waters in the USA.

A further problem is that, as irrigation water is concentrated by evapotranspiration, calcium and magnesium components tend to precipiate as carbonates, leaving sodium ions dominant in the soil solution. The sodium ions tend to be absorbed by colloidal clay particles, deflocculating them and leaving the resultant structureless soil almost impermeable to water and unfavourable to root development.

The death of vegetation in areas of saline patches, due both to poor soil structure and toxicity, creates bare ground which becomes a focal point for erosion by wind and water.

Probably the most serious impact of salination is on plant growth. This takes place partly through its effect on soil structure, but more significantly through its effects on osmotic pressures and through direct toxicity.

When a water solution containing large quantities of dissolved salts comes into contact with a plant cell it causes a shrinkage of the protoplasmic lining. The phenomenon is due to the osmotic movement of the water, which passes from the cell towards the more concentrated soil solution. The cell collapses and the plant succumbs.

The toxicity effect varies with different plants and different salts. Sodium carbonate, by creating highly alkaline soil conditions, may damage plants by a direct caustic effect; while high nitrate may promote undesirable vegetative growth in grapes or sugarbeets at the expense of sugar content. Boron is injurious to many crop plants at solution concentrations of more than 1 or 2 ppm.

The expected yield reduction from salinity build-up is illustrated in table 4.6. The measure of soil salinity utilized is the electrical conductivity in mmho/cm of a saturated soil extract. The tolerances of different plants to salinity vary greatly, but all suffer from increased salinity.

Table 4.6 Yield reduction for selected agricultural crops at different salinity levels, expressed by the electrical conductivity (mmho/cm) of saturated soil extracts

Crop	% yield reduction		
	10	25	50
Barley	11.9	15.8	17.5
Sugarbeet	10.0	13.0	16.0
Cotton	9.9	11.9	16.0
Wheat	7.1	10.0	14.0
Rice	5.1	5.9	8.0
Maize	5.1	5.9	7.0
Alfalfa	3.0	4.9	8.2

Source: from data in Carter, 1975

Reclamation of salt-affected lands

Because of the extent and seriousness of salinity, be the causes natural or anthropogenic, various reclamation techniques have been initiated. These can be divided into three main types: eradication, conversion and control.

Eradication predominantly involves the removal of salt either by improved drainage or by the addition of quantities of fresh water to leach the salt out of the soil. Both solutions involve considerable expense and pose severe technological problems in areas of low relief and limited fresh-water availability. Improved drainage can either be provided by open drains or by the use of tubewells (as at Mohenjo Daro, Pakistan) to reduce ground-water levels and associated salinity and waterlogging. A minor eradication measure, which may have some potential, is the biotic treatment of salinity through the harvesting of salt-accumulating plants such as *Suaeda fruticosa*.

Conversion involves the use of chemical methods to convert harmful salts into less harmful ones. For example, gypsum is frequently added to sodic soils to convert caustic alkali carbonates to soluble sodium sulphate and relatively harmless calcium carbonate:

$$Na_2CO_3 + CaSO_4 \leftrightarrows CaCO_3 + Na_2SO_4 \downarrow \text{ leachable}$$

Some of the most effective ways of reducing the salinity hazard involve miscellaneous control measures, such as less wasteful and lavish application of water through the use of sprinklers rather than traditional irrigation methods; the lining of the canals to reduce seepage; the realignment of canals through less permeable soil; and the use of more salt-tolerant plants. As salinity is a particularly serious threat at the time of germination and for seedlings, various strategies can be adopted during this critical phase of plant growth:

plots can be irrigated lightly each day after seeding to prevent salt build-up; major leaching can be carried out just before planting; and areas to be seeded can be bedded in such a way that salts accumulate at the ridge tops with the seed planted on the slope between the furrow bottom and the ridge top (Carter, 1975). A useful general review of methods for controlling soil salinity is given by Rhoades (1990).

Lateritization

In some parts of the tropics are extensive sheets of a material called laterite, an iron and/or aluminium-rich duricrust (see Maignien, 1966 or Macfarlane, 1976). These iron-rich sheets result naturally, either because of a preferential removal of silica during the course of extensive weathering (leading to a *relative* accumulation of the sesquioxides of iron and aluminium), or because of an *absolute* accumulation of these compounds.

One of the properties of laterites is that they harden on exposure to air and through desiccation. Once hardened they are not favourable to plant growth. One particular way in which exposure may take place is by accelerated erosion, while forest removal may so cause a change in microclimate that desiccation of the laterite surface can take place. Indeed, one of the main problems with the removal of humid tropical rain forest is that soil hardening may occur. The phenomenon may occur in some, but by no means in all so-called tropical soils (Richter and Babbar, 1991). Should hardening occur, it tends to limit the extent of successful soil utilization and severely retards the re-establishment of forest. Although Vine (1968: 90) and Sanchez and Buol (1975) have warned against exaggerating this difficulty in agricultural land use, there are records from many parts of the tropics of accelerated induration brought about by forest removal (Goudie, 1973). In the Cameroons, for example, around 2 m of complete induration can take place in less than a century. In India, foresters have for a long time been worried by the role that plantations of teak (*Tectona grandis*) can play in lateritization. Teak is deciduous, demands light, likes to be well spaced (to avoid crown friction), dislikes competition from undergrowth, and is shallow-rooted. These characteristics mean that teak plantations tend to expose the soil surface to erosive and desiccative forces more than does the native vegetation cover.

One of the main exponents of the role that human agency has played in lateritization in the tropical world has been Gourou (1961: 21–2). Although he may be guilty of exaggerating the extent and significance of laterite, Gourou gives many examples from low latitudes of falling agricultural productivity resulting from the onset of lateritization. It is worth quoting him at length:

On the whole, laterite is hostile to agriculture owing to its sterility and compactness. All tropical countries have not reached the same degree of lateritic 'suicide' but when the evolution has advanced a considerable way, man is placed in very strange conditions ... Laterite is a pedological leprosy. Man's activities aggravate the dangers of laterite and increase the rate of the process of lateritization. To begin with, erosion when started by negligent removal of the forest simply wears away the friable and relatively fertile soil which would otherwise cover the laterite and support forest or crops ... The forest checks the formation of the laterite in various ways. The trees supply plenty of organic matter and maintain a good proportion of humus in the soil. The action of capillary attraction is checked by the loosening of the soil; and the bases are retained through the absorbent capacity of humus. The forest slows down evaporation from the soil ... it reduces percolation and consequently leaching. Lastly the forest may improve the composition of the soil by fixing atmospheric dust.

Accelerated podzolization and acidification

There is an increasing amount of evidence that the introduction of agriculture, deforestation and pastoralism to parts of upland Western Europe promoted some major changes in soil character: notably an increase in the development of acidic and podzolized conditions, associated with the development of peat bogs. Climatic changes of the type envisaged by Blytt and Sernander and later workers (see Goudie, 1972a: 94) may have played a role, as could progressive leaching of Devensian (last Glacial) drifts during the passage of the Holocene. But the association in time and space of human activities with soil deterioration is becoming increasingly clear (Evans et al., 1975).

Replacing the natural forest vegetation with cultivation and pasture, human societies set in train various related processes, especially on base-poor materials. First, the destruction of deep-rooting trees curtailed the enrichment of the surface of the soil by bases brought up from the deeper layers. Second, the use of fire to effect forest clearance may have released nutrients in the form of readily soluble salts, some of which were inevitably lost in drainage, especially in soils poor in colloids (Dimbleby, 1974). Third, the taking of crops and animal products depleted the soil reserves to an extent probably greater than that arising from any of the manuring practices of prehistoric settlements. Fourth, as the soil degraded, the vegetation which invaded — especially bracken and heather — itself tended to produce a more acidic humus type of soil than the original mixed deciduous forest, and so continued the process.

Various workers now attribute much of the podzolization in upland Britain to such processes, and Dimbleby has concluded that 'although a few soils have been podzols since the Atlantic period, the majority are secondary, having arisen as a result of man's assault on the landscape, particularly in the Bronze Age' (see also Bridges, 1978).

The development of podzols, by impeding downward percolating waters, may have accelerated the formation of peats, which tend to develop where there is waterlogging through impeded drainage. Many peat bogs in highland Britain appear to coincide broadly in age with the first major land-clearance episodes (P. D. Moore,

1973; Merryfield and Moore, 1971). Another fact which would have contributed to their development is that when a forest canopy is removed (as by deforestation) the transpiration demand of the vegetation is reduced, less rainfall is intercepted, so that the supply of ground water is increased, aggravating any waterlogging.

However, the role of natural processes must not be totally forgotten, and Ball's assessment would seem judicious (1975: 26):

It seems to be on balance that the highland trends in soil formation due to climate, geology and relief have been clearly running in the direction of leaching, acidity, podzolization, gleying and peat formation. For the British highlands generally, man has only intervened to hasten or slow the rate of these trends, rather than being in a position to alter the whole trend from one pedogenetic trend to another.

Moreover, it would be plainly misleading to stress only the deleterious effects of human actions on European soils. Traditional agricultural systems have often employed laborious techniques to augment soil fertility and to reduce such properties as undesirable acidity. In Britain, for example, the addition of chalk to light sandy land goes back at least to Roman times and the marl pits from which the chalk was dug are a striking feature of the Norfolk landscape, where Prince (1962) has identified at least 27 000 hollows. Similarly, in the Netherlands, Germany and Belgium there are soils which for centuries (certainly more than a thousand years) have been built up (often over 50 cm) and fertilized with a mixture of manure, sods, litter or sand. Such soils are called Plaggen soils (Pope, 1970). Plaggen soils also occur in Ireland, where the addition of sea-sand to peat was carried out in pre-Christian times. Likewise, before European settlement in New Zealand, the Maoris used thousands of tons of gravel and sand, carried in flax baskets, to improve soil structure (Cumberland, 1961).

Another type of soil which owes much to human influence is the category called 'paddy soil'. Long-continued irrigation, levelling and manuring of terraced land in China and elsewhere has changed the nature of the pre-existing soils in the area. Among the most important modifications that have been recognized (Gong, 1983) are an increase in organic matter, an increase in base saturation, and the translocation and reduction of iron and manganese.

Soil structure alteration

One of the most important features of a soil, in terms of both its suitability for plant growth and its inherent erodibility, is its structure. There are many ways in which humans can alter this, especially by compacting it with agricultural machinery, by the use of recreation vehicles and by changing its chemical character through irrigation. Soil compaction tends to increase the resistance of soil to penetration by roots and emerging seedlings, and limits oxygen and carbon dioxide exchange between the root zone and the atmosphere. Moreover, it reduces the rate of water infiltration

Table 4.7 Change to soil properties resulting from the passage of 100 motorcycles in New Zealand

Soil property	Total no. of sites	No. of sites with significant change*	Mean percentage at significant sites	No. of sites with significant increase	No. of sites with significant decrease	Main direction of change	Mean percentage change in direction
Infiltration capacity	16	16 (100%)	84.3	3	13	Decrease	78.1
Bearing capacity	21	10 (48%)	22.8	2	8	Decrease	18.6
Soil moisture	20	14 (70%)	15.5	5	9	Decrease	16.7
Dry bulk density	19	11 (58%)	13.6	9	2	Decrease	13.3

*Change is significant when greater than: 10% for infiltration capacity; 10% for bearing capacity; 5% for soil moisture; 5% for bulk density
Source: Crozier et al., 1978, table 1

into the soil, which may change the soil moisture status and accelerate surface runoff and soil erosion (Chancellor, 1977). For example, the effects of the passage of vehicles on some soil structural properties are shown in table 4.7. Most notable of all is the reduction that is caused in soil infiltration capacity, which may explain why vehicle movements can often lead to gully development. Whether one is dealing with primitive sledges (as in Swaziland), or with the latest recreational toys of leisured Californian adolescents, the effects may be comparable. Grazing is another activity that can damage soil structure through trampling and compaction. Heavily grazed lands tend to have considerably lower infiltration capacities than those found in ungrazed lands. This is indicated for some American examples in table 4.8. The removal of vegetation cover and associated litter also changes infiltration capacity, since cover protects the soil from packing by raindrops and provides organic matter for binding soil particles together in open aggregates. Soil fauna that live on the organic matter assist this process by churning together the organic material and mineral particles. Dunne and Leopold have ranked the relative influence of different land-use types on infiltration (1978, table 6.2 after US Soil Conservation Service):

Highest infiltration   Woods, good
                         Meadows
                         Woods, fair
                         Pasture, good
                         Woods, poor
                         Pasture, fair
                         Small grains, good rotation
                         Small grains, poor rotation
                         Legumes after row crops
                         Pasture, poor
                         Row crops, good rotation
                         (more than one quarter in hay or sod)
                         Row crops, poor rotation (one quarter or less
                         in hay or sod)
Lowest infiltration   Fallow

Table 4.8 Rates of infiltration on grazed and ungrazed lands in America

| | Rate of infiltration ($mm\ h^{-1}$) | |
Site	Ungrazed	Heavily grazed
Montana	2.5–66.0	5.1–15.2
Oklahoma	134.6–309.9	40.6–83.8
Colorado	40.6–83.8	20.3–30.5
Montana	109.2–185.4	20.3–96.5
Wyoming	30.5–38.1	17.8–30.5
Louisiana	45.7	17.8
Kansas	33.0	20.3
Arizona	40.6	30.5

Source: processed by author from data in Gifford and Hawkins, 1978

In general, experiments show that reafforestation improves soil structure, especially the pore volume of the soils (see, for example, Challinor, 1968). Ploughing is also known to produce a compacted layer at the base of the zone of ploughing (Baver et al., 1972). This layer has been termed the 'plough sole'. The normal action of the plough is to leave behind a loose surface layer and a dense subsoil where the soil aggregates have been pressed together by the sole of the plough. The compacting action can be especially injurious when the depth of ploughing is both constant and long term, and when heavy machinery is used on wet ground (Greenland, 1977).

On the other hand, for many centuries farmers have achieved improvements in soil structure by deliberate practices, particularly with a view to developing the all-important crumb structure. In pre-Roman times people in Britain and France added lime to heavy clay soils, while the agricultural improvers of the eighteenth and nineteenth centuries improved the structure of sandy heath soils by adding clay, and of clay soils by adding calcium carbonate (marl). They also compacted such soils and added binding organic matter by breeding sheep and feeding them with turnips and other fodder plants (Russell, 1961).

In the modern era attempts have been made to reduce soil crusting by applying municipal and animal wastes to farm land, by adding chemicals such as phosphoric acid, and by adopting a cultivation system of the no-tillage type. The last practice is based on the idea that the use of herbicides has eliminated much of the need for tillage and cultivation in row crops; seeds are planted directly into the soil without ploughing, and weeds are controlled by the herbicides. With this method less bare soil is exposed and heavy farm machinery is less likely to create soil compaction problems (see Carlson, 1978, for some of these methods).

Soil structures may also be modified to increase water runoff, particularly in arid zones where the runoff obtained can augment the meagre water supply for crops, livestock, industrial and urban reservoirs, and ground-water recharge projects. In the Negev farming was practised in this way, especially in the Nabatean and the Romano-Byzantine periods (about 300 BC to AD 630), and attempts were made to induce runoff by clearing the surface gravel of the soil and heaping it into thousands of mounds. This exposed the finer silty soil beneath, facilitating soil crust formation by raindrop impact, decreasing infiltration capacity and reducing surface roughness, so that runoff increased (Evenari et al., 1971). Today a greater range of techniques are available for the same end: soils can be smoothed and compacted by heavy machinery, soil crusting can be promoted by dispersion of soil colloids with sodium salts, the permeability of the soil surface can be reduced by applying water-repellent materials, and soil pores can be filled with binders (Hillel, 1971).

Soil drainage and its impact

Soil drainage 'has been a gradual process and the environmental changes to which it has led have, by reason of that gradualness, often passed unnoticed' (Green, 1978: 171). To be sure, the most spectacular feats of drainage – arterial drainage – involving the construction of veritable rivers and large dike systems, as seen in the Netherlands and the Fenlands of eastern England, have received attention. However, more widespread than arterial drainage, and sometimes independent of it, is the drainage of individual fields, either by surface ditching or by underdrainage with tile pipes and the like. Green (1978) has attempted to map the areas of drained agricultural land. In Finland, Denmark, Great Britain the Netherlands and Hungary, the majority of agricultural land is drained (figure 4.6).

In Britain underdrainage was promoted by government grants and reached a peak of about 1 000 000 hectares per year in the 1970s in England and Wales. More recently government subsidies have been cut and the uncertain economic future of farming has led to a reduction in farm expenditure. Both tendencies have led to a reduction in field drainage, which may now be at only 40 000 hectares per year (Robinson 1990).

The drainage conditions of the soil have also frequently been altered by the development of ridge and furrow patterns created by ploughing (plate 4.2). Such patterns are a characteristic feature of many of the heavy soils of lowland England where large areas, especially of the Midland lowlands, are striped by long, narrow ridges of soil, lying more or less parallel to each other and usually arranged in blocks of approximately rectangular shape. They were formed by ploughing with heavy ploughs pulled by teams of oxen and the precise mechanism has been described with clarity by Coones and Patten (1986: 154):

The 'heavy' plough cut through the soil and turned the furrow-slice to the right. Consequently, if the field were simply ploughed up and down by working progressively across it from one side to the other, each double run

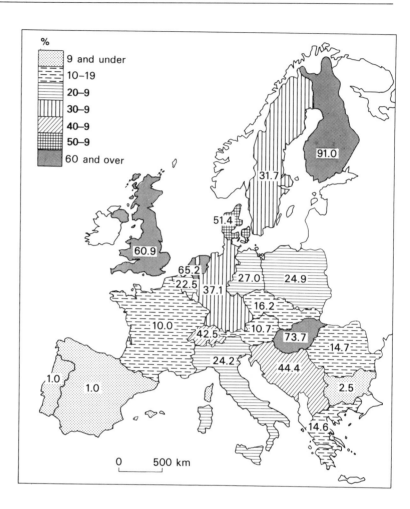

Figure 4.6
Percentage of drained
agricultural land in Europe.
There are no data for the blank
areas (after Green, 1978,
figure 1)

would merely produce a tiny ridge overlying an unploughed ribbon of
ground, the soil being thrown inwards from the two furrows on either side.
This would have been a pointless exercise, so instead the field was
ploughed as a number of separate units, each with its own central axis.
Ploughing started along the centre line and worked gradually away from it,
turning the soil inwards from both edges as the plough went along one side
and back down the other. Every sod on each side was therefore turned the
same way and laid against the previous one. It has been suggested that once
the plough team had progressed to a point a certain distance from the
original centre line, the distance it was obliged to walk along the headlands
to the other side for the return run was such that it was easier to start a new
circuit, based upon the axis of an adjacent unit. Repeated ploughing of
exactly the same units year after year gradually led to the formation of
adjacent ridges, separated from each other by furrows produced where the
furrow-slices were turned away from each other.

Soil drainage has been one of the most successful ways in which
communities have striven to increase agricultural productivity; it
was practised by Etruscans, Greeks and Romans (N. Smith, 1976).

Large areas of marshland and floodplain have been drained to human advantage. By leading water away, the water-table is lowered and stabilized, providing greater depth for the root zone. Moreover, well-drained soils warm up earlier in the spring and thus permit earlier planting and germination of crops. Farming is easier if the soil is not too wet, since the damage to crops by winter freezing may be minimized, undesirable salts carried away, and the general physical condition of the soil improved. In addition, drained land may have certain inherent virtues: tending to be flat it is less prone to erosion and more amenable to mechanical cultivation. It will also be less prone to drought risk than certain other types of land (Karnes, 1971). Paradoxically, by reducing the area of saturated ground, drainage can alleviate flood risk in some situations by limiting the extent of a drainage basin that generates saturation excess overland flow.

Conversely, soil drainage can have quite undesirable or un-planned effects. For example, some drainage systems, by raising drainage densities, can increase flood risk by reducing the distance over which unconcentrated overland flow (which is relatively slow) has to travel before reaching a channel (where flow is relatively fast). In central Wales, for instance, the establishment of drainage

Plate 4.2
Over large tracts of lowland England, the landscape is dominated by systems of ridge and furrow, as shown in this air photograph of Hollingdon, Buckinghamshire
© British Crown copyright 1986/MOD reproduced with the permission of the Controller of Her Britannic Majesty's Stationery Office

ditches in peaty areas to enable afforestation has tended to increase flood peaks in the rivers Wye and Severn (Howe et al., 1966).

Drainage can also cause long-term damage to soil quality. A fall in water level in organic soils can lead to the oxidation and eventual disappearance of peaty materials, which in the early stages of post-drainage use may be highly productive. This has occurred in the English Fenland, and in the Everglades of Florida, where drainage of peat soils has led to a subsidence in the soil of 32 mm per year (Stephens, 1956).

The soil climate of neighbouring areas may also be modified when the water-tables of drained land are lowered. This has been known to create problems for forestry in areas of marginal water availability.

Soil moisture content can also determine the degree to which soils are subjected to expansion and contraction effects, which in turn may affect engineering structures in areas with expansive soils (Holtz, 1983). Soils containing sodium montmorillonite type clays, when drained or planted with large trees, may dry out and cause foundation problems (Driscoll, 1983).

Some of the most contentious effects of drainage are those associated with the reduction of wetland wildlife habitats. In Britain this has become an important political issue, especially in the context of the Somerset Levels and the Halvergate Marshes.

Soil fertilization

The chemistry of soils has been changed deliberately by the introduction of chemical fertilizers. Sometimes these create environmental problems such as water pollution, while their substitution for more traditional fertilizers may accelerate soil structure deterioration and soil erosion. On the other hand, the use of synthetic fertilizers has greatly increased agricultural productivity in many parts of the world, and remarkable increases in yields have been achieved. It is also true that in some circumstances proper fertilizer use can help minimize erosion by ensuring an ample supply of roots and plant residues, particularly on infertile or partially degraded soils (Bøckman et al., 1990).

The employment of chemical fertilizers on a large scale is little more than 150 years old. In the early nineteenth century nitrates were first imported from Chile, and sulphate of ammonia was produced only after the 1820s as a by-product of coal gas manufacture. In 1843 the first fertilizer factory was established at Deptford Creek (London), but for a long time superphosphates were the only manufactured fertilizers in use. In the twentieth century synthetic fertilizers, particularly nitrates, were developed, notably by Scandinavian countries which used their vast resources of water power. Potassic fertilizers came into use much later than the phosphatic and nitrogenous; nineteenth-century farmers hardly knew them (Russell, 1961). The rapid development in the use of the three main

Table 4.9 Quantities of fertilizers used by farmers in the UK expressed in
thousands of imperial tons of plant food

Date	Nitrogen (N)	Phosphoric acid (P₂O₅)	Potash (K₂O)
1900	16	110	7
1913	29	180	23
1929	48	198	52
1939	60	170	75
1946	165	358	120
1956	291	386	305

Source: from Russell, 1961, table 14, p. 185

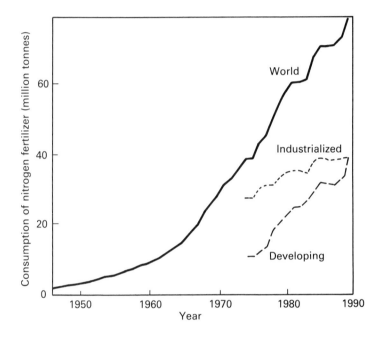

Figure 4.7
Global consumption of nitrogen
fertilizer over the last four
decades (after various sources
in Conway and Pretty, 1991,
figure 4.1)

fertilizers – nitrates, phosphates and potash – in the first half of the
twentieth century is shown in table 4.9, while figure 4.7 shows the
trend in nitrogen fertilizer consumption for the world as a whole
over the last four decades.

Fires and soil

The importance and antiquity of fire as an agency through which
the environment is transformed requires that some attention be
given to the effects of fire on soil characteristics and on soil erosion.
    Fire has often been used intentionally to change soil properties,
and both the release of nutrients by fire and the value of ash have
long been recognized, notably by those involved in shifting agricul-

ture based on slash-and-burn techniques. Following cultivation, the loss of nutrients by leaching and erosion is very rapid (Nye and Greenland, 1964), and this is why after only a few years the shifting cultivators have to move on to new plots. Fire rapidly alters the amount, form and distribution of plant nutrients in ecosystems, and, compared to normal biological decay of plant remains, burning rapidly releases some nutrients into a plant-available form. Indeed, the amounts of P, Mg, K and Ca released by burning forest and scrub vegetation are high in relation to both the total and available quantities of these elements in soils (Raison, 1979). In forests, burning often causes the pH of the soil to rise by three units or more, creating alkaline conditions where formerly there was acidity. Burning also leads to some direct nutrient loss by volatilization and convective transfer of ash, or by loss of ash to water erosion or wind deflation. The removal of the forest causes soil temperatures to increase because of the absence of shade, so that humus is often lost at a faster rate than it is formed (Grigg, 1970).

Soil erosion: general considerations

Loss of soil humus, whether it be a result of fire, drainage, deforestation or ploughing, is an especially serious manifestation of human alteration of soil. As table 4.10 indicates, humus has many beneficial effects on both the chemical and the physical properties of soil. Its removal by human activity can be a potent contributory cause of soil erosion.

The scale of accelerated soil erosion that has been achieved by human activities has been well summarized by Myers (1988: 6):

Since the development of agriculture some 12,000 years ago, soil erosion is said by some to have ruined 4.3 million km^2 of agricultural lands, or an area equivalent to rather more than one-third of today's croplands . . . the amount of agricultural land now being lost through soil erosion, in conjunction with other forms of degradation, can already be put at a minimum of 200,000 km^2 per year.

In the late 1930s, towards the end of the so-called Dust Bowl years, Sauer conducted a campaign against what he called the 'destructive exploitation in modern colonial expansion' (Sauer, 1938: 497), and placed soil erosion squarely into the context of geography:

We may well consider whether the theme of soil erosion should not be moved up to the first category of problems before the geographers of the world. It is very important for the future of mankind. It has critical significance for certain chapters of historical geography. The physical processes involved are poorly observed and generalized and their study will undoubtedly shake somewhat the rather lethargic present position of geomorphology . . . best of all the subject is suited to a 'hologeographic' approach in which the development of surface conditions of specific localization is examined as an interaction of identified physical and economic (i.e. Wirtschaft) processes.

Table 4.10 The beneficial properties of humus

Property	Explanation	Effect
(a) *Chemical properties*		
Mineralization	Decomposition of humus yields CO_2, NH_4^+, NO_3^-, PO_4^{3-}, and SO_4^{2-}	A source of nutrient elements for plant growth
Cation exchange	Humus has negatively charged surfaces which bind cations such as Ca^{++} and K^+	Improved cation exchange capacity (CEC) of soil. From 20 to 70 per cent of the CEC of some soils (e.g. Mollisols) is attributable to humus
Buffer action	Humus exhibits buffering in slightly acid, neutral and alkaline ranges	Helps to maintain a uniform pH in the soil
Acts as matrix for biochemical action in soil	Binds other organic molecules electrostatically or by covalent bonds	Affects bioactivity persistence and biodegradability of pesticides
Chelation	Forms stable complexes with Cu^{2+}, Mn^{2+}, Zn^{2+}, and other polyvalent cations	May enhance the availability of micronutrients to higher plants
(b) *Physical properties*		
Water retention	Organic matter can hold up to 20 times its weight in water	Helps prevent drying and shrinking. May significantly improve the moisture-retaining properties of sandy soils
Combination with clay	Cements soil particles into structural units called aggregates	Improves aeration. Stabilizes structure. Increases permeability
Colour	The typical dark colour of many soils is caused by organic matter	May facilitate warming

Source: Swift and Sanchez, 1984, table 2

That soil erosion is a major and serious aspect of the human role in environmental change is not to be doubted. There is a long history of weighty books and papers on the subject (see, for example, Marsh, 1864; Bennett, 1938; Jacks and Whyte, 1939; Morgan, 1979). Although many techniques have been developed to reduce the intensity of the problem (see Hudson, 1971) it appears to remain intractable. As L. J. Carter (1977: 409) has reported of the USA:

although nearly $15 billion has been spent on soil conservation since the mid-1930s, the erosion of croplands by wind and water . . . remains one of the biggest, most pervasive environmental problems the nation faces. The problem's surprising persistence apparently can be attributed at least in part to the fact that, in the calculation of many farmers, the hope of maximizing short-term crop yields and profits has taken precedence over the longer term advantages of conserving the soil. For even where the loss of topsoil has begun to reduce the land's nautral fertility and productivity, the effect is often masked by the positive response to heavy application of fertilizer and pesticides, which keep crop yields relatively high.

Although construction, urbanization, war, mining and other such activities are often significant in accelerating the problem, the prime causes of soil erosion are deforestation and agriculture. Pimentel (1976) has estimated that in the USA soil erosion on agricultural land operates at a rate of about 30 t ha^{-1} y^{-1}, which is approximately eight times quicker than topsoil is formed. He calculates that water runoff delivers around 4 billion tonnes of soil each year to the rivers of the forty-eight contiguous states, and that three-quarters of this comes from agricultural land. He estimates that another billion tonnes of soil is eroded by the wind, a process which created the Dust Bowl of the 1930s.

One serious consequence of accelerated erosion is the sedimentation that takes place in reservoirs, shortening their lives and reducing their capacity. Many small reservoirs, especially in semi-arid areas, appear to have an expected life of only thirty years or even less (see, for example, Rapp et al., 1972). Soil erosion also has serious implications for soil productivity. A reduction in soil thickness reduces available water capacity and the depth through which root development can occur. The water-holding properties of the soil may be lessened as a result of the preferential removal of organic material and fine sediment. Hardpans and duricrusts may become exposed at the surface, and provide a barrier to root penetration. Furthermore, splash erosion may cause soil compaction and crusting, both of which may be unfavourable to germination and seedling establishment. Erosion also removes nutrients preferentially from the soil. Some damage may be caused by associated excessive sedimentation, while wind erosion may lead to the direct sandblasting of crops. Finally, extreme erosion may lead to wholesale removal of both seeds and fertilizer. Stocking (1984) provides a useful review of these problems.

Soil erosion associated with deforestation and agriculture

Forests protect the underlying soil from the direct effects of rainfall, generating what is generally an environment in which erosion rates tend to be low. The canopy plays an important role by shortening the fall of rain drops, decreasing their velocity and thus reducing kinetic energy. There are some examples of certain types (e.g. beech) in certain environments (e.g. maritime temperate) creating large raindrops, but in general most canopies reduce the erosive effects of rainfalls. Possibly more important than the canopy in reducing erosion rates in forest is the presence of humus in forest soils (Trimble, 1988), for this both absorbs the impact of raindrops and has an extremely high permeability. Thus forest soils have high infiltration capacities. Another reason that forest soils have an ability to transmit large quantities of water through their fabrics is that they have many macropores produced by roots and their rich soil fauna. Forest soils are also well aggregated, making them resistant to both wetting and water drop impact. This superior degree of aggradation is a result of the presence of considerable organic material, which is an important cementing agent in the formation of large water-stable aggregates. Furthermore, earthworms also help to produce large aggregates. Finally, deep-rooted

Plate 4.3
The removal of vegetation in Swaziland creates spectacular gully systems, which in southern Africa are called *dongas*. The smelting of local iron ores in the early nineteenth century required the use of a great deal of fire-wood which may have contributed to the formation of this example

trees help to stabilize steep slopes by increasing the total shear strength of the soils.

It is therefore to be expected that with the removal of forest, for agriculture or for other reasons, rates of soil loss will rise and mass movements will increase in magnitude and frequency. The rates of erosion that result will be particularly high if the ground is left bare; under crops the increase will be less marked. Furthermore, the method of ploughing, the time of planting, the nature of the crop, and size of the fields, will all have an influence on the severity of erosion.

It is seldom that we have reliable records of rates of erosion over a sufficiently long time-span to show just how much human activities have accelerated these effects. Recently, however, techniques have been developed which enable rates of erosion on slopes to be gauged over a lengthy time-span by means of dendrochronological techniques that date the time of root exposure for suitable species of tree. In Colorado, USA, Carrara and Carroll (1979) found that rates over the last 100 years have been about 1.8 mm y^{-1}, whereas in the previous 300 years rates were between 0.2 and 0.5 mm y^{-1}, indicating an acceleration of about sixfold. This great jump has been attributed to the introduction of large numbers of cattle to the area about a century ago.

Another way of obtaining long-term rates of soil erosion is to look at rates of sedimentation on continental shelves and on lake floors. The former method was employed by Milliman et al. (1987) to evaluate sediment removal down the Yellow River in China during the Holocene. They found that, because of accelerated erosion, rates of sediment accumulation on the shelf over the last 2300 years have been ten times higher than those for the rest of the Holocene (i.e. since around 10 000 BP).

Another good example of using long-term sedimentation rates to infer long-term rates of erosion is provided by Hughes et al.'s

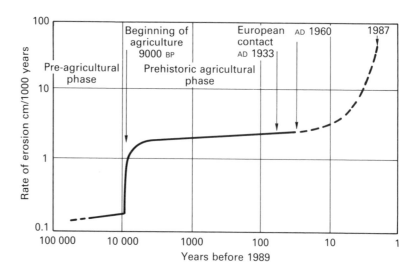

Figure 4.8
Rates of erosion in Papua New Guinea in the Holocene, derived from rates of sedimentation in Kuk Swamp (after Hughes et al., 1991, figure 5, with modifications)

(1991) study of the Kuk Swamp in Papua New Guinea (figure 4.8). They identify low rates of erosion until 9000 BP, when, with the onset of the first phase of forest clearance, erosion rates increased from 0.15 cm 1000 y^{-1} to about 1.2 cm 1000 y^{-1}. Rates remained relatively stable until the last few decades when, following European contact, the extension of anthropogenic grasslands, subsistence gardens and coffee plantations has produced a rate that is very markedly higher: 34 cm 1000 y^{-1}.

Table 4.11, which is based on data from tropical Africa, shows the comparative rates of erosion for three main types of land use: trees, crops and barren soil. It is very evident from these data that under crops, but more especially when ground is left bare or under fallow, soil erosion rates are greatly magnified. At the same time, and causally related, the percentage of rainfall that becomes runoff is increased.

Table 4.11 Runoff and erosion under various covers of vegetation in parts of Africa

Locality	Average annual rainfall (mm)	Slope (%)	Annual runoff (%) A	B	C	Erosion (t ha^{-1} yr^{-1}) A	B	C
Ouagadougou (Burkina Faso)	850	0.5	2.5	2–32	40–60	0.1	0.6–0.8	10–20
Sefa (Senegal)	1300	1.2	1.0	21.2	39.5	0.2	7.3	21.3
Bouake (Ivory Coast)	1200	4.0	0.3	0.1–26	15–30	0.1	1–26	18–30
Abidjan (Ivory Coast)	2100	7.0	0.1	0.5–20	38	0.03	0.1–90	108–170
Mpwapwa* (Tanzania)	c.570	6.0	0.4	26.0	50.4	0	78	146

Note: A = forest or ungrazed thicket
      B = crop
      C = barren soil
* From Rapp et al., 1972, figure 5, p. 259
Source: after Charreau, table 5.5 p. 153 in Greenland and Lal, 1977

In some cases the erosion produced by forest removal will be in the form of widespread surface stripping. In other cases the erosion will occur as more spectacular forms of mass movement, such as mudflows, landslides and debris avalanches. Some detailed data on debris-avalanche production in North American catchments as a result of deforestation and forest road construction are presented in table 4.12. They illustrate the substantial effects created by clear-cutting and by the construction of logging roads. It is indeed probable that a large proportion of the erosion associated with forestry operations is caused by road construction, and care needs to be exercised to minimize these effects. The digging of drainage ditches in upland pastures and peat moors to permit tree-planting in central Wales has also been found to cause accelerated erosion (Clarke and McCulloch, 1979), while the elevated sediment loads can cause reservoir pollution (Burt et al., 1983).

In general, the greater the deforested proportion of a river basin the higher the sediment yield per unit area will be. In the USA the rate of sediment yield appears to double for every 20 per cent loss in forest cover.

Soil erosion resulting from deforestation and agricultural practice is often thought to be especially serious in tropical areas or

Table 4.12 Debris-avalanche erosion in forest, clear-cut and roaded areas

Site	Period of records (years)	Area (%)	Area (km²)	No. of slides	Debris-avalanche erosion (m³ km⁻² yr⁻¹)	Rate of debris-avalanche erosion relative to forested areas
Stequaleho Creek, Olympic Peninsula						
Forest	84	79	19.3	25	71.8	× 1.0
Clear-cut	6	18	4.4	0	0	0
Road	6	3	0.7	83	11 825	× 165
TOTAL	–	–	24.4	108	–	–
Alder Creek, western Cascade Range, Oregon						
Forest	25	70.5	12.3	7	45.3	× 1.0
Clear-cut	15	26.0	4.5	18	117.1	× 2.6
Road	15	3.5	0.6	75	15 565	× 344
TOTAL	–	–	17.4	100	–	–
Selected drainages, Coast Mountains, south-west British Columbia						
Forest	32	88.9	246.1	29	11.2	× 1.0
Clear-cut	32	9.5	26.4	18	24.5	× 2.2
Road	32	1.5	4.2	11	282.5	× 25.2
TOTAL	–	–	276.7	58	–	–
H. J. Andrews Experimental Forest, western Cascade Range, Oregon						
Forest	25	77.5	49.8	31	35.9	× 1.0
Clear cut	25	19.3	12.4	30	132.2	× 3.7
Road	25	3.2	2.0	69	1 772	× 49
TOTAL	–	–	64.2	130	–	–

Source: after Swanston and Swanson, 1976, table 4

Table 4.13 Annual rates of soil loss (tonnes per hectare) under different land-use types in eastern England

Plot	Splash	Overland flow	Rill	Total
1 Bare soil				
Top slope	0.33	6.67	0.10	7.10
Mid-slope	0.82	16.48	0.39	17.69
Lower slope	0.62	14.34	0.06	15.02
2 Bare soil				
Top slope	0.60	1.11	–	1.71
Mid-slope	0.43	7.78	–	8.21
Lower slope	0.37	3.01	–	3.38
3 Grass				
Top slope	0.09	0.09	–	0.18
Mid-slope	0.09	0.57	–	0.68
Lower slope	0.12	0.05	–	0.17
4 Woodland				
Top slope	–	–	–	0.00
Mid-slope	–	0.012	–	0.012
Lower slope	–	0.008	–	0.008

Source: from Morgan, 1977

semi-arid areas (see T. R. Moore, 1979, for a good case study). However, recent measurements by Morgan (1977) on sandy soils in the English East Midlands near Bedford indicate that rates of soil loss under bare soil on steep slopes can reach $17.69 \, \text{ha}^{-1} \, \text{y}^{-1}$, compared with 2.39 under grass and nothing under woodland (table 4.13), and subsequent studies have demonstrated that water-induced soil erosion is a substantial problem, in spite of the relatively low erosivity of British rainfall. Walling and Quine (1991: 123) have identified the following farming practices as contributing to this developing problem:

1 Ploughing up of steep slopes that were formerly under grass, in order to increase the area of arable cultivation.
2 Use of larger and heavier agricultural machinery which has a tendency to increase soil compaction.
3 Removal of hedgerows and the associated increase in field size. Larger fields cause an increase in slope length with a concomitant increase in erosion risk.
4 Declining levels of organic matter resulting from intensive cultivation and reliance on chemical fertilizers, which in turn lead to reduced aggregate stability.
5 Availability of more powerful machinery which permits cultivation in the direction of maximum slope rather than along the contour. Rills often develop along tractor and implement wheelings and along drill lines.
6 Use of powered harrows in seedbed preparation and the rolling of fields after drilling.

7 Widespread introduction of autumn-sown cereals to replace spring-sown cereals. Because of their longer growing season, winter cereals produce greater yields and are therefore more profitable. The change means that seedbeds are exposed with little vegetation cover throughout the period of winter rainfall.

Water is not the only active process creating accelerated erosion in eastern England, though it is important (Evans and Northcliff, 1978). Ever since the 1920s dust storms have been recorded in the Fenlands, the Brecklands, East Yorkshire (Radley and Sims, 1967) and in Lincolnshire (see, for example, Arber, 1946), and they seem to be occurring with increasing frequency. The storms result from changing agricultural practices, including the substitution of artificial fertilizers for farmyard manure, a reduction in the process of 'claying', whereby clay was added to the peat to stabilize it, the removal of hedgerows to facilitate the use of bigger farm machinery, and, perhaps most importantly, the increased cultivation of sugar beet. This crop requires a fine tilth and tends to leave the soil relatively bare in early summer compared with other crops (Pollard and Miller, 1968).

Nevertheless, possibly the most famous case of soil erosion by deflation was the Dust Bowl of the 1930s in the USA (see figure 4.9). In part this was caused by a series of hot, dry years which depleted the vegetation cover and made the soils dry enough to be susceptible to wind erosion. The effects of this drought were gravely exacerbated by years of over-grazing and unsatisfactory farming techniques. However, perhaps the prime cause of the event was the rapid expansion of wheat cultivation in the Great Plains. The number of cultivated hectares doubled during the First World War as tractors (for the first time) rolled out on to the plains by the thousands. In Kansas alone wheat hectarage increased from under 2 million hectares in 1910 to almost 5 million in 1919. After the war wheat cultivation continued apace, helped by the development of the combine harvester and government assistance. The farmer, busy sowing wheat and reaping gold, could foresee no end to his land of milk and honey, but the years of favourable climate were not to last, and over large areas the tough sod which exasperated the earlier homesteaders had given way to friable soils of high erosion potential. Drought, acting on damaged soils, created the 'black blizzards' which have been so graphically described by Coffey (1978: 79–80):

there was something fantastic about a dust cloud that covered 1.35 m. square miles, stood three miles high and stretched from Canada to Texas, from Montana to Ohio – a cloud so colossal it obliterated the sky . . . a four-day storm in May 1934 . . . transported some 300 million tonnes of dirt 1500 miles, darkened New York, Baltimore and Washington for five hours, and dropped dust not only on the President's desk in the White House, but also on the decks of ships some 300 miles out in the Atlantic . . . masses of dust began to billow into huge tumbling clouds ebony black at the base and muddy tan at the top, some so saturated with

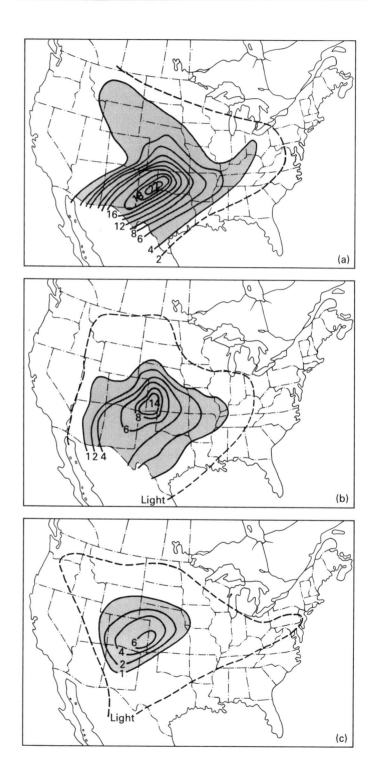

Figure 4.9
The concentration of dust
storms (number of days per
month) in the USA in 1936,
illustrating the extreme
localization over the High
Plains of Texas, Colorado,
Oklahoma and Kansas:
(a) March
(b) April
(c) May
(after Goudie, 1983b)

Plate 4.4
The Dust Bowl created in the 1930s caused severe problems for the inhabitants of Cimarron County, Oklahoma

dust particles that ducks and geese caught in flight, suffocated; some turning the sky so black that chickens, thinking it night, would roost. Oklahoma counted 102 storms in the span of one year; North Dakota reported 300 in eight months . . .

Dust storms are still a serious problem in various parts of the United States: the Dust Bowl was not solely a feature of the 1930s. Thus, for example, in the San Joaquin Valley area of California in 1977 a dust storm caused extensive damage and erosion over an area of about 2000 km^2. More than 25 million tonnes of soil were stripped from grazing land within a 24-hour period. While the combination of drought and a very high wind (as much as 300 km

Plate 4.5
In the High Plains near Lubbock in Texas the effect of soil erosion and drifting was very evident in 1977. Note the vast fields and absence of windbreaks

per hour) provided the predisposing natural conditions for the stripping to occur, over-grazing and the general lack of windbreaks in the agricultural land played a more significant role. In addition, broad areas of land had recently been stripped of vegetation, levelled or ploughed up prior to planting. Other quantitatively less important factors included stripping of vegetation for urban expansion, extensive denudation of land in the vicinity of oilfields, and local denudation of land by vehicular recreation (Wilshire et al., 1981). One interesting observation made in the months after the dust storm was that in subsequent rain storms runoff occurred at an accelerated rate from those areas that had been stripped by the wind, exacerbating problems of flooding and initiating numerous gullies. Elsewhere in California dust yield has been considerably increased by mining operations in dry lake beds (Wilshire, 1980).

A comparable acceleration of dust storm activity has occurred in the CIS. After the 'Virgin Lands' programme of agricultural expansion in the 1950s, dust storm frequencies in the southern Omsk region increased on average by a factor of 2.5 and locally by factors of 5 to 6. Data on trends elsewhere are evaluated by Goudie (1983b: 520) and Goudie and Middleton (1992).

Soil erosion produced by fire

Many fires are started by humans, either deliberately or non-deliberately, and because fires remove vegetation and expose the ground they tend to increase rates of soil erosion.

The burning of forests, for example, can, especially in the first years after the fire event, lead to high rates of soil loss (see table 4.14). Burnt forests often have rates a whole order of magnitude higher than those of protected areas. Comparably large changes in soil erosion rates have been observed to result from the burning of heather in the Yorkshire moors in northern England, and the effects of burning may be felt for the six years or more that may be required to regenerate the heather (*Calluna*) (see table 4.14). In the Australian Alps fire in experimental catchments has been found to lead to a greatly increased flow in the streams, together with a marked surge in the delivery of suspended load. Combining the two effects of increased flow rate and sediment yield, it was found that, after fire, the total sediment load was increased 1000 times (Pereira, 1973). Likewise, watershed experiments in the chaparral scrub of Arizona, involving denudation by a destructive fire, indicated that whereas erosion losses before the fire were only $43 \ t \ km^{-2} \ y^{-1}$ after the fire they were between 50 000 and 150 000 $t \ km^{-2} \ y^{-1}$. The causes of the marked erosion associated with chaparral burning are particularly interesting. There is normally a distinctive 'non-wettable' layer in the soils supporting chaparral. This layer, composed of soil particles coated by hydrophobic substances leached from the shrubs or their litter, is normally associated with the upper part of the soil profile (Mooney and Parsons, 1973), and builds up

Table 4.14 Soil erosion associated with *Calluna* (heather) burning on the North Yorkshire moors

Condition of Calluna or ground surface	Mean rate of litter accumulation (+) or erosion (−) (mm/year)	No. of observations
(a) *Calluna* 30–40 cm high. Complete canopy	+3.81	60
(b) *Calluna* 20–30 cm high. Complete canopy	+0.25	20
(c) *Calluna* 15–20 cm high. 40–100% cover	−0.74	20
(d) *Calluna* 5–15 cm high. 10–100% cover	−6.4	20
(e) Bare ground. Surface of burnt *Calluna*	−9.5	19
(f) Bare ground. Surface of peaty or mineral subsoil	−45.3	25

Source: data in Imeson, 1971

through time in the unburned chaparral. The high temperatures which accompany chaparral fires cause these hydrophobic substances to be distilled so that they condense on lower soil layers. This process results in a shallow layer of wettable soil overlying a non-wettable layer. Such a condition, especially on steep slopes, can result in severe surface erosion.

In chaparral terrain it is possible to envisage a fire-induced sediment cycle (Graf, 1988: 243). It starts with a fire that destroys the scrub and the root net, and changes surface soil properties in the way already discussed. After the fire, a precipitation event of low magnitude (with a return interval of around one or two years) is sufficient to induce extensive sheet and rill erosion, which removes enough soil to retard vegetation recovery. Eventually, a larger precipitation event occurs (with a return interval of around five to ten years) and, because of limited vegetation cover, produces severe debris slides. Slowly the vegetation cover re-establishes itself, and erosion rates diminish. However, in due course enough vegetation grows to create a fire hazard, and the whole process starts again.

Soil erosion associated with construction and urbanization

There are now a number of studies which illustrate clearly that urbanization can create significant changes in erosion rates.

The highest rates of erosion are produced in the construction phase, when there is a large amount of exposed ground and much disturbance produced by vehicle movements and excavations. Wolman and Schick (1967) and Wolman (1967) have shown that the equivalent of many decades of natural or even agricultural erosion may take place during a single year in areas cleared for construction. In Maryland they found that sediment yields during construction reached $55\,000\ \mathrm{t\ km^{-2}\ y^{-1}}$, while in the same area rates under forest were around $80\text{–}200\ \mathrm{t\ km^{-2}\ y^{-1}}$ and those under farm $400\ \mathrm{t\ km^{-2}\ y^{-1}}$. New road cuttings in Georgia were found to have sediment yields up to $20\,000\text{–}50\,000\ \mathrm{t\ km^{-2}\ yr^{-1}}$. Likewise, in Devon, England, Walling and Gregory (1970) found that suspended sediment concentrations in streams draining construction areas were two to ten times (occasionally up to 100 times) higher than those in undisturbed areas. In Virginia, USA, Vice et al. (1969) noted equally high rates of erosion during construction and reported that they were ten times those from agricultural land, 200 times those from grassland and 2000 times those from forest in the same area.

However, construction does not go on for ever, and once the disturbance ceases, roads are surfaced, and gardens and lawns are cultivated. The rates of erosion fall dramatically and may be of the same order as those under natural or pre-agricultural conditions (table 4.15). Moreover, even during the construction phase several techniques can be used to reduce sediment removal, including the excavation of settling ponds, the seeding and mulching of bare surfaces, and the erection of rock dams and straw bales (Reed, 1980).

Table 4.15 Rates of erosion associated with construction and urbanization

Location	Land use	Source	Rate $(t\ km^{-2}\ yr^{-1})$
1 Maryland, USA	Forest	Wolman (1967)	39
	Agriculture		116–309
	Construction		38 610
	Urban		19–39
2 Virginia, USA	Forest	Vice et al. (1969)	9
	Grassland		94
	Cultivation		1876
	Construction		18 764
3 Detroit, USA	General non-urban	Thompson (1970)	642
	Construction		17 000
	Urban		741
4 Maryland, USA	Rural	Fox (1976)	22
	Construction		37
	Urban		337
5 Maryland, USA	Forest and grassland	Yorke and Herb (1978)	7–45
	Cultivated land		150–960
	Construction		1600–22 400
	Urban		830
6 Wisconsin, USA	Agricultural	Daniel et al. (1979)	<1
	Construction		19.2
7 Tama New Town, Japan	Construction	Kadomura (1983)	c.40 000
8 Okinawa, Japan	Construction	Kadomura (1983)	25 000–125 000

Attempts at soil conservation

Because of the adverse effects of accelerated erosion a whole array
of techniques has now been widely adopted to conserve soil
resources (Hudson, 1987), though some of the techniques such as
hill slope terracing may be of some antiquity:

1 *Revegetation*
   (a) Deliberate planting
   (b) Suppression of fire, grazing, etc., to allow regeneration
2 *Measures to stop stream bank erosion*
3 *Measures to stop gully enlargement*
   (a) Planting of trailing plants, etc.
   (b) Weirs, dams, gabions, etc.
4 *Crop management*
   (a) Maintaining cover at critical times of year
   (b) Rotation
   (c) Cover crops
5 *Slope runoff control*
   (a) Terracing
   (b) Deep tillage and application of humus
   (c) Transverse hillside ditches to interrupt runoff
   (d) Contour ploughing
   (e) Preservation of vegetation strips (to limit field width)
6 *Prevention of erosion from point sources like roads, feedlots*
   (a) Intelligent geomorphic location
   (b) Channelling of drainage water to non-susceptible areas
   (c) Covering of banks, cuttings, etc., with vegetation
7 *Suppression of wind erosion*
   (a) Soil moisture preservation
   (b) Increase in surface roughness through ploughing up
       clods or by planting windbreaks

There are some parts of the world where terraces are one of the
most prominent components of the landscape (plate 4.6). This
applies to many wine-growing areas, to some arid zone regions
(such as Yemen and Peru), and to a wide selection of localities in
the more humid tropics (Luzon, Java, Sumatra, Assam, Ceylon,
Uganda, Cameroons, etc.). In areas subject to wind erosion other
strategies may be necessary. Since soil only blows when it is dry,
anything which conserves soil moisture is beneficial. Another
approach to wind erosion is to slow down the wind by physical
barriers, either in the form of an increased roughness of the soil
surface brought about by careful ploughing, or by planted vegetative barriers, such as windbreaks and shelter-belts.

Some attempts at soil conservation have been particularly
successful. For example, in Wisconsin, a study by Trimble and
Lund (1982) showed that in the Coon Creek Basin, erosion rates
declined fourfold between the 1930s and the 1970s. One of the

Plate 4.6
The loess lands of China have been particularly susceptible to the development of large ravines. However, these conservation terraces, built by a production brigade, have effectively reduced the problem and allowed the land to become rehabilitated

main reasons for this was the progressive adoption of contour-strip ploughing (figure 4.10).

Attempts at soil conservation have not always been without their drawbacks. For example, the establishment of ground cover in dry areas to limit erosion may so reduce soil moisture because of accelerated evapotranspiration that the growth of the main crop is adversely affected. On a wider scale major afforestation schemes can cause substantial runoff depletion in river catchments. Likewise, the provision of mulching is sometimes detrimental: in cool climates, reduced soil temperature shortens the growing season, whilst in wet areas, higher soil moisture may induce gleying and anaerobic conditions (Morgan, 1979: 60). Some terrace schemes have also had their shortcomings. They have been known to hold back so much water on hillsides that the soils have become saturated and landsliding has been induced. Similarly strip cropping, because it involves the farming of small areas, is incompatible with highly mechanized agricultural systems, and insect infestation and weed control are additional problems which it has posed.

Soil conservation measures may not always be appropriate, for soil erosion can be a useful phenomenon. Sanchez and Buol (1975), for instance, have pointed out that in recent volcanic areas soil erosion has enabled removal of the more weathered base-depleted material from the soil surface, exposing the more fertile, less weathered, base-rich material beneath. Likewise, in the Nochixtlan area of southern Mexico, soil erosion has been utilized by local farmers to *produce* agricultural land. Severe gullies have cut in to steep valley-side slopes (figure 4.11), and since the Spanish Con-

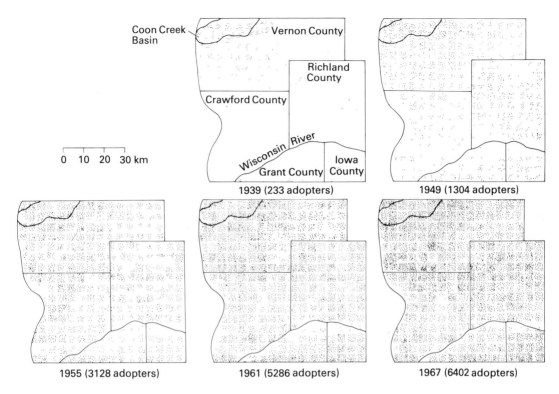

1939 (233 adopters)   1949 (1304 adopters)

1955 (3128 adopters)   1961 (5286 adopters)   1967 (6402 adopters)

quest an average depth of 5 m has been stripped from the entire surface area. The local Mixtec farmers, far from seeing this high rate of erosion as a hazard to be feared, have directed the flow of the eroded material to feed their fields with fertile soil and to extend their land. Over the past 1000 years (see Whyte, 1977), the Mixtec cultivators have managed to use gully erosion to double the width of the main valley floors from 1.5 to 3 km and to infill the narrow tributary valley floors with flights of terraces. Judicious use of the phenomenon of gully erosion has enabled them to convert poor hilltop fields into the rich alluvial farmland below.

None the less it is undoubtedly true that manipulation of the soil is one of the most significant ways in which humans change the environment, and one in which they have had some of the most detrimental effects. Soil deterioration has led to many cases of what W. C. Lowdermilk once termed 'regional suicide', and the overall situation is at least as bleak as it was in the post-Dust Bowl years when Jacks commented:

The organization of civilized societies is founded upon the measures taken to wrest control of the soil from wild Nature, and not until complete control has passed into human hands can a stable super-structure of what we call civilization be erected on the land. (Jacks and Whyte, 1939: 17)

Figure 4.10
The spread of contour-strip soil conservation methods in Wisconsin, USA between 1939 and 1967. One dot represents one adopter (after H. E. Johansen in Trimble and Lund, 1982, figure 22)

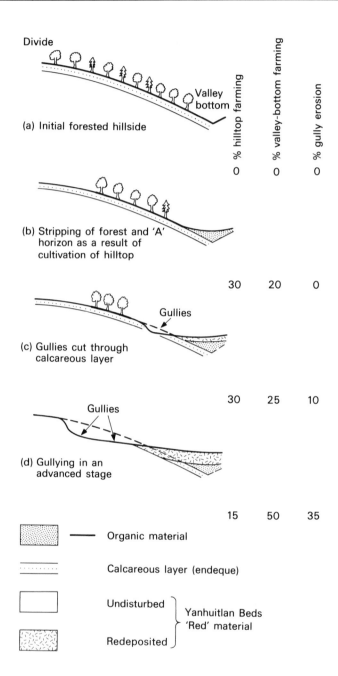

Figure 4.11
Beneficial effect of gullying on production of agricultural land in Nochixtlan, Mexico (after Kirkby in Whyte, 1977, figure 1)

FIVE

The Human Impact on the Waters

Deliberate modification of rivers

Although there are many ways in which humans influence water quantity and quality in rivers and streams – for example, by direct channel manipulation, modification of basin characteristics, urbanization and pollution – the first of these is of particularly great importance (Mrowka, 1974). Indeed, there are a great variety of methods of direct channel manipulation and many of them have a long history. Perhaps the most widespread of these is the construction of dams and reservoirs. The first recorded dam was constructed in Egypt some 5000 years ago, but since that time the adoption of this technique has spread, whether to improve agriculture, to prevent floods, to generate power or to provide a reliable source of water.

In recent decades the demand for water has increased rapidly. Global water use has more than tripled since 1950, and now stands at 4340 km³ per year – equivalent to eight times the annual flow of the Mississippi River. Irretrievable water losses have increased about sevenfold this century (table 5.1a) and the construction of large dams increased markedly, especially between 1945 and the early 1970s (Beaumont, 1978, and figure 5.1). Engineers have now built more than 36000 dams around the world and, as table 5.1b shows, such large dams (i.e. more than 15 m high) are still being constructed at an appreciable rate, especially in Asia. In the late 1980s some 45 very large dams (more than 150 m high) were under construction. Indeed, one of the most striking features of dams and reservoirs is that they have become increasingly large (Beckinsale, 1969). Thus in the 1930s the Hoover or Boulder Dam in the USA (221 m high) was by far the tallest in the world and it impounded the biggest reservoir, Lake Mead, which contained 38 billion m³ of water. By the 1980s it was exceeded in height by at least 18 others, and some of these impounded reservoirs with four times the volume of Lake Mead.

Table 5.1 Major changes in the hydrological environment

(a) *Irretrievable water losses (km³/yr)*

Users	1900	1940	1950	1960	1970	1980	1990	2000
Agriculture	409	679	859	1180	1400	1730	2050	2500
Industry	3.5	9.7	14.5	24.9	38.0	61.9	88.5	117
Municipal supply	4.0	9.0	14	20.3	29.2	41.1	52.4	64.5
Reservoirs	0.3	3.7	6.5	23.0	66.0	120	170	220
Total	417	701	894	1250	1540	1950	2360	2900

(b) *Number of large dams (>15m high) constructed or under construction, 1950–86*

Continent	1950	1982	1986	Under construction 31.12.86
Africa	133	665	763	58
Asia	1,562	22,789	22,389	613
Australasia/Oceania	151	448	492	25
Europe	1,323	3,961	4,114	222
North and Central America	2,099	7,303	6,595	39
South America			884	69
Total	5,268	35,166	36,327	1,026
of which in China	8	18,595	18,820	183

Source: data provided by UNEP

Such large dams are capable of causing almost total regulation of the streams they impound but, in general, the degree to which peak flows are reduced depends on the size of the dam and the impounded lake in relation to catchment characteristics. In Britain, as table 5.2 shows, peak flow reduction downstream from selected reservoirs varies considerably, with some tendency for the greatest degree of reduction to occur in those catchments where the reservoirs cover the largest percentage of the area. When considering the magnitude of floods of different recurrence intervals before and after dam construction, it is clear that dams have much less effect on rare events of high magnitude (Petts and Lewin, 1979), and this is brought out in table 5.3. None the less, most dams achieve their aim: to regulate river discharge. They are also highly successful in fulfilling the needs of surrounding communities: millions of people depend upon them for survival, welfare and employment.

However, dams may have a whole series of environmental consequences that may or may not have been anticipated (figure 5.2, and Makkaveyev, 1972). Some of these are dealt with in greater detail elsewhere, such as subsidence, p. 254, earthquake triggering, (p. 296), the transmission and expansion in the range of organisms, (p. 74) the build-up of soil salinity, (p. 140), changes in ground-water levels creating slope instability, (p. 273), and water-

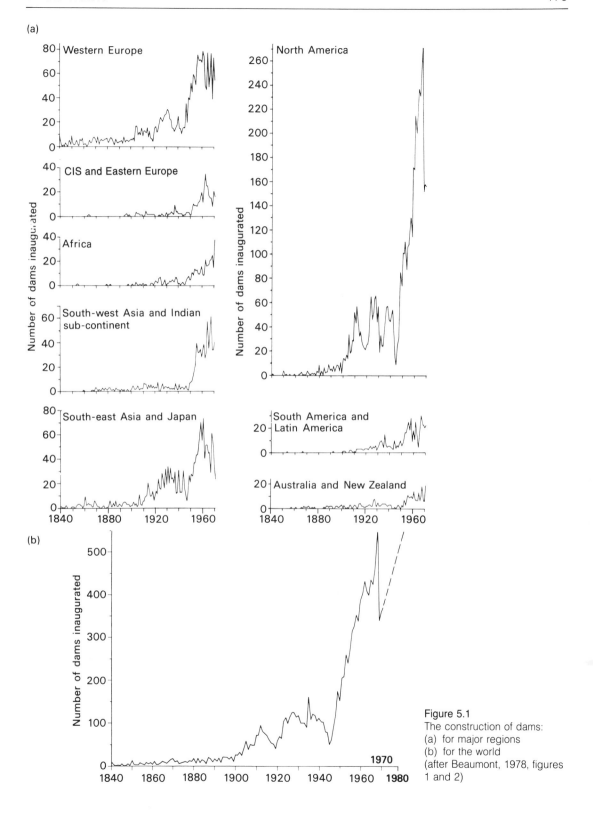

Figure 5.1
The construction of dams:
(a) for major regions
(b) for the world
(after Beaumont, 1978, figures
1 and 2)

Plate 5.1
The Thames Barrier. Its
objective is to remove the
threat of flooding by controlling
the movement of high tidal
surges up the river

Table 5.2 Peak flow reduction downstream from selected British reservoirs

Reservoir	% of catchment inundated	% of peak flow reduction
Avon, Dartmoor	1.38	16
Fernworth, Dartmoor	2.80	28
Meldon, Dartmoor	1.30	9
Vyrnwy, mid-Wales	6.13	69
Sutton Bingham, Somerset	1.90	35
Blagdon, Mendip	6.84	51
Stocks, Forest of Bowland	3.70	70
Daer, Southern Uplands	4.33	56
Camps, Southern Uplands	3.13	41
Catcleugh, Cheviots	2.72	71
Ladybower, Peak District	1.60	42
Chew Magna, Mendips	8.33	73

Source: after Petts and Lewin, 1979, table 1, p. 82

Table 5.3 The ratios of post- to pre-dam discharges for flood magnitudes of selected frequency

Reservoir	Recurrence interval (years)			
	1.5	2.3	5.0	10.0
Avon, R. Avon	0.90	0.89	0.93	1.02
Stocks, R. Hodder	0.83	0.86	0.84	0.95
Sutton Bingham, R. Yeo	0.52	0.61	0.69	0.79

Source: after Petts and Lewin, 1979, table 2, p. 84

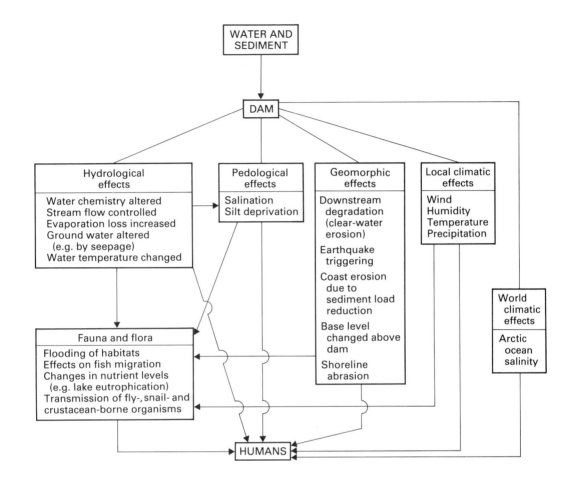

Figure 5.2
Generalized representation of the possible effects of dam construction on human life and various components of the environment

logging, (p. 141). Several of these processes may in turn affect the viability of the scheme for which the dam was created.

A particularly important consequence of impounding a reservoir behind a dam is the reduction in the sediment load of the river downstream. A clear demonstration of this effect has been given for the South Saskatchewan River in Canada (table 5.4) by Rasid (1979). During the pre-dam period, typified by 1962, the total annual suspended loads at Saskatoon and Lemsford Ferry were remarkably similar. As soon as the reservoir began to fill late in 1963, however, some of the suspended sediment began to be trapped, and the transitional period was marked by a progressive reduction in the proportion of sediment which reached Saskatoon. In the four years after the dam was fully operational the mean annual sediment load at Saskatoon was only 9 per cent that at Lemsford Ferry.

Even more dramatic are the data for the Colorado River in the USA (figure 5.3). Prior to 1930 it carried around 125–150 m t of suspended sediment per year to its delta at the head of the Gulf of

Table 5.4 Total yearly suspended load (imperial tons) of the South Saskatchewan River at Lemsford Ferry and Saskatoon, 1962–70

Period of record		Lemsford Ferry	Saskatoon	Difference at Saskatoon (%)
Pre-dam				
1962		1 813 305	1 873 278	+3
Transitional				
1963		4 892 159	446 788	−8
1964		7 711 006	4 145 553	−46
1965		9 731 837	2 721 021	−72
1966		5 228 474	1 675 142	−68
	MEAN	6 890 869	3 254 626	−53
Post-dam				
1967		12 618 814	445 967	−96
1968		2 660 643	101 412	−96
1969		10 562 413	2 146 496	−80
1970		5 643 278	117 847	−98
	MEAN	7 871 287	702 930	−91

Source: Rasid, 1979, table 1

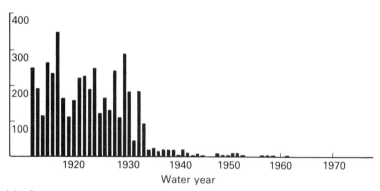

(a)   Suspended-sediment discharge (millions of tons/yr)

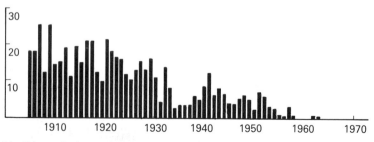

Figure 5.3
Historical (a) sediment and (b) water discharge trends for the Colorado River, USA (after the United States Geological Survey in Schwarz et al., 1991)

(b)   Water discharge (millions of acre-feet/yr)

Table 5.5 Silt concentrations (in parts per million) in the Nile at Gaafra before and after the constructiion of the Aswan High Dam

BEFORE (averages for the period 1958–63)											
Jan.											*Dec.*
64	50	45	42	43	85	674	2702	2422	925	124	77
AFTER											
44	47	45	50	51	49	48	45	41	43	48	47
RATIO											
1.5	1.1	1.0	0.8	0.8	1.7	14.0	60.0	59.1	21.5	2.58	1.63

Source: Abu-Atta, 1978, p. 199

California. Following a series of dams the Colorado now discharges neither sediment nor water to the sea (Schwarz et al., 1991).

Sediment retention is also well illustrated by the Nile (table 5.5), both before and after the construction of the great Aswan High Dam. Until its construction the late summer and autumn period of high flow was characterized by high silt concentrations, but since it has been finished the silt load is rendered lower throughout the year and the seasonal peak is removed. Petts (1985, table XVIII) indicates that the Nile now only transports 8 per cent of its natural load below the Aswan High Dam, although this figure seems to be exceptionally low. Other rivers for which data are available carry between 8 and 50 per cent of their natural suspended loads below dams.

Plate 5.2
The Aswan High Dam in Egypt largely controls the flow of the Nile, providing protection both against floods and water shortages. The massive structure does, however, have some undesirable consequences including the trapping of fertile silts

Sediment removal in turn has various possible consequences, including a reduction in flood-deposited nutrients on fields, less nutrients for fish in the south-east Mediterranean Sea, accelerated erosion of the Nile Delta, and accelerated riverbed erosion since less sediment is available to cause bed aggradation. The last process is often called 'clear-water erosion' (see Beckinsale, 1972), and in the case of the Hoover Dam it affected the river channel of the Colorado for 150 km downstream by causing incision. In turn, such channel incision may initiate a rejuvenation of headward erosion in tributaries and may cause the lowering of ground-water tables and the undermining of bridge piers and other structures downstream of the dam. On the other hand, in regions such as northern China, where modern dams trap silt, the incision of the river channel downstream may alleviate the strain on levées and lessen the expense of levée strengthening or heightening.

However, clear-water erosion does not always follow from silt retention in reservoirs. There are examples of rivers where, before impoundment, floods carried away the sediment brought into the main stream by steep tributaries. Reduction of the peak discharge after the completion of the dam leaves some rivers unable to scour away the sediment that accumulates as large fans of sand or gravel below each tributary mouth (Dunne and Leopold, 1978). The bed of the main stream is raised and if water-intakes, towns or other structures lie alongside the river they can be threatened again by flooding or channel shifting across the accumulating wedge of sediment. Rates of aggradation of a metre a year have been observed, and tens of kilometres of channel have been affected by sedimentation. One of the best documented cases of aggradation concerns the Colorado River below Glen Canyon Dam in the USA. Since dam closure the extremes of river flow have been largely eliminated so that the ten years' recurrence interval flow has been reduced to less than one-third. The main channel flow is no longer capable of removing sediment provided by flash-flooding tributaries, and deposits up to 2.6 m thick have accumulated within the upper Grand Canyon (Petts, 1985: 133).

Some landscapes in the world are dominated by dams, canals and reservoirs. Probably the most striking example of this (figure 5.4) is the 'tank' landscape of south-east India where myriads of little streams and areas of overland flow have been dammed by small earth structures to give what Spate (Spate and Learmonth, 1967: 778) has likened to 'a surface of vast overlapping fish-scales'.

In the northern part of the subcontinent, in Sind, the landscape changes wrought by hydrology are no less striking, with the mighty snow-fed Indus being controlled by large embankments (*bunds*) and interrupted by great barrages. Its waters are distributed over thousands of square kilometres by a canal network that has evolved over the past 4000 years (figure 5.5).

Another landscape where equally far-reaching changes have been wrought is the Netherlands. Coates (1976) has calculated that, before 1860, reclamation of that country from the sea, in the

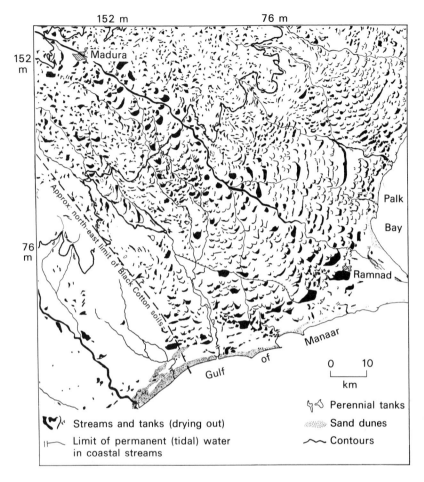

Figure 5.4
The Madurai–Ramanathapuram
tank country in south India
(after Spate and Learmonth,
1967, figure 25.12)

extension of drainage lines, involved the movement of 1000 million
m³ of material. The area is dominated by human constructions:
canals, rivers, drains and lakes. As C. T. Smith (1978: 506) has put
it:

It can be said of the Netherlands with more truth than of any other country
that it is a creation of man, for it owes its existence to the long continued
struggle to drain and reclaim new land and to protect existing lands from
the invasion of the sea.

Particularly in the sixteenth, seventeenth and eighteenth centuries
the Dutch applied their remarkable expertise in hydrological management to countries other than their own and drained many areas
both on coasts and inland (figure 5.6), creating major river
networks.

Another direct means of river manipulation is channelization.
This involves the construction of embankments, dikes, levées and
floodwalls to confine floodwaters; and improving the ability of

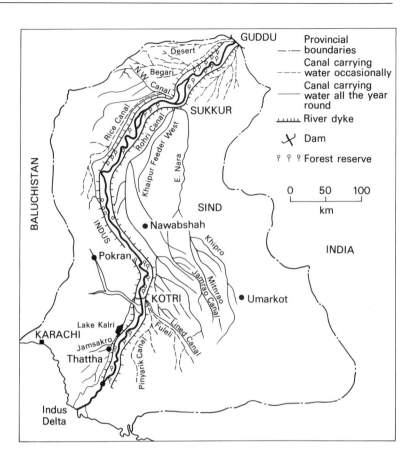

Figure 5.5
The irrigated areas in Sind
(Pakistan) along the Indus
Valley (after Manshard, 1974,
figure 5.7)

channels to transmit floods by enlarging their capacity through straightening, widening, deepening or smoothing (table 5.6).

Some of the great rivers of the world are now lined by extensive embankment systems such as those that run for more than 1000 km alongside the Nile, 700 km along the Hwang Ho, 1400 km by the Red River in Vietnam, and over 4500 km in the Mississippi Valley (Ward, 1978). Like dams, embankments and related structures often fulfil their purpose but they may also create some environmental problems and have some disadvantages. For example, they reduce natural storage for floodwaters, both by preventing water from spilling on to much of the floodplain, and by stopping bank storage in cases where impermeable floodwalls are used. Likewise the flow of water in tributaries may be constrained. In addition, embankments may occasionally exacerbate the flood problem they were designed to reduce by preventing floodwaters downstream of a breach from draining back into the channel once the peak has passed.

Channel improvement, designed to improve water flow, may also have unforseen or undesirable effects. For example, the more rapid movement of water along improved channel sections can aggravate

Plate 5.3
Polders have transformed the
nature of the low-lying
coastline of the Netherlands.
Large areas have been
reclaimed from the sea. The
Afsluitdijk separates the
Wazenezee and the IJsselmeer

flood peaks further downstream and cause excessive erosion. The
lowering of water-tables in the 'improved' reach may cause over-
drainage of adjacent agricultural land so that sluices must be
constructed in the channel to maintain levels. On the other hand,
lined channels may obstruct soil water movement (interflow) and
shallow ground water and so cause surface saturation. Brookes
(1985) and Gregory (1985a) provide useful reviews on the impact
of channelization.

Channelization may also have miscellaneous effects on fauna
through the increased velocities of water flow, reductions in the
extent of shelter in the channel bed, and by reduced nutrient inputs
due to the destruction of overhanging bank vegetation (see Keller,
1976). In the case of large swamps, like those of the Sudd in Sudan
or the Okavango in Botswana, the channelization of rivers could

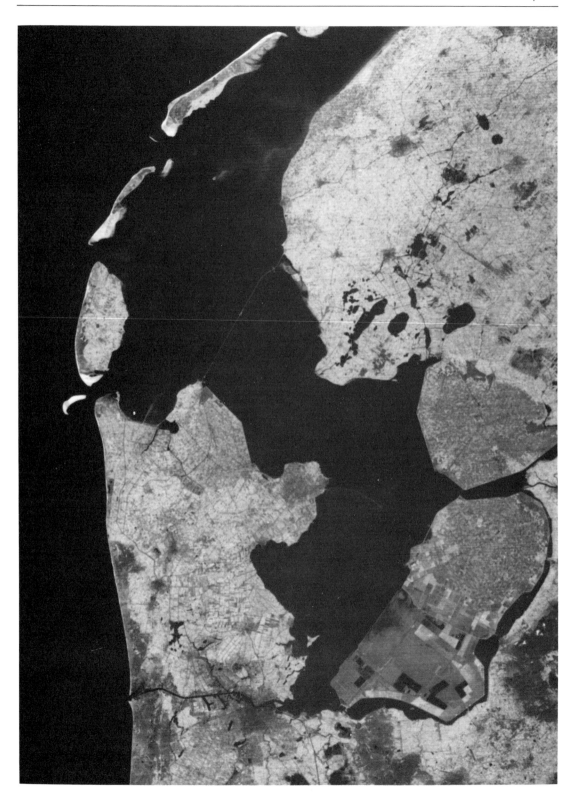

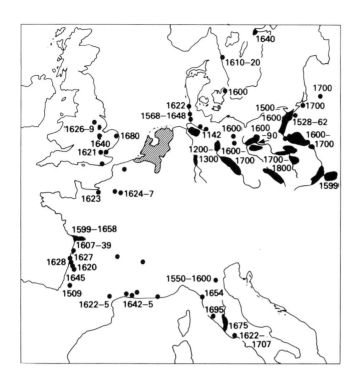

Figure 5.6
Areas where the great Dutch
engineers were involved in
wetland reclamation schemes
(after Van Veen in C. T. Smith,
1978, figure 9.12)

Table 5.6 Selected terminologies for the methods of river channelization in the USA and UK

American term	British equivalent	Method involved
Widening	Resectioning	Increase of channel capacity by manipulating width
Deepening	Resectioning	and/or depth variable
Straightening	Realigning	Increasing velocity of flow by steepening the gradient
Diking	Embanking	Raising of channel banks to confine floodwaters
Bank stabilization	Bank protection	Methods to control bank erosion e.g. gabions and concrete structures
Clearing and snagging	Pioneer tree clearance	Removal of obstructions from a watercourse, thereby
	Control of aquatic plants	decreasing the resistance
	Dredging of sediments	and increasing the velocity of
	Urban clearing	flow

Source: Brookes, 1985, table 1

completely transform the whole character of the swamp environment. Figure 5.7 illustrates some of the differences between natural and artificial channels.

Another type of channel modification is produced by the construction of bypass and diversion channels to carry excess floodwater or to enable irrigation to take place. Such channels may be as

Plate 5.4
The Landsat image of part of
the Netherlands shows the
Ijsselmeer, the polders and the
coastal barrier

NATURAL CHANNEL

Suitable water temperatures:
  adequate shading; good cover for fish
  life; minimal variation in temperatures;
  abundant leaf material input

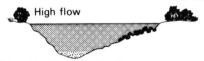

Pool-riffle sequence

Pool
(silt, sand
and fine
gravel)

Riffle
(coarse gravel)

Sorted gravels provide diversified habitats
for many stream organisms

ARTIFICIAL CHANNEL

Increased water temperatures:
  no shading; no cover for fish life;
  rapid daily and seasonal fluctuations
  in temperatures; reduced leaf material
  input

Mostly riffle

Unsorted gravels:
  reduction in habitats; few organisms

POOL ENVIRONMENT

High flow

Diversity of water velocities:
  high in pools, lower in riffles. Resting areas
  abundant beneath undercut banks or behind
  large rocks, etc.

Low flow

Sufficient water depth to support fish and
other aquatic life during dry season

High flow

May have stream velocity higher than
some aquatic life can withstand. Few
or no resting places

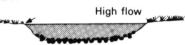

Low flow

Insufficient depth of flow during dry
seasons to support diversity of fish
and aquatic life. Few if any pools
(all riffle)

Figure 5.7
Comparison of the natural
channel morphology and
hydrology with that of a
channelized stream,
suggesting some possible
ecological consequences (after
Keller, 1976, figure 4)

old as irrigation itself. They may contribute to the salinity problems
encountered in many irrigated areas.

Deliberate modification of a river regime can also be achieved by
long-distance inter-basin water transfers (Shiklomanov, 1985),
transfers necessitated by the unequal spatial distribution of water
resources and by the increasing rates of water consumption. At
present, the world water consumption for all human needs is
$3500 \text{ km}^3 \text{ y}^{-1}$: at the start of the century it was eight times less than
this figure, and by the end of the century it is expected to be
$6000 \text{ km}^3 \text{ y}^{-1}$. The total volume of water in the various transfer
systems in operation and under construction on a global scale is

about 300 km^3 y^{-1}, with the largest countries in terms of volume of transfers being Canada, the CIS, the USA and India.

In future decades it is likely that many even greater schemes will be constructed (see figure 5.8); route lengths of some hundreds of kilometres will be common, and the water balances of many rivers and lakes will be transformed. This is already happening in the CIS (figure 5.9), where the operation of various anthropogenic activities of this type have caused runoff in the most intensely cultivated central and southern areas to decrease by 30–50 per cent compared to normal natural runoff. At the same time, inflows into the Caspian and Aral Seas have declined sharply, so that their levels have fallen and their areas decreased.

When one turns to the coastal portions of rivers, to estuaries, the possible effects of another human impact, dredging, can be as complex as the effects of dams and reservoirs upstream (La Roe, 1977). Dredging and filling are certainly widespread and often desirable. Dredging may be performed to create and maintain canals, navigation channels, turning basins, harbours and marinas; to lay pipelines; and to obtain a source of material for fill or construction. Filling is the deposition of dredged materials to create new land. There are miscellaneous ecological effects of such actions. In the first place, filling directly disrupts habitats. Second, the generation of large quantities of suspended silt tends physically to smother bottom-dwelling plants and animals; tends to smother fish by clogging their gills; reduces photosynthesis through the effects of turbidity; and tends to lead to eutrophication by an increased nutrient release. Likewise, the destruction of marshes, mangroves and sea grasses by dredge and fill can result in the loss of these natural purifying systems. The removal of vegetation may also cause erosion. Moreover, as silt deposits stirred up by dredging accumulate elsewhere in the estuary they tend to create a 'false bottom'. Characterized by shifting, unstable sediments, the dredged bottom, fill deposits or spoil areas are slowly – if at all – recolonized by fauna and flora. Furthermore, dredging tends to change the configuration of currents, the rate of fresh-water drainage and may provide avenues for salt-water intrusion.

Urbanization and its effects on river flow

The process of urbanization has a considerable hydrological impact, in terms of controlling rates of erosion and the delivery of pollutants to rivers, and in terms of influencing the nature of runoff and other hydrological characteristics. An attempt to generalize some of these impacts using a historical model of urbanization is summarized usefully by Savini and Kammerer (1961) and reproduced here in table 5.7.

One of the most important effects is the way in which urbanization affects flood runoff. Research both in the United States and in Britain has shown that, because urbanization produces extended

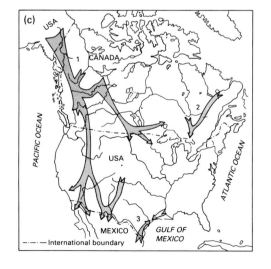

Figure 5.8
Some major schemes
proposed for large-scale
inter-basin water transfers:
(a) Projected water transfer
    systems in the CIS:
    1. from the Onega River,
    and in future from Onega
    Bay 2. from the Sukhona
    and Northern Dvina Rivers
    3. from the Svir River and
    Lake Onega 4. from the
    Pechora River 5. from the
    Ob River 6. from the
    Danube delta
(b) Projected systems for
    water transfers in India:
    1. scheme of the national
    water network
    2. scheme of Grand Water
    Garland
(c) Some major projects for
    water transfers in North
    America: 1. NAWAPA
    2. Grand Canal 3. Texas
    . River basins
    (after Shiklomanov, 1985,
    figures 12.6, 12.9 and
    12.11, in *Facets of
    Hydrology II*, ed. J. C.
    Rodda, by permission of
    John Wiley and Sons Ltd)

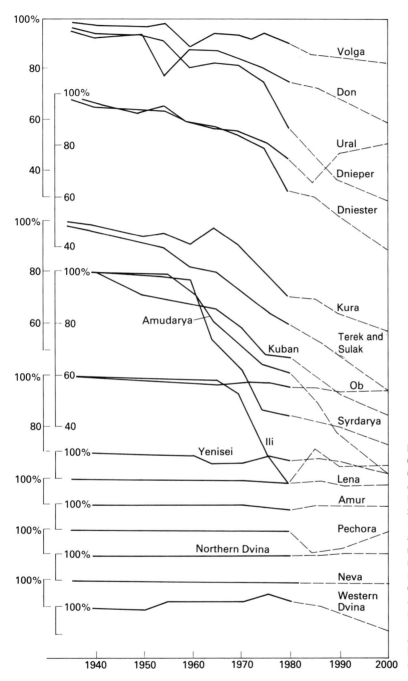

Figure 5.9
Changes in annual runoff in the CIS due to human activity during 1936–2000 (for the Kura, Terek and Sulak, Syrdarya, Amudarya, Ili Rivers, 100 per cent has been accepted as the norm at the mouths for a conventionally natural period; for the other rivers, natural water resources of the basins have been accepted as the norm of 100 per cent) (from Shiklomanov 1985, figure 12.7 in *Facets of Hydrology II*, ed. J. C. Rodda, by permission of John Wiley and Sons Ltd)

impermeable surfaces of tarmac, tiles and concrete, there is a tendency for flood runoff to increase in comparison with rural sites. City drainage densities may be greater than those in natural conditions (Graf, 1977) and the installation of sewers and storm drains accelerates runoff, as illustrated in figure 5.10. The greater

Plate 5.5
The desiccation of the Aral
Sea, perhaps the greatest
environmental disaster of the
CIS, has in large measure
resulted from the transfer of
water from rivers flowing into it.
The fishing industry has
suffered greatly

the area that is sewered the greater is the discharge for a particular
recurrence level (figure 5.11). Peak discharges are higher and occur
sooner after runoff starts in basins that have been affected by
urbanization and the installation of sewers. Some runoff may be
generated in urban areas because low vegetation densities mean
that evapotranspiration is limited.

However, in many cases the effect of urbanization is greater on
small floods; as the size of the flood and its recurrence interval
increase, so the effect of urbanization diminishes (Martens, 1968;
Hollis, 1975). A probable explanation for this is that, during a
severe and prolonged storm event, a non-urbanized catchment may
become so saturated and its channel network so extended that it
begins to behave hydrologically as if it were an impervious catchment with a dense surface-water drain network. Under these
conditions, a rural catchment produces floods of a type and size
similar to those of its urban counterpart. Moreover, a further
mechanism probably operates in the same direction, for in an urban
catchment it seems probable that some throttling of flow occurs in
surface-water drains during intense storms, tending to attenuate the
very highest discharges. Thus, Hollis believes, whilst the size of
small frequent floods is increased many times by urbanization,
large, rare floods (the ones likely to cause extreme damage) are not
significantly affected by the construction of surburban areas within
a catchment area (figure 5.12). None the less, a whole series of

Table 5.7 Stages of urban growth and their miscellaneous hydrological impacts

Stage	Impact
1	*Transition from pre-urban to early-urban stage:*	
(a) Removal of trees or vegetation	Decrease in transpiration and increase in storm flow
(b) Construction of scattered houses with limited sewerage and water facilities	
(c) Drilling of wells	Some lowering of water-table
(d) Construction of septic tanks, etc.	Some increase in soil moisture and perhaps some contamination
2	*Transition from early-urban to middle-urban stage:*	
(a) Bulldozing of land	Accelerated land erosion
(b) Mass construction of houses, etc.	Decreased infiltration
(c) Discontinued use and abandonment of some shallower wells	Rise in water-table
(d) Diversion of nearby streams for public supply	Decrease in runoff between points of diversion of disposal
(e) Untreated or inadequately treated sewerage into streams and wells	Pollution of streams and wells
3	*Transition from middle-urban to late-urban stage:*	
(a) Urbanization of area completed by addition of more buildings	Reduced infiltration and lowered water-table; higher flood peaks and lower low flows
(b) Larger quantitites of untreated waste into local streams	Increased pollution
(c) Abandonment of remaining shallow wells because of pollution	Rise in water-table
(d) Increase in population requiring establishment of new water supply and distribution systems	Increase in local stream flow if supply is from outside basin
(e) Channels of streams restricted at least in part to artificial channels and tunnels	Higher stage for a given flow (therefore increased flood damage); changes in channel geometry and sediment load
(f) Construction of sanitary drainage system and treatment plant for sewage	Removal of additional water from area
(g) Improvement of storm drainage system	
(h) Drilling of deeper, large-capacity industrial wells	Lowered water pressure, some subsidence, salt-water encroachment
(i) Increased use of water for air-conditioning	Overloading of sewers and other drainage facilities
(j) Drilling of recharge wells	Raising of water pressure surface
(k) Waste-water reclamation and utilization	Recharge to ground-water aquifers; more efficient use of water resources

Source: modified after Savini and Kammerer, 1961

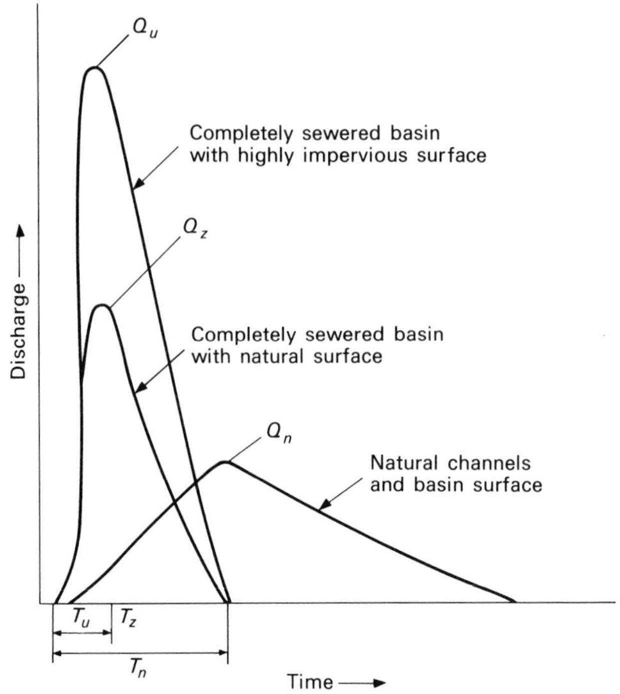

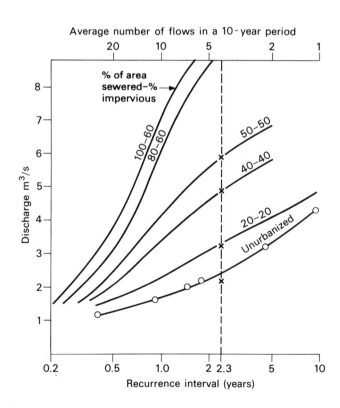

Figure 5.10
Effect of urban development
on flood hydrographs. Peak
discharges (Q) are higher and
occur sooner after runoff starts
(T) in basins that have been
developed or sewered (after
Fox, 1979, figure 3)

Figure 5.11
Flood frequency curves for a 1
square mile basin in various
states of urbanization (after US
Geological Society in Viessman
et al., 1977, figure 11.33)

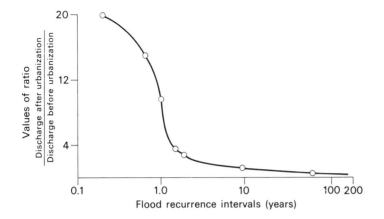

Figure 5.12
Effects on flood magnitude of
paving 20 per cent of a basin
(after Hollis, 1975)

techniques have been developed in an attempt to reduce and delay
urban storm runoff (table 5.8), for Hollis's findings may not be
universally applicable. For example, K. V. Wilson (1967) working
in Jackson, Mississippi, found that the fifty-year flood for an
urbanized catchment was three times higher than that of a rural one
(plate 5.6).

Water transfers for industrial and municipal purposes are having
an increasing impact on river regimes in urbanized societies
(Richards and Wood, 1977). Under dry weather flow conditions,
stream flow in urban catchments in the Midlands and south-east
England is mainly imported water. This is illustrated by the River
Tame which contains about 90 per cent effluent in its 95 per cent
duration flow. This proportion is increasing as more water is
imported (figure 5.13), reduced in quality and added to low natural
base-flows reduced by lack of recharge.

Deforestation and its effect on river flow

As we have already noted, one of the first major indications that
humans could inadvertently adversely affect the environment was
the observation that deforestation could create torrents and floods.
The deforestation that gives rise to such flows can be produced
both by felling and by fire.

The first experimental study, in which a planned land-use change
was executed to enable observation of the effects of stream flow,
began at Wagon Wheel Gap, Colorado, in 1910 (Pereira, 1973).
Here stream flow from two similar watersheds of about 80 hectares
each were compared for eight years. One valley was then clear-felled
and the records were continued. After the clear-felling the annual
water yield was 17 per cent above that predicted from the flows of
the unchanged control valley. Peak flows are also increased.
Studies on two small basins in the Australian Alps (Wallace's
Creek – 41 km^2; Yarrango Billy River – 224 km^2), which were

Table 5.8 Measures for reducing and delaying urban storm runoff

Area	Reducing runoff	Delaying runoff
Large flat roof	Cistern storage Rooftop gardens Pool storage or fountain storage Sod roof cover	Ponding on roof by constricted drainpipes Increasing roof roughness (1) rippled roof (2) gravelled roof
Car parks	Porous pavement (1) gravel car parks (2) porous or punctured asphalt Concrete vaults and cisterns beneath car parks in high value areas Vegetated ponding areas around car parks Gravel trenches	Grassy strips on car parks Grassed waterways draining car parks Ponding and detention measures for impervious areas (1) rippled pavement (2) depressions (3) basins
Residential	Cisterns for individual homes or groups of homes Gravel drives (porous) Contoured landscape Ground-water recharge (1) perforated pipe (2) gravel (sand) (3) trench (4) porous pipe (5) dry wells Vegetated depressions	Reservoir or detention basin Planting a high-delaying grass (high roughness) Grassy gutters or channels Increased length of travel of runoff by means of gutters, diversions and so
General	Gravel alleys Porous pavements	Gravel alleys

Source: after US Department of Agriculture, Soil Conservation Service, 1972 in Viessman et al., 1977: 569

burned over, showed that rain storms, which from previous records would have been expected to give rise to flows of $60–80 \text{ m}^3 \text{ s}^{-1}$, produced a peak of $370 \text{ m}^3 \text{ s}^{-1}$ (a five- or sixfold increase). Likewise catchment experiments in Arizona have shown that when chaparral scrub is burned there is a tenfold increase in water yield.

Experiments with tropical catchments have shown typical maximum and mean annual stream-flow increases of 400–450 mm per year on clearance with increases in water yield of up to 6 mm per year for each percentage reduction in forest area above a 15 per cent change in cover characteristics (Anderson and Spencer, 1991: 49).

Additional data for the effects of land-use change on annual runoff levels are presented for tropical areas in table 4.11.

Plate 5.6
Flood risk in rivers like the
Mississippi may be increased
by human activities. These
floods occurred at Cape
Giradean, Missouri, 1973.
Among major causes are
deforestation and urbanization

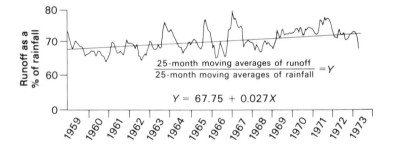

Figure 5.13
Trends in 25-month moving
averages of runoff as a
percentage of 25-month
moving averages of rainfall for
the River Tame, Lee Marston,
England, 1957–74. The trend
reflects increasing amounts of
imported 'runoff' (after
Richards and Wood, 1977,
figure 24.6)

As vegetation regenerates after a forest has been cut or burned, so stream flow tends to revert to normal, though the process may take some decades. This is illustrated in figure 5.14 which shows the dramatic effects produced on the Coweeta catchments in North Carolina by two spasms of clear-felling, together with the gradual return to normality in between.

The substitution of one forest type for another may also affect stream flow. This can again be exemplified from the Coweeta catchments, where two experimental catchments were converted from a mature deciduous hardwood forest cover to a cover of white pine (*Pinus strobus*). Fifteen years after the conversion, annual stream flow was found to be reduced by about 20 per cent (Swank and Douglass, 1974). The reason for this notable change is that the interception and subsequent evaporation of rainfall is greater for pine than it is for hardwoods during the dormant season.

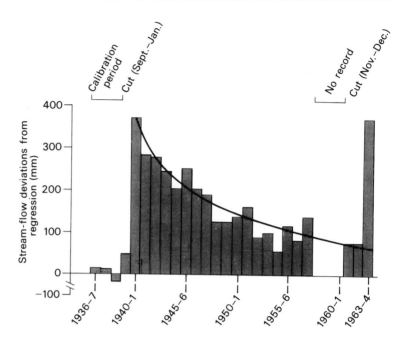

Figure 5.14
The increase of water yield after clear-felling a forest: a unique confirmation from the Coweeta catchment in North Carolina, USA (after Pereira, 1973, figure 8)

Fears have also been expressed that the replacement of tall natural forests by eucalpts will produce a decline in stream-flow. However, most current research does not support this contention, for transpiration rates from eucalpyts are similar to those from other tree species (except in situations with a shallow ground-water table) while their interception losses are if anything generally rather less than those from other tree species of similar height and planting density (Bruijnzeel, 1990).

The reasons why the removal of a forest cover and its replacement with pasture, crops or bare ground have such important effects on stream flow are many. A mature forest probably has a higher rainfall interception rate, a tendency to reduce rates of overland flow, and probably generates soils with a higher infiltration capacity and better general structure. All these factors will tend to produce both a reduction in overall runoff levels and less extreme flood peaks. However, with careful management the replacement of forest by other land-use types need not be detrimental in terms of either sediment loss or flood generation. In Kenya, for example, the tea plantations with shade trees, protective grass covers, and carefully designed culverts were found to be 'a hydrologically effective substitute for natural forest' (Pereira, 1973: 127).

In many studies the runoff from clean-tilled land tends to be greater than that from areas under a dense crop cover, but tilling the soil surface does not always increase runoff (Gregory and Walling, 1973: 345). There are reports from the CIS suggesting that the reverse may be the case, and that autumn ploughing can decrease surface runoff, presumably because of its effect on surface detention and on soil structure.

Grazing practices also influence runoff, for heavy grazing can both compact the soil and cause vegetation removal. In general it tends to lead to an increase in runoff.

Changes in river-bank vegetation may have a particularly strong influence on river flow. In the south-west USA, for instance, many streams are lined by the salt cedar (*Tamarix pentandra*). With roots either in the water-table or freely supplied by the capillary fringe, these shrubs have full potential transpiration opportunity. The removal of such vegetation can cause large increases in stream flow. It is interesting to note that the salt cedar itself is an alien, native to Eurasia, which was introduced by humans. It has spread explosively in the south-west USA, increasing from about 4000 hectares in 1920 to almost 400 000 hectares in the early 1960s (Harris, 1966). In the Upper Rio Grande valley in New Mexico, the salt cedar was introduced to try and combat anthropogenic soil erosion, but it spread so explosively that it came to consume approximately 45 per cent of the area's total available water (Hay, 1973).

Reforestation of abandoned farmlands reverses the effects of deforestation: increased interception and evapotranspiration can cause a decline in water yield. In parts of the eastern USA, farm abandonment and recolonization of the land by pines, spruce and cedar has been occurring throughout the twentieth century, and this has reduced stream flow by important amounts at a time when water supplies for some eastern cities were becoming critically short (Dunne and Leopold, 1978).

A process which is often associated with afforestation is peat drainage. The hydrological effects of this are the subject of controversy, since there are cases of both increased and decreased flood peaks after drainage. It has been suggested that differences in peat type alone might account for the different effects. Thus it is conceivable that the drainage of a *Sphagnum* catchment would lead to increased flooding since *Sphagnum* compacts with drainage, reducing its storage volume and its permeability. On the other hand, in the case of non-*Sphagnum* peat there would be relatively less change in structure, but there would be a reduction in moisture content and an increase in storage capacity, thereby tending to reduce flood flows. The nature of the peat is, however, but one feature to consider (Robinson, 1979). The intensity of the drainage works (depth, spacing etc.) may also be important. In any case, there may be two (sometimes conflicting) processes operating as a result of peat drainage: the increased drainage network will facilitate rapid runoff, and the drier soil conditions will provide greater storage for rainfall. Which of these two tendencies is dominant will depend on local catchment conditions.

However, the impact of land drainage upon downstream flood incidence has long been a source of controversy. Much depends on the scale of study, the nature of land management and the character of the soil that has been drained. After a detailed review of experience in the UK, Robinson (1990) found that the drainage of heavy clay soils that are prone to prolonged surface saturation in their undrained state generally led to a reduction of large and medium flow peaks. He attributed this to the fact that their natural response is 'flashy' (with limited soil water storage available), whereas their drainage largely eliminates surface saturation. By contrast, the drainage of permeable soils, which are less prone to such surface saturation, improves the speed of subsurface flow, thereby tending to increase peak flow levels.

It is not always easy to determine whether an increase in flood frequency or intensity is the result of land-use changes of the type we have been discussing, or whether some natural changes in rainfall have played a role. In central and southern Wales there is some clear evidence of changes in flood magnitude and frequency over recent decades, and this has sometimes been attributed to the increasing amount of afforestation that has been carried out since the First World War, and to the drainage of upland areas that this has necessitated. While in the Severn catchment this appears to be a partial explanation (Howe et al., 1967), in other basins the main cause of more frequent and intense floods appears to have been a marked increase in the magnitude and frequency of heavy daily rainfalls. For example, in the case of the Tawe valley near Swansea, of 17 major floods since 1875, 14 occurred from 1929–81 and only 3 during the 1875–1928 period. Of 22 notable widespread heavy rainfalls in the Tawe catchment since 1875, only 2 occurred during 1875–1928, but 20 from 1929–81 (Walsh et al., 1982).

The human impact on lake levels

One of the results of human modification of river regimes is that lake levels have suffered some change, though it is not always possible to distinguish between the part played by humans and that played by natural climatic changes.

A lake basin for which there are particularly long records of change is the Valencia Basin in Venezuela (Böckh, 1973). It was the declining level of the waters in this lake which so struck the great German geographer, von Humboldt, in 1800. He recorded its level as being about 422 m above sea-level, and some previous observations on its level were made by Manzano in 1727, which established it as being at 426 m. The 1968 level was about 405 m, representing a fall of no less than 21 m in about 240 years (figure 5.15). Humboldt believed that the cause of the declining level was the deforestation brought about by humans, and this has been supported by Böckh (1973), who points also to the abstraction of water for irrigation. This remarkable fall in level meant that the lake ceased to have an overflow into the River Orinoco. It has as a consequence become subject to a build-up in salinity, and is now eight times more saline than it was two-and-a-half centuries ago.

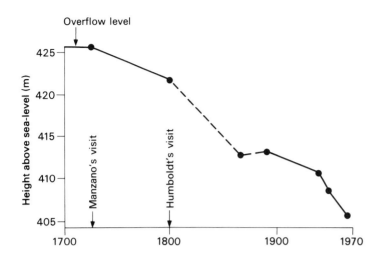

Figure 5.15
Variations in the level of Lake
Valencia, Venezuela, to 1968
(after Böckh, 1973, figure 18.2)

Even the world's largest lake, the Caspian, has been modified by
human activities. The most important change has been the fall of
3 m in its level since 1929 (see figure 5.16). This decline is
undoubtedly partly the product of climatic change (Micklin 1972),
for winter precipitation in the northern Volga Basin, the chief
flow-generating area of the Caspian, has been generally below
normal since 1930 because of a reduction in the number of moist
cyclones penetrating into the Volga Basin from the Atlantic. None
the less, human actions have contributed to this fall, particularly
since the 1950s because of reservoir formation, irrigation, muni-

Figure 5.16
The changing levels of the Aral
and Caspian Seas in the CIS
(a) The nature of the changes
    until 1973, illustrating the
    very rapid decline in levels
    since the Revolution.
(b) The past and predicted
    contraction of the Aral Sea
    as its level falls (modified
    after Hollis, 1978, p. 63)

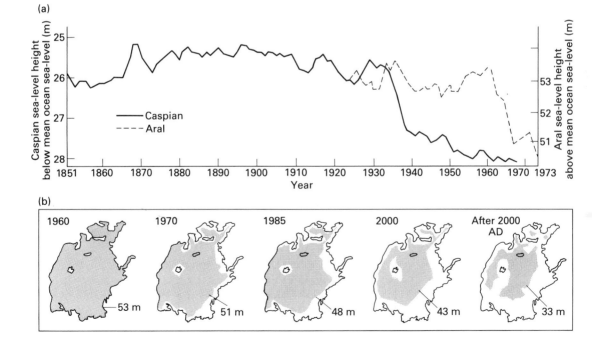

cipal and industrial withdrawals, and agricultural practices. In addition to the fall in level, salinity in the northern Caspian has increased by 30 per cent since the early 1930s. A secondary effect of the changes in level has been a decline in fish numbers due to the disappearance of the shallows. These are biologically the most productive zones of the lake, providing a food base for the more valuable types of fish and also serving as spawning grounds for some species. There are plans to divert some water from northward-flowing rivers in Siberia towards the Volga to correct the decline in Caspian levels, but the possible climatic impacts of such action have caused some concern.

Similar declines in inflow to those evident in the Caspian apply to the Sea of Azov (Mote and Zumbrunne, 1977). The reduction of freshwater inputs from inflowing rivers has permitted an increase in the ingress of saline water from the Black Sea. Salinity increased at the rate of 30–40 per cent in five years in the early 1970s (figure 5.17). As a consequence, the valuable sturgeon and carp fisheries have declined, while the numbers of unwelcome stinging Black Sea jellyfish have multiplied.

Perhaps the most severe change to a major inland sea is that taking place in the Aral Sea of the CIS (figure 5.18). Since 1960 the Aral Sea has lost more than 40 per cent of its area and about 60 per cent of its volume, and its level has fallen by more than 14 m (Kotlyakov, 1991). This has lowered the artesian water table over a band 80–170 km in width, has exposed 24 000 km^2 of former lake bed to desiccation, and has created salty surfaces from which salts are deflated to be transported in dust storms, to the detriment of soil quality. The mineral content of what remains has increased almost threefold over the same period. It is probably the most dire ecological tragedy to have afflicted the CIS, and as with the Caspian's decline much of the blame rests with excessive use of water which would otherwise replenish the sea.

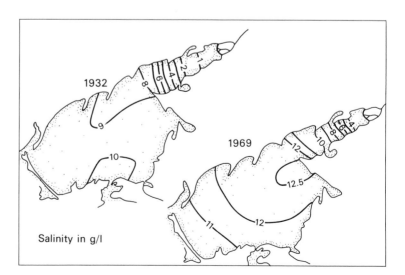

Figure 5.17
The changing salinity of the Sea of Azov (after Mote and Zumbrunne, 1977, figure 2)

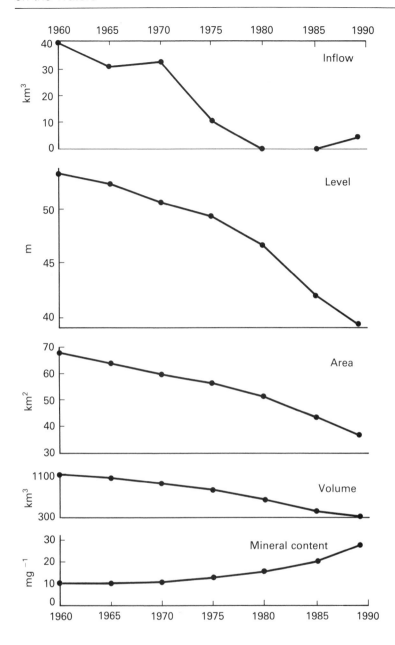

Figure 5.18
Changes in the Aral Sea,
1960–1989 (Kotlyakov, 1991)

   Water abstraction from the Jordan River for irrigation purposes has caused a decline in the level of the Dead Sea. Under natural conditions fresh water from the Jordan constantly fed the less salty layer of the sea which occupied roughly the top 40 m of the 320 m deep body of water. Because the amount of water entering the lake via the Jordan was more or less equal to the quantity lost by evaporation, the lake maintained its stable, stratified state, with less salty water resting on the waters of high salinity. However, with the recent human-induced diminution in Jordan discharge, the sea's upper layer has receded because of the intense levels of evaporation. Its salinity has approached that of the older and deeper waters. It has now been established that as a consequence the layered structure has collapsed, creating a situation where there is increased precipitation of salts. Moreover, now that circulating waters carry oxygen to the bottom, the characteristic hydrogen-sulphide smell has largely disappeared (Maugh, 1979).

From time to time there have been proposals for major augmentation of lake volumes, either by means of river diversions or by allowing the ingress of sea water through tunnels or canals. Among such plans have been those to flood the salt lakes of the Kalahari by transferring water from the rivers of central Africa such as the Zambezi; the scheme to transfer Mediterranean water to the Dead Sea and to the Quattara Depression; and the Zaire–Chad scheme. Perhaps the most ambitious idea has been the so-called 'Atlantropa' project, whereby the Mediterranean would be empoldered, a dam built at the Straits of Gibraltar, and water fed into the new lake from the Zaire River (Cathcart, 1983).

Changes in ground-water conditions

In many parts of the world humans obtain water supplies by pumping from ground water (see, for example, Drennan, 1979). This has two main effects: the reduction in the levels of water-tables and the replacement, in coastal areas, of fresh water by salt water. Environmental consequences of these two phenomena include ground subsidence and soil salinization. Increasing population levels and the adoption of new exploitation techniques (for example, the replacement of irrigation methods involving animal or human power by electric and diesel pumps) has increased these problems.

Some of the reductions in ground-water levels that have been caused by abstraction are considerable. Figure 5.19 shows the rapid increase in the number of wells tapping ground water in the London area from 1850 until after the Second World War, while figure 5.20 illustrates the widespread and substantial changes in ground-water conditions that resulted. The piezometric surface in the confined chalk aquifer has fallen by more than 60 m over hundreds of square kilometres. Likewise, beneath Chicago, Illinois, pumping since the late nineteenth century has lowered the piezometric head by some 200 m. The drawdown that has taken place in the Great Artesian Basin of Australia locally exceeds 80–100 m (Lloyd, 1986).

The reductions in water levels that are taking place in the High Plains of Texas are some of the most serious, and threaten the long-term viability of irrigated agriculture in that area. Before irrigation development started in the 1930s, the High Plains ground-water system was in a state of dynamic equilibrium, with long-term recharge equal to long-term discharge. However, the

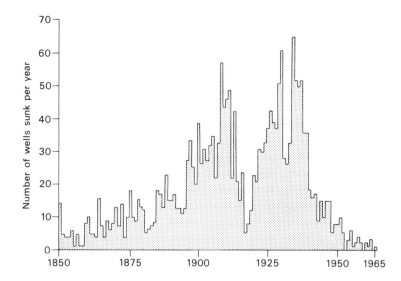

Figure 5.19
Construction of wells tapping
the confined aquifer below
London, 1850–1965

ground-water is now being mined at a rapid rate to supply centre-pivot and other schemes. In a matter of only fifty years or less, the water level has declined by 30 to 50 m in a large area to the north of Lubbock. The thickness of the aquifer has decreased more than 50 per cent in large parts of Castro, Crosby, Floyd, Hale, Lubbock, Parmer and Swisher counties, and already a decrease in the area irrigated by each well is taking place as a consequence of decreasing well yields (*The Cross Section*, Lubbock, 1982: 1).

However, there are situations where humans deliberately endeavour to increase the natural supply of ground water by attempting artificial recharge of ground-water basins (figure 5.21). Where the materials containing the aquifer are permeable (as in some alluvial fans, coastal sand dunes or glacial deposits) the technique of water-spreading is much used. In relatively flat areas river water may be diverted to spread evenly over the ground so that infiltration takes place. Alternative water-spreading methods may involve releasing water into basins which are formed by excavation, or by the construction of dikes or small dams. On alluvial plains water can also be encouraged to percolate down to the water-table by distributing it to a series of ditches or furrows. In some situations natural channel infiltration can be promoted by building small check dams down a stream course. In irrigated areas surplus water can be spread by irrigating with excess water during the dormant season. When artificial ground-water recharge is required in sediments with impermeable layers such water-spreading techniques are not effective and the appropriate method may then be to pump water into deep pits or into wells. This last technique is used on the coastal plain of Israel, both to replenish the ground-water reservoirs when surplus irrigation water is available, and to attempt to diminish the problems associated with salt-water intrusion.

(a)

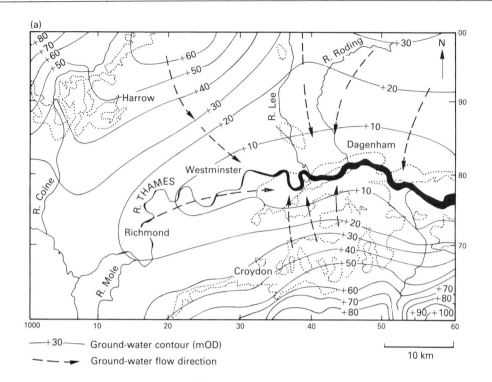

─+30─── Ground-water contour (mOD)

── ─ ➤ Ground-water flow direction

10 km

(b)

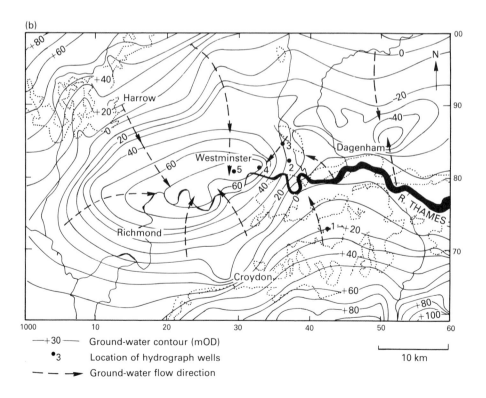

─+30 ── Ground-water contour (mOD)

•3   Location of hydrograph wells

── ─ ➤ Ground-water flow direction

10 km

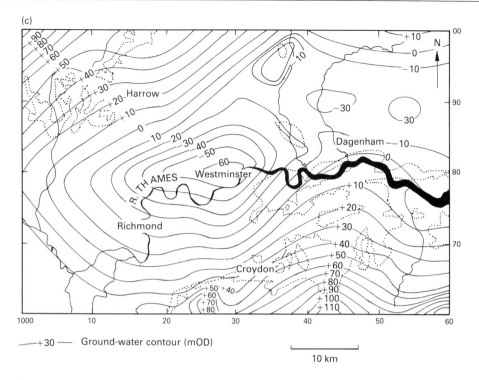

+30 —— Ground-water contour (mOD)

|——————————| 10 km

Figure 5.20
Ground-water levels in the London area, (a) prior to major development, (b) in 1965 and (c) in 1985 (from Wilkinson and Brassington, 1991, figures 4.5, 4.6, 4.7)

As figure 5.22 shows, ground-water extraction in the Tel Aviv area greatly exceeded the sustained yield of the aquifer during the 1950s and water levels declined below sea-level over about 60 km². Sea-water intrusion became a problem. Completion of a new inter-basin water transfer scheme in the mid-1960s allowed the injection of fresh water into a line of wells 8 km long, in an attempt to provide a fresh-water 'barrier' along the coast. Other boreholes were injected to the landward of this 'barrier' to help fill the aquifer that had been depressed by over-pumping. This operation proved to be effective almost immediately (Mandel, 1977), and by 1969 the situation had definitely stabilized.

In some industrial areas, reductions in industrial activity have caused a recent reduction in ground-water abstraction, and as a consequence ground-water levels have begun to rise, a trend that is exacerbated by considerable leakage losses from ancient, deteriorating pipe and sewer systems. In cities like London, Liverpool and Birmingham an upward trend has already been identified (Brassington and Rushton, 1987; Price and Reed, 1989). In London, because of a 46 per cent reduction in ground-water abstraction, the water table in the Chalk and Tertiary beds has risen by as much as 20 m. Such a rise has numerous implications, which are listed in table 5.9.

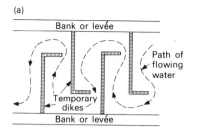

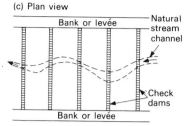

(c) Elevation view

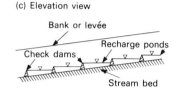

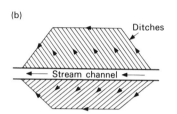

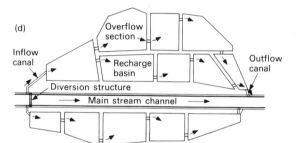

Plate 5.7 (opposite)
In Nebraska, America, fields are irrigated by centre-pivot irrigation schemes which use ground water. Ground-water levels have fallen rapidly in many areas because of the adoption of this type of irrigation technology

Figure 5.21
Examples of ways of recharging ground water by means of water-spreading devices:
(a) temporary earth dikes
(b) ditch networks
(c) check dams and ponds
(d) connected recharge basins adjoining a stream channel (after Todd, 1964, figures 13.46 and 13.48)

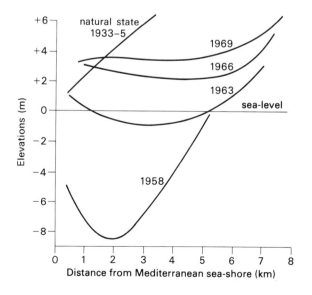

Figure 5.22
Profiles of the water-table in the Tel Aviv area, Israel, perpendicular to the sea coast showing its natural state, its state after overpumping, and its state after the aquifer had been injected with fresh water brought in by inter-basin water transfer. (After Bachmat in Mandel, 1977, figure 3.3, from *Arid Zone Development*, ed. Mundlak and Singer. Copyright © 1977, the American Academy of Arts and Sciences. Reprinted by permission of the Ballinger Publishing Company)

Table 5.9 Possible effects of a rising ground-water level

Increase in spring and rivers flows
Re-emergence of 'dry springs'
Surface water flooding
Pollution of surface waters and spread of underground pollution
Flooding of basements
Increased leakage into tunnels
Reduction of slope and retaining wall stability
Reduction in bearing capacity of foundations and piles
Increased hydrostatic uplift and swelling pressures on foundations and
structures
Swelling of clays
Chemical attack on foundations

Source: modified after Wilkinson and Brassington, 1991, table 4.3, p. 42

Water pollution

Water pollution is not new. During the reign of King George III a
Member of Parliament reportedly wrote a letter to the Prime
Minister complaining about the odour and appearance of the
Thames. He wrote the letter, it is said, not with ink, but with water
from the Thames itself (Strandberg, 1971).

Water pollution is frequently undesirable: it causes disease
transmission through infection; it may poison humans and animals;
it may create objectional odours and unsightliness; it may be the
cause of the unsatisfactory quality even of treated water; and it may
affect economic activities like shellfish culture.

The causes and forms of water pollution created by humans are
many and can be classified into groups as follows (after Strandberg, 1971):

1 Sewage and other oxygen-demanding wastes
2 Infectious agents
3 Organic chemicals
4 Other chemical and mineral substances
5 Sediments (turbidity)
6 Radioactive substances
7 Heat (thermal pollution)

Moreover, many human activities can contribute to changes in
water quality, including agriculture, fire, urbanization, industry,
mining, irrigation and many others. Of these, agriculture is probably the most important. Some pollutants merely have local
effects, while others, such as acid rain or DDT, may have continental or even planetary implications.

It is not a simple matter to recognize trends in pollution. In
general long-term records are sparse, gauging stations are often
moved, analytical methods change, and some trends in water
chemistry may be due to natural factors (including climatic
change) rather than to people. However, where long-term data are

available some of the changes in water chemistry are striking. In their study of the information collated on Lake Michigan, the Mississippi River, the Ohio River and the Illinois River, Ackerman et al. (1970) found that since the start of the twentieth century dissolved solids concentrations had risen markedly. For chlorides the increase has varied from 100 per cent in 52 years to 300 per cent in 70 years, while for sulphates it has increased from 80 per cent in 43 years to 155 per cent in 62 years.

Figure 5.23 shows an historical data set, albeit imperfect, for some rivers in Europe and North America, of changes in sulphur concentrations. These data show an increase in the European rivers by a factor of about 3, while that for the North American rivers is about 1.5 (Husar and Husar, 1991).

It is also plainly not a simple matter to try to estimate the global figure for the extent of water pollution caused by humans. For one thing we know too little about the natural long-term levels of dissolved materials in the world's rivers. None the less, Meybeck (1979) has calculated that about 500 million tonnes of dissolved salts reach the oceans each year as a result of human activity. These inputs have increased by more than 30 per cent the natural

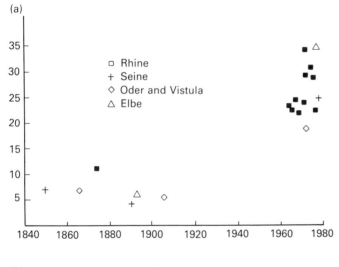

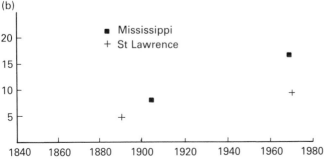

Figure 5.23
Trends in sulphur concentrations (expressed as mg/l) for (a) selected European rivers and (b) selected North American rivers (from Husar and Husar, 1991, figures 24.15, 24.16)

values for sodium, chloride and sulphate, and have created an overall global augmentation of river mineralization by about 12 per cent. Likewise, Peierls et al. (1991) have demonstrated that the quantity of nitrates in world rivers now appears to be closely correlated to human population density. Using published data for 42 major world rivers they found a highly significant correlation between annual nitrate concentration and human population density that explained 76 per cent of the variation in nitrate concentration for the 42 rivers. They maintain that 'human activity clearly dominates nitrate export from land.'

Chemical pollution by agriculture and other activities

Agriculture may be one, if not the most, important cause of pollution, either by the production of sediments or by the generation of chemical wastes (Ministry of Agriculture, Fisheries and Food, 1976). With regard to the latter, it has been suggested that denitrification processes in the environment are incapable of keeping pace with the rate at which atmospheric nitrogen is being mobilized through industrial fixation processes and being introduced into the biosphere in the form of commercial fertilizers (Manners, 1978). Nitrogen, with phosphorus, tends to regulate the growth of aquatic plants and therefore the eutrophication of inland waters. Excess nitrates can also cause health hazards to humans and animals, including methaemo-globinaemia in infants.

Eutrophication is the enrichment of waters by nutrients (Ryding and Rast, 1989). The process occurs naturally during, for example, the slow ageing of lakes, but it can be accelerated both by runoff from fertilized agricultural land and by the discharge of domestic sewage and industrial effluents (Lund, 1972). This process, often called 'cultural eutrophication', commonly leads to excessive growths of algae followed in some cases by a serious depletion of dissolved oxygen as the algae decay after death. Oxygen levels may become too low to support fish life, resulting in fish kills. Changes in diatom assemblages can also occur (see, particularly, Battarbee, 1977). The nature of these changes is summarized in table 5.10.

As agriculture has, in the developed world, become of an increasingly specialized and intensive nature, so the pollution impact has increased. The traditional mixed farm tended to be a more or less closed system which generated relatively few external impacts. This was because crop residues were fed to livestock or incorporated in the soil; and manure was returned to the land in amounts that could be absorbed and utilized. Many farms have become more specialized with the separation of crop and livestock activities, large numbers of stock may be kept on feedlots, silage may be produced in large silos, and synthetic fertilizers may be applied to fields in large quantities (Conway and Pretty, 1991).

Increases in the application of nitrogenous fertilizers have almost amounted to a revolution (Green, 1976). The long-term

Table 5.10 Characteristics of lakes experiencing 'cultural eutrophication'

Biological factors
(a) *Primary productivity*: usually much higher than in unpolluted water and is manifest as extensive algal blooms
(b) *Diversity of primary produces*: initially green algae increase, but blue-green algae rapidly become dominant and produce toxins. Similarly, macrophytes (e.g. reed maces) respond well initially but due to increased turbidity and anoxia (see below) they decline in diversity as eutrophication proceeds
(c) *Higher tropic level productivity*: overall decrease in response to factors given in this table
(d) *Higher tropic level diversity*: decreases due to factors given in this table. The species of macro- and micro-invetebrates which tolerate more extreme conditions increase in numbers. Fish are also adversely affected and populations are dominated by surface dwelling coarse fish such as pike and perch

Chemical factors
(a) *Oxygen content of bottom waters (hypolimnion)*: this is usually low due to algal blooms restricting oxygen exchange between the water and atmosphere. Oxygen-deficient conditions (anoxia) develop, especially at night when algae are not photosynthesizing. Thus seasonal and diurnal patterns of oxygen availability occur. The decay of algal blooms also produces anoxia
(b) *Salt content of water*: this can be very high and a further restriction on floral and faunal diversity

Physical factors
(a) *Mean depth of water body*: as infill occurs the depth decreases
(b) *Volume of hypolimnion*: varies
(c) *Turbidity*: this increases, as sediment input increases, and restricts the depth of light penetration which can become a limiting factor for photosynthesis. It is also increased if boating is a significant activity

Source: modified after Mannion, 1992, table 11.4

trends in the UK are illustrated in figure 5.24a. The growth in fertilizer usage since the Second World War has been exponential, and in spite of the increasing costs of energy supplies and hydro-carbons in the 1970s world fertilizer production has continued to rise inexorably (table 5.11).

When one examines the evidence to see whether increasing nitrate fertilizer applications are indeed leading to widespread upward trends in nitrate concentrations in river and ground waters, it is often conflicting (Manners, 1978; Rodda et al., 1976). This may be because soil and vegetation have a large storage capacity for nitrogen, while a great deal probably depends on local soil, climate and land-management conditions. Further problems are presented by the inadequacy of long-term data. This complexity in the evidence is brought out strongly in Tomlinson's (1970) analysis of trends of nitrate concentrations in English rivers summarized in table 5.12. In six rivers concentrations increased significantly with time, in eight the correlation coefficients were positive

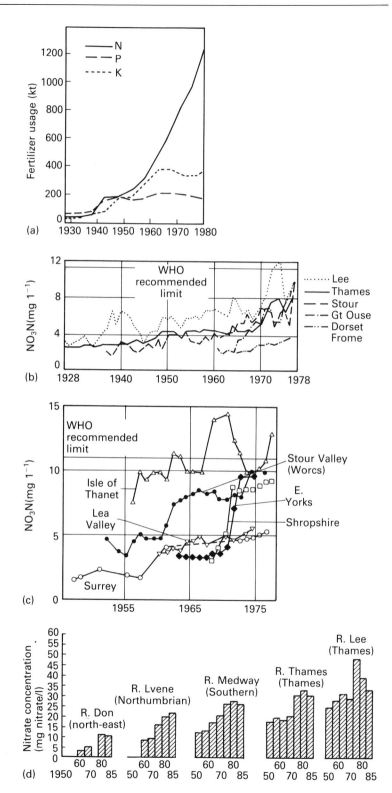

Figure 5.24
Nitrates in surface and ground
waters in the UK:
(a) The trends in annual
    fertilizer usage in the UK
    during the period 1928–80
(b) Trends in mean and
    annual nitrate
    concentration in five rivers
    for which long-term data
    are available
(c) Nitrate concentrations in
    selected public water
    supply abstraction
    boreholes in the Chalk and
    Triassic sandstone aquifers
    of the UK (Royal Society
    Study Group, 1983, figures
    4, 18 and 31)
(d) Changing nitrate
    concentrations in five UK
    rivers. The averages are
    five-year means (from
    Department of the
    Environment statistics in
    Conway and Pretty, 1991,
    figure 4.8)

Table 5.11 World fertilizer production

| | | | (index 1970 = 100) | | | Tonnes/km² |
	1970	1975	1980	1985	1988	1988
Canada	100	184	301	403	371	2.6
USA	100	128	147	129	132	5.1
Japan	100	102	101	112	109	13.7
France	100	114	142	156	166	13.3
Germany	100	109	139	136	138	20.6
Italy	100	145	198	206	182	7.6
Netherlands	100	116	122	121	101	46.7
Spain	100	134	161	171	201	5.5
Sweden	100	115	111	111	103	7.6
UK	100	121	143	179	169	20.9
North America	100	130	152	138	140	4.6
OECD Pacific	100	98	99	112	114	2.1
OECD Europe	100	124	153	169	172	9.9
OECD	100	125	148	149	150	5.8
World	100	136	185	212	242	5.4

Source: OECD, 1991, figure 27

Table 5.12 Trends in nitrate concentrations in English rivers in relation to fertilizer use

River	Place	Period	Correlation coefficient with time
Dee	Chester	1961–7	−0.77
Great Ouse	Bedford	1953–67	−0.22
Kennett	Reading	1953–67	+0.34
Rother	Horsham	1957–67	+0.90
Severn	Tewkesbury	1953–67	+0.78
Stour	Langham	1953–67	+0.73
Tees	Darlington	1958–67	−0.34
Thames	Hampton	1953–64	+0.22
Blithe	Newton	1953–67	+0.56
Derwent	Hathersage	1953–67	+0.26
Devon	Hawton	1953–67	+0.47
Dove	Monks Bridge	1953–67	+0.68
Dove	Stratton	1961–7	+0.47
Leen	Newstead Abbey	1957–63	−0.16
Manifold	Iham	1956–67	+0.25
Poulter	Elkesley	1953–67	+0.41
Tyne	Wylam	1956–67	+0.08
Wensum	Norwich	1953–67	+0.05

Source: from data in Tomlinson, 1970

but not significant, in three cases the correlations were negative and not significant while in one case the correlation was negative and significant. However, the situation since Tomlinson has now become rather clearer (Royal Society Study Group, 1983) and, although there are considerable fluctuations from year to year, the trend in nitrate levels in English rivers is clearly apparent (figure

5.24b). They are 50 to 400 per cent higher than they were 20 years ago. The River Thames, which provides the major supply of London's water, has increased its mean annual nitrate concentration from around 11 mg per litre in 1928 to 35 mg per litre in the 1980s.

Also causing concern is the trend in nitrate concentrations in British ground waters. Investigations have revealed a large quantity of nitrate in the unsaturated zone of the principal aquifers (mainly Chalk and Triassic sandstone), and this is slowly moving down towards the main ground-water body. The slow transit time means that in many water supply wells increased nitrate concentrations will not occur for 20 to 30 years, but they will then be above acceptable levels for human health (figure 5.24c).

However, even if an increase in nitrate levels is evident this may not necessarily be because of the application of fertilizers (Burt and Haycock, 1992). Some nitrate pollution may be derived from organic wastes. There has also been some anxiety in Britain over a possible decline in the amount of organic matter present in the soil, which could limit its ability to assimilate nitrogen. Moreover, the pattern of tillage may have affected the liberation of nitrogen. The increased area and depth of modern ploughing accelerates the decay of residues and may change the pattern of water movement in the soil. Finally, tile drainage has also expanded in area very greatly in recent decades in England (figure 5.25). This has affected the movement of water through the soil and hence the degree of leaching of nitrates and other materials (Edwards, 1975).

Given the uncertainties surrounding the question of nitrate pollution, and bearing in mind the major advances in agricultural

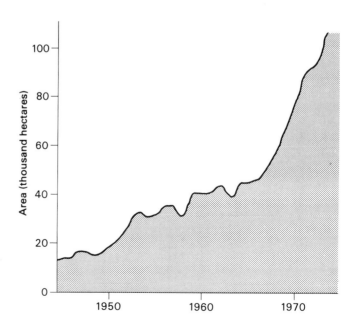

Figure 5.25
Annual total area of field
drainage by tile drains in
England and Wales (after
Green, 1976, figure 3)

productivity that they have permitted, attempts at controlling their use have not been received with complete favour. Indeed, as Viets (1971) has pointed out, if fertilizer use were curtailed there would be less vegetation cover and increased erosion and sediment delivery to rivers. This, he suggests, would necessitate an increase in land hectarage to maintain production levels which would also cause greater erosion; at the same time the increased ploughing would lead to greater nitrate loss from grasslands. Nevertheless some farmers are inefficient and wasteful in their use of fertilizers and there is scope for economy (Cooke, 1977).

Pesticides are another source of chemical pollution brought about by agriculture. There is now a tremendous range of pesticides and they differ greatly in their mode of action, in the length of time they remain in the biosphere, and in their toxicity.

Much of the most adverse criticism of pesticides has been directed against the chlorinated hydrocarbon group of insecticides, which includes DDT and Dieldrin. These insecticides are toxic not only to the target organism but to other insects too (that is, they are non-specific). They are also highly persistent. Appreciable quantities of the original application may survive in the environment in unaltered form for years. This can have two rather severe effects: global dispersal and the 'biological magnification' of these substances in food chains (Manners, 1978). This last problem is well illustrated for DDT in figure 5.26.

In addition to the influence of nitrate fertilizers, the confining of animals on feedlots results in tremendously concentrated sources of nutrients and pollutants. Animal wastes in the USA are estimated to be as much as 1.6 billion tonnes per year, with 50 per cent of this amount originating from feedlots. It needs to be

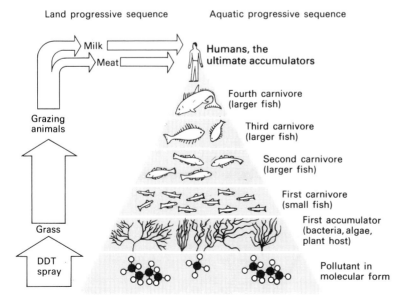

Figure 5.26
Biological concentration occurs when relatively indestructible substances (DDT for example) are ingested by lesser organisms at the base of the food pyramid. An estimated 1000 kg of plant plankton are needed to produce 100 kg of animal plankton. These, in turn, are consumed by 10 kg of fish, the amount needed by a person to gain 1 kg. The ultimate consumer (man or woman) then takes in the DDT taken in by 1000 kg of the lesser creatures at the base of the food pyramid, when he or she ingests enough fish to gain 1 kg (after Strandberg, 1971, figure 2)

remembered that 1000 head of beef produce the equivalent organic load of 6000 people (Sanders, 1972). The number of cattle fed
on feedlots in the USA has increased rapidly (see figure 5.27),
partly because *per capita* consumption of meat has moved sharply
upward, and partly because the confinement of cattle on small
lots, on which they are fed a controlled selection of feeds, leads to
the greatest possible weight gains with the least possible cost and
time (Bussing, 1972). Cattle feedlot runoff is a high-strength
organic waste, high in oxygen-demanding material, and has caused
many fish kills in rivers in states such as Kansas.

There has been a marked increase in Britain in the phosphorus
content of some rivers since the early 1950s, an increase related to
the use of detergents (figure 5.28). Other changes in surface-water
chemistry may be produced by the washing out of acids that pollute
the air in precipitation. This is particularly clear in the heavily
industrialized area of north-western Europe, where there is a
marked zonation of acid rain with pH values that often fall below 4
or 5. A broadly comparable situation exists in North America
(Johnson, 1979).

Very considerable fears have been expressed about the possible
effects that acid deposition may have on aquatic ecosystems, and

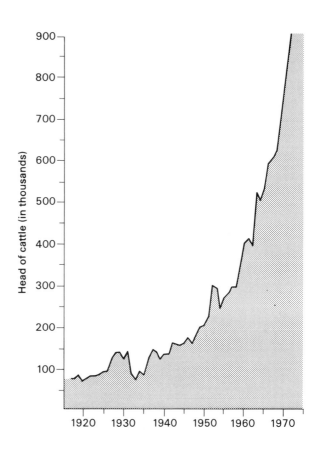

Figure 5.27
Cattle on feed in feedlots in
Colorado, 1917–71 (after
Bussing, 1972, figure 2)

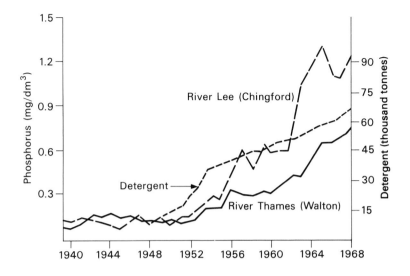

Figure 5.28
Comparison of the increasing
phosphate content of the
Rivers Thames and Lee with
the annual consumption of
detergents (after Royal
Commission on Environmental
Pollution in Rodda et al., 1976,
figure 7.11)

particularly on fish populations. It is now generally accepted that
increasing lake and river acidification has caused fish kills and
stock depletion. Fishless lakes now occur in areas like the Adiron-
dacks in the north-east USA. Species of fish vary in their tolerance
to low pH, with rainbow trout being largely intolerant to values of
pH below 6.0, while salmon, brown and brook trout are
somewhat less sensitive. Values of 4.0–3.5 are lethal to salmonids,
while values of 3.5 or less are lethal to most fish.

Numerous lacustrine molluscs and crustaceans are not found
even at weakly acid pH values of 5.8–6.0, and crayfish rapidly lose
their ability to recalcify after moulting as the pH drops from 6.0 to
5.5 (Committee on the Atmosphere and the Biosphere, 1981).

Fish declines and deaths in areas afflicted by acid rain are not
solely caused by water acidity *per se*. It has become evident that
under conditions of high acidity certain metal ions, including
aluminium, are readily mobilized, leading to their increasing concentration in fresh waters. Some of these metal ions are highly
toxic and many mass mortalities of fish have been attributed to
aluminium poisoning rather than high acidity alone. The aluminium adversely affects the operation of fish gills, causing mucus to
collect in large quantities; this eventually inhibits the ability of the
fish to take in necessary oxygen and salts. In addition, aluminium's
presence in water can reduce the amount of available phosphates,
an essential food for phytoplankton and other aquatic plants. This
decreases the available food for fish higher in the food chain,
leading to population decline (Park, 1987).

Mining can also create serious chemical pollution. In the strip-
mining area of Kentucky, for example. Collier (1964) found that
oxidation of the iron sulphide in the spoil banks and other parts of
the mining area produced an excess of hydrogen ions. The ions
then reacted with the mineral matter of the spoil bank, releasing

soluble products at a rate ten to fifteen times greater than in the non-mined area. Such pollution was not entirely undesirable in its effects. Collier found that this release of minerals was good for tree growth, increasing the rate of development.

In Britain, coal seams and mudstones in the coal measures contain pyrite and marcasite (ferrous sulphide). In the course of mining the water-table is lowered, air gains access to these minerals and they are oxidized. Sulphuric acid may be produced.

As a consequence, mine drainage waters may have pH values as low as 2 or 3. They have high sulphate and iron concentrations, they may contain toxic metals, and because of the reaction of the acid on clay and silicates in the rocks, they may also contain appreciable amounts of calcium, magnesium, aluminium and manganese. Following reactions with sediments or mixing with alkaline river waters, or when chemical or bacterial oxidation of ferrous compounds occurs, the iron may precipitate as ferric hydroxide. This may discolour the water and leave unsightly deposits (Rodda et al., 1976: 299). Some of the reactions involved in the development of acid mine drainage have been set down by Down and Stocks as follows (1977: 110–11):

1 Oxidation of the sulphide (usually FeS_2) in a wet environment producing ferrous sulphate and sulphuric acid.
$$2FeS_2 + 2H_2O + 7O_2 = 2FeSO_4 + 2H_2SO_4$$
2 Ferrous sulphate, in the presence of sulphuric acid and oxygen, can oxidize to produce ferric sulphate, especially in the presence of the bacterium *Thiobacillus ferro-oxidans*.
$$4FeSO_4 + 2H_2SO_4 + O_2 = 2Fe_2(SO_4)_3 + 2H_2O$$
3 The ferric iron so produced combines with the hydroxyl $(OH)^-$ ions of water to form ferric hydroxide, which is insoluble in acid and precipitates.
$$Fe_2(SO_4)_3 + 6H_2O = 2Fe(OH)_3 + 3H_2SO_4$$

The use of salt for de-icing has increased the chloride content of runoff in some areas. For example, the chloride content of Lough Neagh, Northern Ireland, almost doubled between 1958 and 1969 because of salt application to roads in its catchment. In the USA some 6 million tonnes of salt are applied to the roads each year, and in those states where it is applied the quantities amount to some 32 tonnes km^{-1}. Sugar maples seem susceptible to such high levels of salt and ancient specimens are dying along rural roads all over New England (Wagner, 1974: 99).

Similarly, storm-water runoff from urban areas may contain large amounts of contaminants, derived from litter, garbage, car-washings, horticultural treatments, vehicle drippings, industry, construction, animal droppings and the chemicals used for snow and ice clearance. Comparison of contaminant profiles for urban runoff and raw domestic sewage, based on surveys throughout the USA, indicates the importance of pollution from urban runoff sources (table 5.13). Further data for a variety of world cities are listed in table 5.14.

Table 5.13 Comparison of contaminant profiles for the urban surface runoff and raw domestic sewage, based on surveys throughout the USA

Constituent*	Urban surface runoff	Raw domestic sewage
Suspended solids	250–300	150–250
BOD †	10–250	300–350
Nutrients		
(a) Total nitrogen	0.5–5.0	25–85
(b) Total phosphorus	0.5–5.0	2–15
Coliform bacteria (MPN/100 ml)	10^4–10^6	10^6 or greater
Chlorides	20–200	15–75
Miscellaneous substances		
(a) Oil and grease	yes	yes
(b) Heavy metals	(10–100) times sewage conc.	traces
(c) Pesticides	yes	seldom
(d) Other toxins	potential exists	seldom

* All concentrations are expressed in mg l^{-1} unless stated ortherwise
† Biochemical oxygen demand
MPN – Most probable number
Source: Burke, 1972, table 7.36.1

In New York City some half million dogs leave up to 20 000 tonnes of pollutant faeces and up to 3.8 million litres of urine in the city streets each year, all of which is flushed by gutters to storm-water sewers. Taking Britain as a whole, dogs deposit 1000 tons of excrement and three million gallons of urine on the streets every day (Ponting, 1991). Sewage is still a major pollutant. The

Table 5.14 Comparison of storm-water quality from various sources for selected parameters

Study location	BOD (mg/l)	Total solids (mg/l)	Suspended solids (mg/l)	Phosphate (mg/l)	Chloride (mg/l)	Faecal coliforms (MPN/100 ml)
Durham, North Carolina	2–232	194–8860	27–7340	0.15–2.5	3–390	7000–86 000
Cincinnati, Ohio	1–173	–	5–1200	0.02–7.3	5–705	500–76 000
Coshocton, Ohio	0.05–23	–	5–2074	0.08	–	2–56 000
Detroit, Michigan	96–234	310–914	–	–	–	25 000–930 000
Seattle, Washington	10	–	–	4.3	–	16 000
Stockholm, Sweden	18–80	300–3000	–	–	–	4000–200 000
Pretoria, South Africa	30	–	–	–	–	240 000
Oxhey, Herts, UK	100	–	2 045	–	–	–
Moscow, Russia	18–285	–	100–3500	–	–	–
Morristown, New Jersey	3–17	–	56–550	0.02–4.3	–	–
Ann Arbor, Michigan	28	–	2 080	0.8	–	100

MPN = Most probable number
Source: Ellis, 1975

New York Metropolitan area alone produces 6.8 billion litres of sewage per day of which about 16 per cent is raw. Much of this enters the Hudson and East Rivers around New York. In addition, the city produces many tons of sewage sludge, fly ash and dredge waste per day, and 8.6 million tonnes of waste are dumped each year into the New York bight, off the mouth of the Hudson River. In an area of about 100 km² there is a black toxic sludge which smothers all forms of marine life (Southwick, 1976: 16).

None the less, it would be unfair to create the impression that water pollution is either a totally insoluble problem or that levels of pollution must inevitably rise. It is true that some components of the hydrological system – for example, lakes, because they may act as sumps – will prove relatively intractable to improvement, but in many rivers and estuaries striking developments have occurred in water quality in recent decades. This is clear in the case of the River Thames, which suffered a serious decline in its quality and in its fish as London's pollution and industries expanded. However, after about 1950, because of more stringent controls on effluent discharge, the downward trend in quality was reversed (see figure 5.29), and many fish, long absent from the Thames, are now returning (Gameson and Wheeler, 1977).

A similar picture emerges if one considers the trend in the state of Lake Washington near Seattle, USA. Studies there in 1933, 1950 and 1952 showed steady increases in the nutrient content and in associated growth of phytoplankton (algae). These resulted from the increasing levels of untreated sewage pumped into the lake from the Seattle area. A sewage diversion project, started in 1963 and

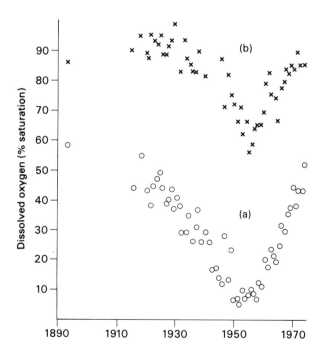

Figure 5.29
The average dissolved oxygen content of the River Thames at half-tide in the July–September quarter since 1890
(a) 79 km below Teddington weir
(b) 95 km below Teddington weir (after Gameson and Wheeler, 1977, figure 4)

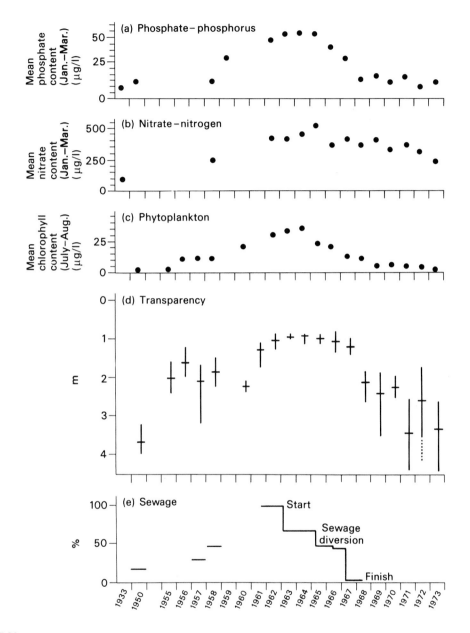

Figure 5.30
Changes in the state of Lake Washington, USA, associated with levels of untreated sewage from 1933 to 1973. The relative amount of treated sewage entering the lake is shown as a percentage of the maximum rated capacity of the treatment plants, 76 million litres per day (after Edmonson, 1975, figure 3)

completed in 1968 (see figure 5.30), resulted in a significant decrease in the sewage input to the lake and was reflected in a rapid decline in phosphates, plankton and turbidity, and a less marked decline in nitrate levels. Likewise, a long-term study of the levels of many heavy metals in sea-water off the British Isles suggests that levels are generally low and that concentrations do not seem to be increasing (Preston, 1973).

Deforestation and its effects on water quality

The removal of a tree cover not only may affect river flow and stream-water temperature, it may also cause changes in stream-water chemistry. The reason for this is that in many forests large quantities of nutrients are cycled through the vegetation, and in some cases (notably in humid tropical rain forest) the trees are a great store of the nutrients. If the trees are destroyed the cycle of nutrients is broken and a major store is disrupted. The nutrients thus released may become available for crops, and shifting cultivators, for example, may utilize this fact and gain good crop yields in the first year after burning and felling forest. Some of the nutrients may, however, be leached out of the soils and appear as dissolved load in streams. Large increases in dissolved load may have undesirable effects, including eutrophication (Hutchinson, 1973), soil salination and a deterioration in public water supply (Conacher, 1979).

A classic, but, it must be added, extreme exemplification of this is the experiment that was carried out in the Hubbard Brook catchments in New Hampshire, USA (Bormann et al., 1968). These workers monitored the effects of 'savage' treatment of the forest on the chemical budgets of the streams. The treatment was to fell the trees, leave them in place, kill lesser vegetation and prevent regrowth by the application of herbicide. The effects were dramatic: the total dissolved inorganic material exported from the basin was about fifteen times larger after treatment, and, in particular, there was a very substantial increase in stream nitrate levels. Their concentration increased by an average of fifty times.

It has been argued, however, that the Hubbard Brook results are atypical (Sopper, 1975) because of the severity of the treatment. Most other forest clear-cutting experiments in the US (table 5.15) indicate that nitrate–nitrogen additions to stream water are not so greatly increased. Under conventional clear-cutting the trees are harvested rather than left, all saleable material is removed, and encouragement is given to the establishment of a new stand of trees. Aubertin and Patric (1974), using such conventional clear-cutting methods rather than the drastic measures employed in Hubbard Brook, found that clear-cutting in their catchments in West Virginia had a negligible effect on stream temperatures (they left a forest strip along the stream), pH and the concentration of most dissolved solids.

Even if clear-cutting is practised, there is some evidence of a rapid return to steady-state nutrient cycling because of quick regeneration. As Marks and Bormann (1972) have put it:

Because terrestrial plant communities have always been subjected to various forms of natural disturbances, such as wind storms, fires, and

Table 5.15 Nitrate–nitrogen losses from control and disturbed forest ecosystems

Site	Nature of distrubance	Nitrate–nitrogen loss ($kg\ ha^{-1}\ y^{-1}$)	
		Control	Disturbed
Hubbard Brook	Clear-cutting without vegetation removal, herbicide inhibition of re-growth	2.0	97
Gale River (New Hampshire)	Commercial clear-cutting	2.0	38
Fernow (West Virginia)	Commercial clear-cutting	0.6	3.0
Coweeta (North Carolina)	Complex	0.05	7.3*
H. J. Andrews' Forest (Oregon)	Clear-cutting with slash-burning	0.08	0.26
Alsea River (Oregon)	Clear-cutting with slash burning	3.9	15.4
	MEAN	1.44	26.83

* This value represents the second year of recovery after a long-term disturbance. All other results for disturbed ecosystems reflect the first year after disturbance.
Source: modified after Vitousek et al., 1979

insect outbreaks, it is only reasonable to consider recovery from distur-bance as a normal part of community maintenance and repair.

Likewise Hewlett et al. (1984) found no evidence in the Piedmont region of the USA that clear-cutting would create such extreme nitrate loss that soil fertility would be impaired or water eutrophi-cation caused. They point out that rapid growth of vegetation minimizes nutrient losses from the ecosystem by three main mechanisms. It channels water from runoff to evapotranspiration, thereby reducing erosion and nutrient loss; it reduces the rates of decomposition of organic matter through moderation of the microclimate so that the supply of soluble ions for loss in drainage water is reduced; and it causes the simultaneous incorporation of nutrients into the rapidly developing biomass so that they are not lost from the system.

In Britain, a comparison of water chemistry in forested and clear-cut areas in the Plynlimon area of mid-Wales has revealed certain trends (figure 5.31). The most notable of these is the increase in nitrate levels, which impacts on catchment acidification processes by causing a decrease in pH and releasing toxic metals such as aluminium. Also important is the increase in dissolved organic carbon, which causes discoloration of stream water (Institute of Hydrology Report for 1990/1).

Thermal pollution

The pollution of water by increasing its temperature is called thermal pollution. Many fauna are affected by temperature, so this environmental impact has some significance (Langford, 1990).

In industrial countries probably the main source of thermal pollution is from condenser cooling water released from electricity generating stations. Water discharged from power stations has been

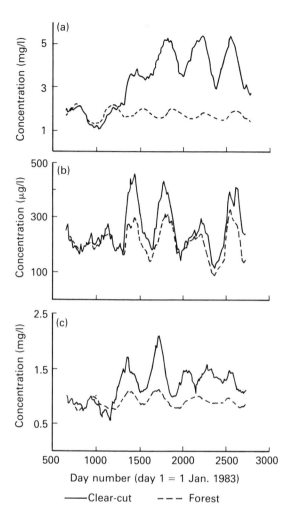

Figure 5.31
Changes in the chemical
composition of stream water
from clear-cut and forested
catchments in mid-Wales,
1983–1990
(a) nitrate (NO_3)
(b) aluminium (Al)
(c) dissolved organic carbon
     (DOC)
(modified after Institute of
Hydrology Report 1990/1991,
figure 20)

heated some 6–9°C, but usually has a temperature of less than 30°C. The extent to which such water affects river temperature depends very much on the state of flow. For example, below the Ironbridge power station in England, the Severn River undergoes a temperature increase of only 0.5°C during floods, compared with an 8°C increase at times of low flow (Rodda et al., 1976).

Over the past two decades or more (see Royal Commission on Environmental Pollution, 1972), both the increase in the capacity of individual electricity power-generating units and the improvements in their thermal efficiency have led to a diminution in the heat rejected in relation to the amount of cooling water per unit of production (table 5.16). Economic optimization of the generating plant has cut down the flow of cooling water per unit of electricity, though it has raised its temperature. This increased temperature rise of the cooling water is more than offset by the reduction achieved in the volume of water utilized. None the less, the expansion of

Table 5.16 Cooling water used per unit of power generation in the UK

Year	Temperature rise (°C)	Specific water circulation per unit of power generation (m³)
(a) Coal and oil-fired		
1950	7	0.21
1960	9	0.15
1970	10	0.12
1980 (estimated)	12–14	0.08–0.09
(b) Nuclear		
1962	8	0.33
1967	9	0.24
1980 (estimated)	12–14	0.09–0.14

Source: Royal Commission on Environmental Pollution, 1972, table 7, p. 21

generating capacity has meant that the total quantity of heat discharged has increased, even though it is much less than would be expected on a simple proportional basis.

Thermal pollution of streams may also follow from urbanization (Pluhowski, 1970). This results from various sources: changes in the temperature regime of streams brought about by reservoirs according to their size, their depth and the season; changes produced by the urban heat island effect; changes in the configuration of urban channels (for example, their width–depth ratio); changes in the degree of shading of the channel, either by covering it over or by removing a natural vegetation cover; changes in the volume of storm runoff; and changes in the ground-water contribution. Pluhowski found that the basic effects of cities on river-water temperatures on Long Island, New York, was to raise temperatures in summer (by as much as 5–8°C) and to lower then in winter (by as much as 1.5–3.9°C).

Reservoir construction can also affect stream-water temperatures. Crisp (1977) found, for example, that as a result of the construction of a reservoir at Cow Green (Upper Teesdale, northern England) the temperature of the river downstream was modified. He noted a reduction in the amplitude of annual water-temperature fluctuations, and a delay in the spring rise of water temperature by 20–50 days, and in the autumn fall by up to 20 days. In rural areas human activities can also cause significant modifications in river-water temperature. Deforestation (table 5.17) is especially important (Lynch et al., 1984). Swift and Messer (1971), for example, examined stream temperature measurements during six forest-cutting experiments on small basins in the southern Appalachians, USA. They found that with complete cutting, because of shade removal, the maximum stream temperatures in summer went up from 19 to 23°C. They believed that such temperature increases were detrimental to temperature-sensitive fish like trout (see figure 5.32). On the other hand, the temporary

Table 5.17 Effect of forest-cutting on mean monthly maximum river-water temperature in the USA

Location	Temperature increases (°C)
Coweeta (North Carolina)	4
Fernow (West Virginia)	5.5
Penn State (Pennsylvania)	6
Hubbard Brook (New Hampshire)	6
H. J. Andrews (Oregon)	6.7
Coastal Range (Oregon)	8
MEAN	6.03

Source: from data in Sopper, 1975, table 2

Figure 5.32
Maximum temperatures for the spawning and growth of fish. Heated waste water may be up to 5–10°C warmer than receiving waters and consequently the local fish populations cannot reproduce or grow properly. (Figure 13–2 (after Giddings, *Chemistry, Man and environmental change*, Harper & Row) from *Encounter with the Earth* by Leo F. Laporte. Copyright © 1975 by permission of Harper & Row, Publishers, Inc.)

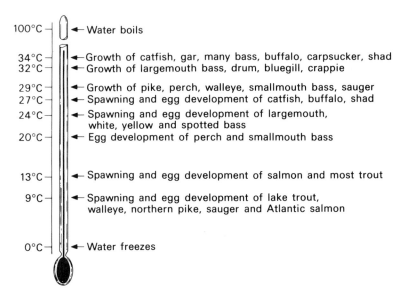

shutdown of power plants may create severe cold-shock kills of fish in discharge-receiving waters in winter (J. R. Clark, 1977). More-over, an increase in water temperature causes a decrease in the solubility of oxygen which is needed for the oxidation of bio-degradable wastes. At the same time, the rate of oxidation is accelerated, imposing a faster oxygen demand on the smaller supply and thereby depleting the oxygen content of the water still further. Temperature also affects the lower organisms, such as plankton and crustaceans. In general, the more elevated the temper-ature is, the less desirable the types of algae in water. In cooler waters diatoms are the predominant phytoplankton in water that is not heavily eutrophic; with the same nutrient levels, green algae begin to become dominant at higher temperatures and diatoms decline; at the highest water temperatures, blue-green algae thrive and often develop into heavy blooms. One further ecological consequence of thermal pollution is that the spawning and migra-

tion of many fish are triggered by temperature and this behaviour
can be disrupted by thermal change.

Pollution with suspended sediments

Probably the most important effect that humans have had on water
quality is the contribution to levels of suspended sediments in
streams. This is a theme which is intimately tied up with that of soil
erosion, on the one hand, and with channel manipulation by
activities such as reservoir construction, on the other. The clearance
of forest, the introduction of ploughing and grazing by domestic
animals, the construction of buildings, and the introduction of spoil
materials into rivers by mineral extraction industries, have all led to
very substantial increases in levels of stream turbidity. Frequently
sediment levels are a whole order of magnitude higher than they
would have been under natural conditions. However, the introduction of soil conservation measures, or a reduction in the intensity of
land use, or the construction of reservoirs, can cause a relative (and
sometimes absolute) reduction in sediment loads (see, for example,
table 5.4 and figure 5.3). All three of these factors have contributed
to the observed reduction in turbidity levels monitored in the rivers
of the Piedmont region of Georgia in the USA (figure 5.33).

Marine pollution

It is not within the scope of this book to discuss human impacts on
the oceans to any great extent. However, it is worth making a few
points on this subject. At first sight, as Jickells et al. (1991: 313)
point out, two contradictory thoughts may cross our minds on this
issue:

The first is the observation of ocean explorers, such as Thor Heyerdahl, of
lumps of tar, flotsam and jetsam, and other products of human society
thousands of kilometers from inhabited land. An alternative, vaguer feeling
is that given the vastness of the oceans (more than 1,000 billion billion
litres of water!), how can man have significantly polluted them?

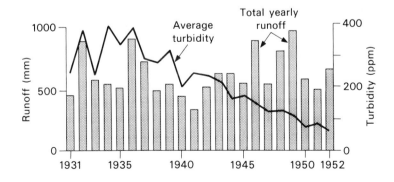

Figure 5.33
Decline of turbidity in the
Chattahoochee Basin, Atlanta,
Georgia, USA, 1931–52 (after
Albert and Spector in Trimble,
1974, figure 27)

What is the answer to this conundrum? Jickells et al. (p. 330) draw a clear distinction between the open oceans and regional seas and in part come up with an answer:

The physical and chemical environment of the open oceans has not been greatly affected by events over the past 300 years, principally because of their large diluting capacity . . . Material that floats and is therefore not diluted, such as tar balls and litter, can be shown to have increased in amount and to have changed in character over the past 300 years.

In contrast to the open oceans, regional areas in close proximity to large concentrations of people show evidence of increasing concentrations of various substances that are almost certainly linked to human activities. Thus the partially enclosed North Sea and Baltic show increases in phosphate concentrations as a result of discharges of sewage and agriculture.

Likewise it is clear that pollution in the open ocean is, as yet, of limited biological significance. GESAMP (1990), an authoritative review of the state of the marine environment for the United Nations Environment Programme, reported (p. 1):

The open sea is still relatively clean. Low levels of lead, synthetic organic compounds and artificial radionuclides, though widely detectable, are biologically insignificant. Oil slicks and litter are common along sea lanes, but are, at present, of minor consequence to communities of organisms living in open-ocean waters.

On coastal waters it reported (p. 1):

The rate of introduction of nutrients, chiefly nitrates but sometimes also phosphates, is increasing, and areas of eutrophication are expanding, along with enhanced frequency and scale of unusual plankton blooms and excessive seaweed growth. The two major sources of nutrients to coastal waters are sewage disposal and agricultural run-off from fertilizer-treated fields and from intensive stock raising.

Attention is also drawn to the presence of synthetic organic compounds – chlorinated hydrocarbons, which build up in the fatty tissues of top predators such as seals which dwell in coastal waters. Levels of contamination are decreasing in northern temperate areas but rising in tropical and subtropical areas due to continued use of chlorinated pesticides there.

The world's oceans have been greatly contaminated by oil. The effects of this pollution on animals have been widely studied. The sources of pollution include tanker collisions with other ships, the explosion of individual tankers because of the build-up of gas levels, the wrecking of tankers on coasts through navigational or mechanical failure, seepage from offshore oil installations, and the flushing of tanker holds (figure 5.34a). There is no doubt that the great bulk of the oil and related materials polluting the oceans results from human action (Blumer, 1972), though humans should not be attributed with all the blame, for natural seepages are reasonably common (Landes, 1973). Paradoxically, human actions

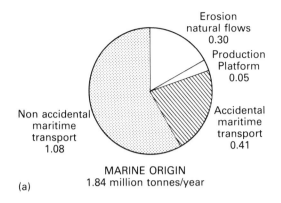

(a)

MARINE ORIGIN
1.84 million tonnes/year

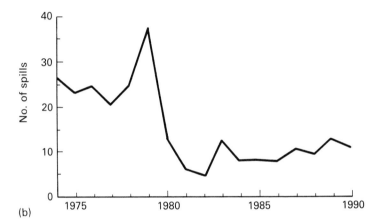

(b)

Figure 5.34
Causes and trends of oil
pollution in the oceans
(a) the origin of oil pollution
(b) the number of accidental
    oil spills over 700 tonnes,
    1974–88
(from OECD Report, 1991,
figure 15, with modifications)

mean that many natural seepages have diminished as wells have drawn down the levels of hydrocarbons in the oil-bearing rocks. Likewise, the flow of asphalt that once poured from the Trinidad Pitch Lake into the Gulf of Paria has also ceased because mining of the asphalt has lowered the lake below its outlet.

Of the various serious causes of oil pollution, a more important mechanism than the much-publicized role of tanker accidents is the discharge of the ballast water taken into empty tankers to provide stability on the return voyage to the loading terminal (Pitt, 1979). It is in this field, however, that technical developments have led to most amelioration. For example, the quantity of oil in tanker ballast water has been drastically reduced by the LOT (load on top) system, in which the ballast water is allowed to settle so that the oil rises to the surface. The tank is then drained until only the surface oil in the tank remains, and this forms part of the new cargo. In the future large tankers will be fitted with separate ballast tanks not used for oil cargoes, called segregated or clean ballast systems. Given such advances it appears likely that an increasing proportion

Plate 5.8

of oil pollution will be derived from accidents involving tankers of ever increasing size, such as the spillage of 300 million litres, which came from the wreck of the *Amoco Cadiz* in 1978. Even here, however, some progress has been made. Since 1980 there has been a reduction in the number of major oil spills, partly as a consequence of diminished long distance oil transport to western Europe (figure 5.34b).

SIX

Human Agency in Geomorphology

Introduction

The human role in creating landforms and modifying the operation of geomorpholical processes such as weathering, erosion and deposition is a theme of great importance though one that, particularly in the Western world, has not received the attention it deserves. Recently there have been some useful general surveys (see, for example, Brown, 1970; Jennings, 1966; Haigh, 1978; Nir, 1983) and, with the increasing concern for the relevance and application of geomorphological studies to human problems (see, for example, Cooke and Doornkamp, 1990; Hails, 1977), the situation is likely to change. Moreover, the development of process studies in post-Second World War geomorphology has led to a relative reduction in the importance of evolutionary and historical geomorphology within the discipline. Examination of processes, and in particular of their rate of operation, has served to highlight the part played by humans in modifying the physical landscape.

The range of the human impact on both forms and process is considerable. For example, in table 6.1 there is a list of some anthropogenic landforms together with some of the causes of their creation. There are very few spheres of human activity which do not, even indirectly, create landforms. It is, however, useful to recognize that some features are produced by direct anthropogenic processes. These tend to be more obvious in their form and origin and are frequently created deliberately and knowingly. They include land-forms produced by constructional activity (such as tipping), by excavation, by hydrological interference, and by farming (Spencer and Hale, 1961). Hillsides have been terraced in many parts of the world for many centuries, notably in the arid and semi-arid highlands of the New World (Donkin, 1979). Haigh (1978) has classified anthropogenic landforming processes as follows:

Table 6.1 Some anthropogenic landforms

Feature	Cause
Pits and ponds	Mining, marling
Broads	Peat extraction
Spoil heaps	Mining
Terracing, lynchets	Agriculture
Ridge and furrow	Agriculture
Cuttings	Transport
Embankments	Transport, river and coast management
Dikes	River and coast management
Mounds	Defence, memorials
Craters	War, *qanat* construction
City mounds (*tells*)	Human occupation
Canals	Transport, irrigation
Reservoirs	Water management
Subsidence depressions	Mineral and water extraction
Moats	Defence
Banks along roads	Noise abatement

1 *Direct anthropogenic processes*
  1.1 Constructional
      tipping: loose, compacted, molten
      graded: moulded, ploughed, terraced
  1.2 Excavational
      digging, cutting, mining, blasting of cohesive or
      non-cohesive materials
      cratered
      tramped, churned
  1.3 Hydrological interference
      flooding, damming, canal construction
      dredging, channel modification
      draining
      coastal protection
2 *Indirect anthropogenic processes*
  2.1 Acceleration of erosion and sedimentation
      agricultural activity and clearances of vegetation
      engineering, especially road construction and urbani-
      zation
      incidental modifications of hydrological regime
  2.2 Subsidence: collapse, settling
      mining
      hydraulic
      thermokarst
  2.3 Slope failure: landslide, flow, accelerated creep
      loading
      undercutting
      shaking
      lubrication
  2.4 Earthquake generation
      loading (reservoirs)
      lubrication (fault plane).

Landforms produced by indirect anthropogenic processes are often less easy to recognize, not least because they tend to involve, not the operation of a new process or processes, but the acceleration of natural processes. They are the result of environmental changes brought about inadvertently by human technology. None the less, it is probably this indirect and inadvertent modification of process and form which is the most crucial aspect of anthropogeomorphology. By removing natural vegetation covers – through the agency of cutting, burning and grazing – humans have accelerated erosion and sedimentation. Sometimes the results will be obvious, for example when major gully systems rapidly develop; other results may have less immediate effect on landform but are, nevertheless, of great importance. By other indirect means humans may create subsidence features and problems, trigger off mass movements like landslides, and even influence the operation of phenomena like earthquakes.

Finally there are situations where, through a lack of understanding of the operation of processes and the links between different processes and phenomena, humans may deliberately and directly alter landforms and processes and thereby set in train a series of events which were not anticipated or desired. There are, for example, many records of attempts to reduce coast erosion by important and expensive engineering solutions, which, far from solving erosion problems, only exacerbated them.

Landforms produced by excavation

Of the landforms produced by direct anthropogenic processes those resulting from excavation are widespread, and may have some antiquity. For example, Neolithic peoples in the Breckland of East Anglia in England used antler picks and other means to dig a remarkable cluster of deep pits in the chalk. The purpose of this was to obtain good-quality non-frost-shattered flint to make stone tools. The cratered area, called Grimes Graves, is well displayed on aerial photographs, and is illustrated in figure 6.1. In many parts of Britain chalk has also been excavated to provide marl for improving acidic, light, sandy soils, and Prince (1962, 1964) has made a meticulous study of the 27 000 pits and ponds in Norfolk that have resulted mainly from this activity – an activity particularly prevalent in the eighteenth century. It is often difficult in individual cases to decide whether the depressions are the results of human intervention for solutional and periglacial depressions are often evident in the same area, but pits caused by human action do tend to have some distinctive features: irregular shape, a track leading into them, proximity to roads, etc. (see the debate between Prince, and Sperling et al., 1979).

Difficulties of identifying the true origin of excavational features were also encountered in explaining the Broads, a group of twenty-five fresh-water lakes in the county of Norfolk (plate 6.1).

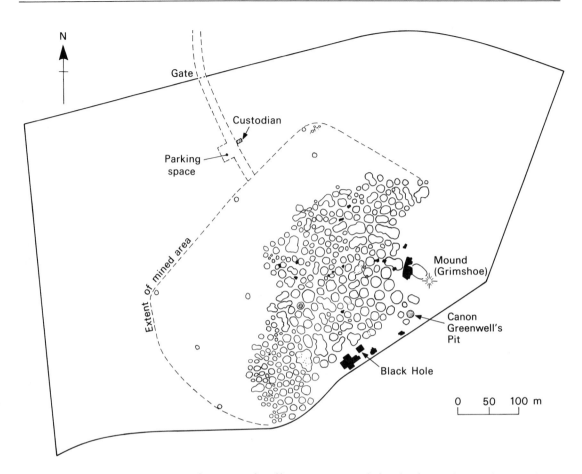

N

Gate

Custodian

Parking
space

Extent of mined area

Mound
(Grimshoe)

Canon
Greenwell's
Pit

Black Hole

0   50   100 m

Figure 6.1
Flint mines at Grimes Graves,
Norfolk, England, dug during
Neolithic times (around 2000
BC) (after Clark, 1963)

They are of sufficient area, and depth, for early workers to have precluded a human origin. Later it was proposed instead that they were natural features caused by uneven alluviation and siltation of river valleys which were flooded by the rapidly rising sea-level of the Holocene (Flandrian) transgression. It was postulated by Jennings (1952) that the Broads were initiated as a series of discontinuous natural lakes, formed beyond the limits of a thick estuarine clay wedge laid down in Romano-British times by a transgression of the sea over earlier valley peats. The waters of the Broads were thought to have been ponded back in natural peaty hollows between the flanges of the clay and the marginal valley slopes, or in tributary valleys whose mouths were blocked by the clay.

It is now clear, however, that the Broads are the result of human work (Lambert et al., 1970). Some of them have rectilinear and steep boundaries, most of them are not mentioned in early topographic books, and archival records indicate that peat-cutting (turbary) was widely practised in the area. On these and other grounds it is believed that peat-diggers, before AD 1300, excavated 25.5 million m^3 of peat and so created the depressions in which the lakes have formed. The flooding may have been aided by sea-level

change. Comparably extensive peat excavation was also carried on in the Netherlands, notably in the fifteenth century.

Other excavational features result from war, especially craters caused by bomb or shell impact. Regrettably, human power to create such forms is increasing. It has been calculated (Westing and Pfeiffer, 1972) that between 1965 and 1971, 26 million craters, covering an area of 171 000 hectares, were produced by bombing in Indo-China. This represents a total displacement of no less than 2.6 billion m^3 of earth, a figure much greater than calculated as being involved in the peaceable creation of the Netherlands.

Some excavation is undertaken on a large scale for purely aesthetic reasons, when Nature offends the eye. This can be illustrated by brief reference to the work of some English landscape gardeners. In 1676 at Cassiobury Park, Watford, Moses Cook the gardener says that he was 'forced to cut through one Hill, thirty rod, most of the hill two feet deep, into sharp gravel', to produce the landscape he desired. At nearby Moor Park in 1720 the intended view across the Vale of St Albans from a new house was found to be obscured by the brow of a hill, so that the hilltop was cut off by a depth of 10 m (Prince, 1959).

In many countries where land is scarce, whole hills are levelled and extensive areas stripped to provide fill for harbour reclamation. One of the most spectacular examples of this kind was the deliberate removal of steep-sided hills in the centre of Brazil's Rio

Plate 6.1
The Norfolk Broads of eastern England were originally thought to be natural lakes. However, recent researches by historical geographers have shown that for the most part they result from the activities of medieval peat diggers

de Janeiro, for housing development. In Bahrain, an island in the Arabian–Persian Gulf, a substantial part of the desert surface has been stripped to provide fill for land reclamation in the commercial and industrial area in the north of the island.

An excavational activity of a rather specialized kind is the removal of limestone pavements – areas of exposed limestone which were stripped by glaciers and then moulded into bizarre shapes by solutional activity – for ornmental rock gardens. These pavements which consist of arid, bare rock surfaces (clints) bounded by deep, humid fissures (grikes) have both an aesthetic and a biological significance (Ward, 1979). They occur in northern England, notably in the Pennines and in the Morecombe Bay area, and provide a habitat for various rare plants (including *Actaea spicata*, *Ribes spicatum* and *Dynopteris villarii*). A recent survey indicates that the pavements, which only cover about 2150 hectares in Britain, have been severely damaged. On only 3 per cent of 537 pavement units recorded could no damage be detected, while the proportion considered to be 95 per cent or more intact was only 13 per cent.

The most important cause of excavation is still mineral extraction, producing open-pit mines, strip mines, quarries for structural materials, borrow pits along roads, and similar features (Doerr and

Plate 6.2
The Bingham Canyon open pit copper mine in Utah, USA is one of the largest excavational landforms on the face of the Earth. It has involved the removal of 3355 million tonnes over an area of 7.21 km² to a depth of 774 m, seven times the amount of material moved to build the Panama Canal

Guernsely, 1956). Of these, 'without question, the environmental devastation produced by strip mining exceeds in quantity and intensity any of the other varied forms of man-made land destruction' (Strahler and Strahler, 1973: 284). This form of mining (plate 6.3) is a particular environmental problem in the states of Pennsylvania, Ohio, West Virginia, Kentucky and Illinois. If, because of increasing pressure on available energy resources, the production of coal is expanded then this particular geomorphological impact may also increase. The same applies to the probable exploitation of oil shales, a potential source of oil that at present is relatively untapped. This can be exploited by open-pit mining, by traditional room and pillar mining, and by underground *in situ* pyrolysis. Vast reserves exist but the amount of excavation required will probably be about three times the amount of oil produced (on a volume basis), suggesting that the extent of both the excavation and subsequent dumping of overburden and waste will be considerable (Routson et al., 1979). Some of the waste, that produced by the retorting of the shale to release the oil, contains soluble salts and potentially harmful trace elements, which limit the speed of ground reclamation (Petersen, 1981).

An attempt to provide a general picture of the importance of excavation in the creation of the landscape of Britain was given by Sherlock (1922). He estimated (see table 6.2) that up until the time in which he wrote, human society had excavated around 31 billion m^3 of material in the pursuit of its economic activities. That figure must now be a gross underestimate, partly because Sherlock himself was not in a position to appreciate the anthropogenic role in creating features like the Norfolk Broads, and partly because, since his time, the rate of excavation has greatly accelerated. Sherlock's 1922 study covers a period when earth-moving equipment was still ill-developed. None the less, on the basis of his calculations, he was able to state that 'at the present time, in a densely peopled country like England, Man is many times more powerful, as an agent of denudation, than all the atmospheric denuding forces combined' (p. 333). The most notable change since Sherlock wrote has taken place in the production of aggregates for concrete which are, in descending order of importance, sand and gravel, crushed limestone, artificial or manufactured aggregates, and crushed sandstone. Demand for these materials in the UK grew from 20 million tonnes per annum in 1900, to 50 million tonnes in 1948 and 276 million tonnes in 1973, an increase in *per capita* consumption from 0.6 tonnes per year to about 5 tonnes per year (Jones, 1983).

For the world as a whole the annual movement of soil and rock resulting from mineral extraction may be as high as 3000 billion tonnes (Holdgate et al., 1982: 186). By comparison it has been estimated that the amount of sediment carried into the ocean by the world's rivers each year amounts to 24 billion tonnes per year (Judson, 1968).

Plate 6.3
Strip-mining for coal in
Missouri, USA, is a particularly
unsightly example of the
production of landforms by
human agency

Landforms produced by construction and dumping

The process of constructing mounds and embankments and the
creation of dry land where none previously existed is longstanding.
In Britain the mound at Silbury Hill (plate 6.4) dates back to

Table 6.2 Total excavation of material in Great Britain until 1992

Activity	Approximate volume (m³)
Mines	15 147 000 000
Quarries and pits	11 920 000 000
Railways	2 331 000 000
Manchester Ship Canal	41 154 000
Other canals	153 800 000
Road cuttings	480 000 000
Docks and harbours	77 000 000
Foundations of buildings and street excavation	385 000 000
TOTAL	30 534 954 000

Source: Sherlock, 1922, p. 86

prehistoric times, and the pyramids of central America, Egypt and the Far East are even more spectacular early feats of landform creation. Likewise hydrological management has involved, over many centuries, the construction of massive banks and walls; the ultimate result being the landscape of the present-day Netherlands. Transport developments have also required the creation of large constructional landmarks, but probably the most important features are those resulting from the dumping of waste materials, especially those derived from mining (see figure 6.2). It has been calculated that there are at least 2000 million tonnes of shale lying in pit heaps in the coalfields of Britain (Richardson, 1976). In the Middle East and other areas of long-continued human urban settlement the accumulated debris of life has gradually raised the level of the land surface, and occupation mounds (tells) are a fertile source of information to the archaeologist. Today, with the technical ability to build that humans have, even estuaries may be converted from ecologically productive environments into suburban sprawl (figure 6.3) by the processes of dredging and filling. Indeed one of the striking features of the distribution of the world's population is the tendency for large human concentrations to occur near vast expanses of water. Of those cities with more than 4 million inhabitants, more than three-quarters are situated on a lake or ocean shore. Many of these cities have extended out on to land that has been reclaimed from the sea (for example, Hong Kong, figure 6.4, plate 6.5), thereby providing valuable sites for development, but sometimes causing the loss of rich fishing grounds and ecologically valuable wetlands (Hudson, 1979).

The ocean floors are also being affected because of the vast bulk of waste material that humankind is creating. Waste-solid disposal by coastal cities is now sufficiently large to modify shorelines, and it covers adjacent ocean bottoms with characteristic deposits on a scale large enough to be geologically significant. This has been brought out dramatically by Gross (1972: 3174), who has undertaken a quantitative comparison of the amount of solid wastes dumped into the Atlantic by humans in the New York metropolitan

Plate 6.4
In prehistoric times some striking constructional landforms were created. Top: Silbury Hill, Wiltshire, England. Bottom: a tell in the Middle East produced by the accumulation of the debris of years of human occupation

region with the amount of sediment brought into the ocean by rivers:

The discharge of waste solids exceeds the suspended sediment load of any single river along the U.S. Atlantic coast. Indeed, the discharge of wastes from the New York Metropolitan region is comparable to the estimated suspended-sediment yield (6.1 megatons per year) of all rivers along the Atlantic coast between Maine and Cape Hatteras, North Carolina.

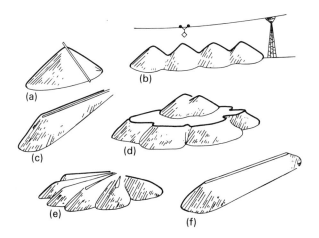

Figure 6.2
Some shapes produced by shale tipping:
(a) conical resulting from MacClane tipping
(b) multiple cones tipped from aerial ropeways
(c) high 'fan-ridges' by tramway tipping over slopes
(d) high plateau mounds topped with cones
(e) low multiple 'fan-ridges' by tramway tipping
(f) lower ridge by tramway tipping
(after Haigh, 1978, figure 2.1)

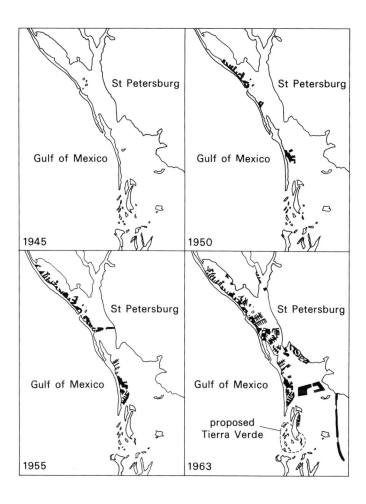

Figure 6.3
Boca Ciega Bay on the Gulf of Mexico near St Petersburg is rapidly being converted by dredging and filling into a typical suburban sprawl (after Lauf, in Wagner, 1974, figure 7.4)

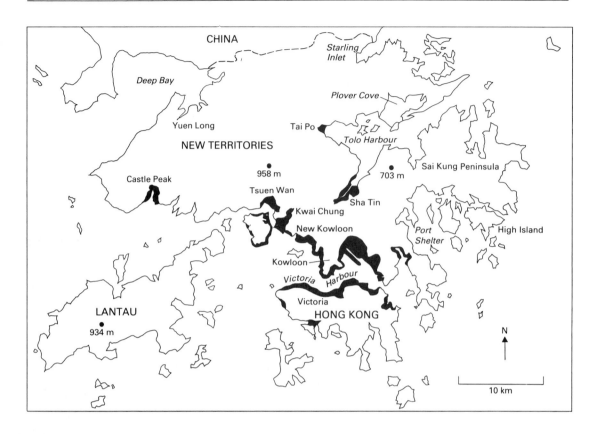

Figure 6.4
Map of Hong Kong showing
main urban reclamation areas
shaded black (reprinted from
Hudson, 1979, figure 1, in
Reclamation Review, 2, 3–16,
permission of Pergamon Press
Ltd © Pergamon Press Ltd)

Not only are the rates of sedimentation high, but the anthropogenic sediments tend to contain abnormally high contents of such substances as carbon and heavy metals (Goldberg et al., 1978) (table 6.3).

Many of the features created by excavation in one generation are filled in by another, since phenomena like water-filled hollows

Table 6.3 Fluxes of heavy metals to coastal environments in the USA ($\mu g \ cm^{-2} \ y^{-1}$)

Chemical	Santa Barbara Basin California	Narangansett Bay Rhode Island	Chesapeake Bay	
			Cove 1411	Cove 1314
Lead				
anthropogenic	2.1	124	7.6	1.9
natural	1	2.6	5.3	3.9
Zinc				
anthropogenic	2.2	230	34	4.1
natural	9.7	14	19	15
Copper				
anthropogenic	1.4	193	4.5	0.8
natural	2.6	3.1	4	3.6

Source: Goldberg et al., 1978

produced by mineral extraction are often both wasteful of land and suitable locations for the receipt of waste. The same applies to natural hollows such as karstic or ground-ice depressions. Watson (1976), for example, has mapped the distribution of hollows, which were largely created by marl diggers in the lowlands of south-west Lancashire and north-west Cheshire, as they were represented on mid-nineteenth-century topographic maps (figure 6.5a). When this distribution is compared with the present-day distribution for the same area (figure 6.5b), it is evident that a very substantial proportion of the holes have been infilled and obliterated by humans, with only 2114 out of 5380 remaining. Hole densities have fallen from 121 km^{-2} to 47 km^{-2}.

Plate 6.5
Hong Kong is one of the most densely populated territories on earth, and its landscapes have been transformed by coastal reclamation and by carving out ledges on steep slopes. Debris flows and landslips may result

Accelerated sedimentation

An inevitable consequence of the accelerated erosion produced by human activities has been accelerated sedimentation. This has been heightened by the deliberate addition of sediments to stream

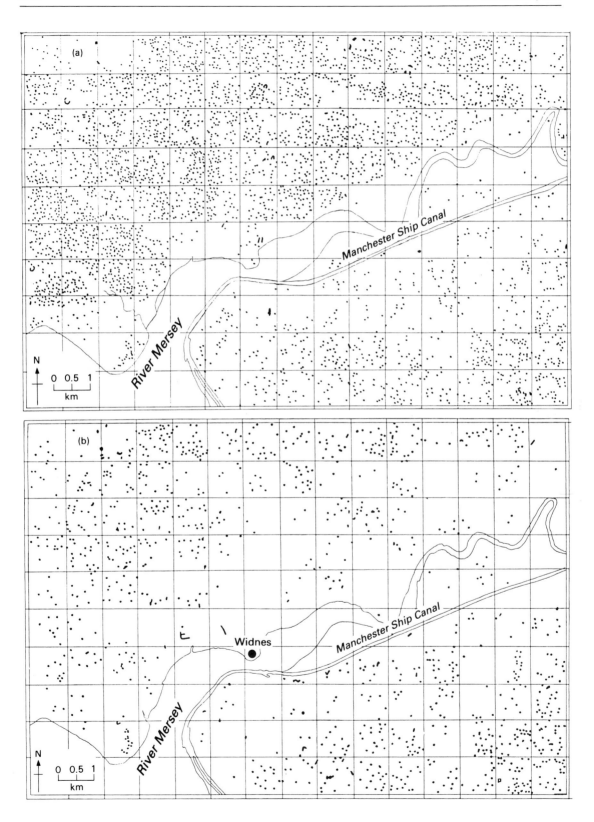

Plate 6.6
Gold-washing in a long sluice
at Yale, British Columbia, in the
1890s. Such activities caused
rivers to become choked with
debris

channels as a result of the need to dispose of mining and other wastes.

In a classic study, G. K. Gilbert (1917) demonstrated that hydraulic mining in the Sierra Nevada mountains of California led to the addition of vast quantities of sediments into the river valleys draining the range. This in itself raised their bed levels, changed their channel configurations and caused the flooding of lands that had previously been immune. Of even greater significance was the

Figure 6.5 (opposite)
The distribution of pits and ponds in a portion of north-western England:
(a) in the mid-nineteenth century
(b) in the mid-twentieth century (after Watson, 1976)

fact that the rivers transported increased quantities of debris into the estuarine bays of the San Francisco system, and caused extensive shoaling which in turn diminished the tidal prism of the bay. Gilbert calculated the volume of shoaling produced by hydraulic mining since the discovery of gold to be 846 million m^3. Comparably serious sedimentation of bays and estuaries has also been caused by human activity on the eastern coast of America. As Gottschalk (1945: 219) wrote:

Both historical and geological evidence indicates that the preagricultural rate of silting of eastern tidal estuaries was low. The history of sedimentation of ports in the Chesapeake Bay area is an epic of the effects of uncontrolled erosion since the beginning of the wholesale land clearing and cultivation more than three centuries ago.

He has calculated that at the head of the Chesapeake Bay, 65 million m^3 of sediment were deposited between 1846 and 1938. The average depth of water over an area of 83 km^2 was reduced by 0.76 m. New land comprising 318 hectares was added to the state of Maryland and, as Gottschalk remarked, 'the Susquehanna River is repeating the history of the Tigris and Euphrates'. Much of the material entrained by erosive processes on upper slopes as a result of agriculture in Maryland, however, was not evacuated as far as the coast. Costa (1975) has suggested, on the basis of the study of sedimentation, that only about one-third of the eroded material left the river valleys. The remainder accumulated on floodplains as alluvium and colluvium at rates of up to 1.6 cm per year. Similarly, Happ (1944), working in Wisconsin, carried out an intensive augering survey of floodplain soil and established that, since the development of agriculture, floodplain aggradation had proceeded at a rate of approximately 0.85 cm per year. He noted that channel and floodplain aggradation had caused the flooding of low alluvial terraces to be more frequent, more extensive and deeper. The rate of sedimentation has since declined (Trimble, 1976).

Such valley sedimentation is by no means restricted to the newly settled terrains of North America. There is increasing evidence to suggest that silty valley fills in Germany, France and Britain, many of them dating back to the Bronze Age and the Iron Age, are the result of accelerated slope erosion produced by the activities of early farmers (Bell, 1982). Indeed, in recent years, various studies have been undertaken with a view to assessing the importance of changes in sedimentation rate caused by humans at different times in the Holocene in Britain. Among the formative events that have been identified are initial land clearance by Mesolithic and Neolithic peoples; agricultural intensification and sedentarization in the late Bronze Age; the widespread adoption of the iron plough in the early Iron Age; settlement by the Vikings and the introduction of sheep farming (table 6.4).

A core from Llangorse Lake in the Brecon Beacons of Wales (Jones et al., 1985) provides excellent long-term data on changing sedimentation rates:

Period (years BP)	Sedimentation rate (cm 100 y^{-1})
9000–7500	3.5
7500–5000	1.0
5000–2800	13.2
2800–AD 1840	14.1
c.AD 1840–present	59.0

The thirteenfold increase in rates after 5000 BP seems to have occurred rapidly and is attributed to initial forest clearance. The

Table 6.4 Accelerated sedimentation in Britain in prehistorical and historical times

Location	Source	Evidence and date
Howgill Fells	Harvey et al. (1981)	Debris cone production following tenth century AD introduction of sheep farming
Upper Thames basin	Robinson and Lambrick (1984)	River alluviation in Late Bronze Age and early Iron Age
Lake District	Pennington (1981)	Accelerated lake sedimentation at 5000 BP as a result of Neolithic agriculture
Mid-Wales	Macklin and Lewin (1986)	Floodplain sedimentation (Capel Bangor unit) on Rheidol as a result of early Iron Age sedentary agriculture
Brecon Beacons	Jones et al. (1985)	Lake sedimentation increase after 5000 BP at Llangorse due to forest clearance
Weald	Burrin (1985)	Valley alluviation from Neolithic onwards until early Iron Age
Bowland Fells	Harvey and Renwick (1987)	Valley terraces at 5000–2000 BP (Bronze or Iron Age settlement) and after 1000 BP (Viking settlement)
Southern England	Bell (1982)	Fills in dry valleys Bronze Age and Iron Age
Callaly Moor (Northumberland)	Macklin et al. (1991)	Valley fill sediments of late Neolithic to Bronze Age

second dramatic increase of more than fourfold took place in the last 150 years and is a result of agricultural intensification.

The work of Binford et al. (1987) on the lakes of the Peten region of northern Guatemala (Central America), an area of tropical lowland dry forest, is also instructive with respect to early agricultural colonization. Combining studies of archaeology and lake sediment stratigraphy, they were able to reconstruct the diverse environmental consequences of the growth of Mayan civilization (figure 6.6). This civilization showed a dramatic growth after 3000 years BP, but collapsed in the ninth century AD. The hypotheses put forward to explain this collapse include warfare, disease, earth-

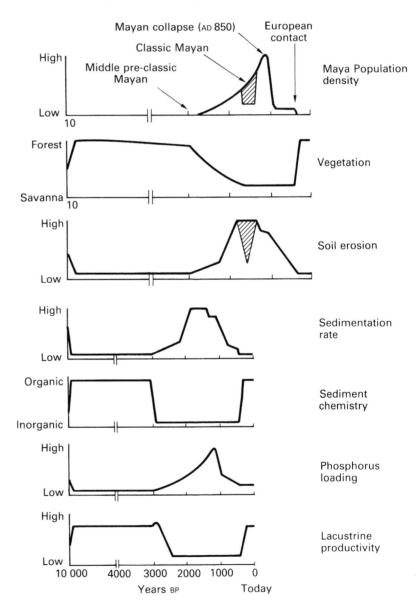

Figure 6.6
The human and environmental history of the Peten Lakes, Guatemala. The hatched areas indicate a phase of local population decline (modified after Binford et al., 1987)

quakes and soil degradation. The population has remained relatively low ever since, and after the first European contact (AD 1525) the region was virtually depopulated. The period of Mayan success saw a marked reduction in vegetation cover, an increase in lake sedimentation rates and in catchment soil erosion, an increased supply of inorganic silts and clays to the lakes, a pulse of phosphorus derived from human wastes, and a decrease in lacustrine productivity caused by high levels of turbidity.

In the last two centuries rates of sedimentation in lake basins have changed in different ways in different basins according to the differing nature of economic activities in catchments.

Some data from various sources are listed for comparison in table 6.5. In the case of the Loe Pool in Cornwall (south-west England) rates of sedimentation were high while mining industry was active, but fell dramatically when mining was curtailed. In the case of Seeswood Pool in Warwickshire, a dominantly agricultural catchment area in central England, the highest rates have occurred since 1978 in response to various land management changes, such as larger fields, continuous cropping and increased dairy herd size.

In other catchments, preafforestation ploughing may have caused sufficient disturbance to cause accelerated sedimentation. For example, Battarbee et al. (1985b) looked at sediment cores in the Galloway area of south-west Scotland and found that in Loch

Table 6.5 Data on rates of erosion and sedimentation in lakes in the last two centuries

Dates	Activity	Rates of erosion in catchment (R. Cober) as determined from lake sedimentation rates ($t\ km^{-2}\ y^{-1}$)
(a) Loe Pool (Cornwall) from O'Sullivan et al., 1982)		
1860–1920	Mining and agriculture	174
1930–6	Intensive mining and agriculture	421
1937–8	Intensive mining and agriculture	361
1938–81	Agriculture	12
(b) Seeswood Pool (Warwickshire) (from Foster et al., 1986)		
1765–1853		7.0
1854–80		12.2
1881–1902		8.1
1903–19		9.6
1920–5		21.6
1926–33		16.1
1934–47		12.7
1948–64		12.0
1965–72		13.9
1973–7		18.3
1978–82		36.2

Grannoch the introduction of ploughing in the catchment caused an increase in sedimentation from 0.2 cm y^{-1} to 2.2 cm y^{-1}.

Sedimentation has severe economic implications because of the role it plays in reducing the effective lifetime of reservoirs. In the tropics capacity depletion through sedimentation is commonly around 2 per cent per annum (Myers, 1988: 14), so that the expected useful life of the Paute hydroelectric project in Ecuador (cost US$ 600 million) is 32 years, that of the Mangla Dam in Pakistan (cost US$ 600 million) is 57 years, and that of the Tarbela Dam, also in Pakistan, is just 40 years.

Ground subsidence

Not all ground subsidence is caused by humans. For example, limestone solutional processes can, in the absence of humankind, create a situation where a cavern collapses to produce a surface depression, and permafrost will sometimes melt to produce a thermokarst depression without human intervention. None the less, ground subsidence can be caused or accelerated by humans in a variety of ways: by the transfer of subterranean fluids (such as oil, gas and water); by the removal of solids through underground mining or by dissolving solids and removing them in solutions (for example, sulphur and salt); by the disruption of permafrost; and by the compaction or reduction of sediments because of drainage and irrigation (Johnson, 1991).

Some of the most dangerous and dramatic collapses have occurred in limestone areas because of the dewatering of limestone caused by mining activities. In the Far West Rand of South Africa, gold-mining has required the abstraction of water to such a degree that the local water-table has been lowered by more than 300 m. The fall of the water-table caused miscellaneous clays and other materials filling the roofs of large caves to dry out and shrink so that they collapsed into the underlying void. One collapse created a depression 30 m deep and 55 m across, killing twenty-nine people. In Alabama, USA, water-level decline consequent upon pumping has had equally serious consequences in a limestone terrain; and Newton (1976) has estimated that since 1900 about 4000 induced sink-holes or related features have been formed, while fewer than fifty natural sink-holes have been reported over the same interval. Sink-holes that may result from such human activity are also found in Georgia, Florida, Tennessee, Pennsylvania and Missouri.

In some limestone areas, however, a reverse process can operate. The application of water to overburden above the limestone may render it more plastic so that the likelihood of collapse is increased. This has occurred beneath reservoirs, such as the May Reservoir in central Turkey, and as a result of the application of waste water and sewerage to the land surface.

The process can be accelerated by the direct solution of suscep-tible rocks. For example, collapses have occurred in gypsum

Plate 6.7
In recent years the city of
Venice has suffered from
increasingly severe winter
floods. The pumping of ground
water for industrial purposes
has caused ground
subsidence to occur

bedrock because of solution brought about by the construction of a
reservoir. In 1893 the MacMillan Dam was built on the Pecos River
in New Mexico, but within twelve years the whole river flowed
through caves which had developed since construction. Both the
San Fernando and Rattlesnake Dams in California suffer severe
leakage for similar reasons.

Subsidence produced by oil abstraction is an increasing problem
in some parts of the world. The classic area is Los Angeles, where
9.3 m of subsidence occurred as a result of exploitation of the
Wilmington oilfield between 1928 and 1971. The Inglewood

Plate 6.8
A gaping chasm in the
Transvaal, South Africa,
became the world's biggest
grave in August, 1964. The
sink-hole opened because
ground-water conditions had
been altered by mining. It
swallowed up a family of five,
their servants and homes

oilfield displayed 2.9 m of subsidence between 1917 and 1963.
Some coastal flooding problems occurred at Long Beach because of
this process. Similar subsidence has been recorded from the Lake
Maracaibo field in Venezuela (Prokopovich, 1972) and from some
Russian fields (Nikonov, 1977).

A more widespread problem is posed by ground-water abstraction for industrial, domestic and agricultural purposes. Table 6.6
presents some data for such subsidences from various parts of the
world. The ratios of subsidence to water-level decline are strongly
dependent on the nature of the sediment composing the aquifer, so
that ratios range from 1:7 for Mexico City, to 1:80 for the Pecos in
Texas, and to less than 1:400 for London, England (Rosepiler and
Reilinger, 1977).

Table 6.6 Ground subsidence

Location	Amount (m)		Rate (mm/y)
(a) Ground subsidence produced by oil and gas abstraction			
Azerbaydzhan, USSR	2.5	(1912–62)	50
Atravopol, USSR	1.5	(1956–62)	125
Wilmington, USA	9.3	(1928–71)	216
Inglewood, USA	2.9	(1917–63)	63
Maracaibo, Venezuela	5.03	(1929–90)	84
(b) Ground subsidence produced by ground-water abstraction			
London, England	0.06–0.08	(1865–1931)	0.91–1.21
Savannah, Georgia (USA)	0.1	(1918–55)	2.7
Mexico City	7.5	–	250–300
Houston, Galveston, Texas	1.52	(1943–64)	60–76
Central Valley, California	8.53	–	–
Tokyo, Japan	4	(1892–1972)	500
Osaka, Japan	>2.8	(1935–72)	76
Niigata, Japan	>1.5	–	–
Pecos, Texas	0.2	(1935–66)	6.5
South-central Arizona	2.9	(1934–77)	96
Bangkok, Thailand	0.5	–	100
Shanghai, China	2.62	(1921–65)	60

Source: data in Cooke and Doornkamp, 1974; Prokopovich, 1972; Rosepiler and Reilinger, 1977; Holzer, 1979; Nutalaya and Rau, 1981; Johnson, 1991

The extent of subsidence that has taken place in the United States as a result of ground-water abstraction has recently been reassessed on the basis of geodetic evidence (Chi and Reilinger, 1984). To the major areas previously identified:

south-central Arizona
Savannah (Georgia)
Pecos and Houston–Galveston (Texas)
Denver (Colorado)
San Joaquin Valley, Santa Clara Valley, Saugus Basin, Los
    Angeles Basin, Bunker Hill–San Timoteo (California)
Milford (Utah)
Raft River Valley (Idaho)
Baton Rouge and New Orleans (Louisiana)
Las Vegas (Nevada)

they have added:

Ventura, Ontario and San Pedro–Santa Monica (California)
Monroe and Alexandria (Louisiana)
Jackson (Mississippi).

In Japan subsidence has now emerged as a major problem (Nakano and Matsuda, 1976). In 1960 only 35.2 km² of the Tokyo

lowland was below sea-level, but continuing subsidence meant that by 1974 this had increased to 67.6 km^2, exposing a total of 1.5 million people to major flood hazard.

The subsidence caused by mining (which led to court cases in England as early as the fifteenth century as a consequence of associated damage to property) is perhaps the most familiar, though its importance varies according to such factors as the thickness of seam removed, its depth, the width of working, the degree of filling with solid waste after extraction, the geological structure, and the method of working adopted (Wallwork, 1974). In general terms, however, the vertical displacement by subsidence is less than the thickness of the seam being worked, and decreases with an increase in the depth of mining. This is because the overlying strata collapse, fragment and fracture, so that the mass of rock fills a greater space than it did when naturally compacted. Consequently the surface expression of deep-seated subsidence may be equal to little more than one-third of the thickness of the material removed. Subsidence associated with coal-mining may disrupt surface drainage and the resultant depressions then become permanently flooded. In England, the valleys of the Aire and Calder in the Castleford–Knottingley area, the Anker Valley in east Warwickshire, and the Wigan–Leigh area of Lancashire, provide examples of subsidence lakes formed on stream floodplains.

Coal-mining regions are not the only areas where subsidence problems are serious. In Cheshire, north-west England, salt is extracted from two major seams, each about 30 m in thickness. Moreover, these seams occur at no great depth – the uppermost being at about 70 m below the surface. A further factor to be considered is that the rock salt is highly soluble in water, so the

Plate 6.9 (opposite)
The Japanese city of Tokyo occupies a low-lying coastal position that is well demonstrated on this satellite image. Ground-water extraction is creating a major subsidence problem, and large areas of the city are becoming more prone to coastal flooding

Plate 6.10
Dramatic subsidence that occurred in the 1930s around Hodbarrow coal mine, Cumbria, north-west England

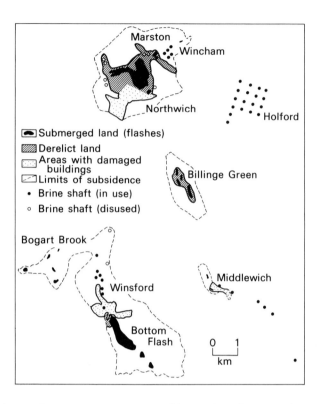

Figure 6.7
Subsidence in the salt area of
mid-Cheshire, England, in
1954 (after Wallwork, 1956,
figure 3)

flooding of mines may cause additional collapse. These three
conditions – thick seams, shallow depth and high solubility – have
produced optimum conditions for subsidence and many subsidence
lakes called 'flashes' have developed (Wallwork, 1956). Some of
these are illustrated in figure 6.7. The main railway line between
Crewe and Manchester at Elton subsided no less than 4.9 m
between 1892 and 1956 (Wallwork, 1960). Likewise, salt extraction at Windsor, Canada, is known to have produced a water-filled
depression 150 m in diameter and about 8 m deep (Martinez,
1971).

Some subsidence is created by a process called hydrocompaction,
which is explained thus. Moisture-deficient, unconsolidated, low-
density sediments tend to have sufficient dry strength to support
considerable effective stresses without compacting. However, when
such sediments, which may include alluvial fans or loess, are
thoroughly wetted for the first time (for example, by percolating
irrigation water) the inter-granular strength of the deposits is
diminished, rapid compaction takes place, and ground surface
subsidence follows. Unequal subsidence can create problems for
irrigation schemes.

Land drainage can promote subsidence of a different type,
notably in areas of organic soils. The lowering of the water-table
makes peat susceptible to oxidation and deflation so that its volume
decreases. One of the longest records of this process, and one of the

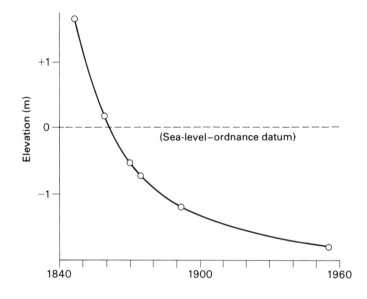

clearest demonstrations of its efficacy, has been provided by the measurements at Holme Fen Post in the English Fenlands. Approximately 3.8 m of subsidence occurred between 1848 and 1957 (Fillenham, 1963), with the fastest rate occurring soon after drainage had been initiated (figure 6.8). The present rate averages about 1.4 cm y^{-1} (Richardson and Smith, 1977). At its maximum natural extent the peat of the English Fenland covered around 1750 km^2. Now only about one-quarter (430 km^2) remains.

A further type of subsidence, sometimes associated with earthquake activity, results from the effects on the earth's crust of large masses of water impounded behind reservoirs. Seismic effects can be generated in areas with susceptible fault systems and this may account for earthquakes recorded at Koyna (India) and elsewhere. The process whereby a mass of water causes coastal depression is called hydro-isostasy, and the degree of subsidence that has been monitored for various large reservoirs is listed in table 6.7.

In permafrost areas ground subsidence is associated with thermokarst development, thermokarst being irregular, hummocky terrain

Table 6.7 Hydro-isostatic subsidence caused by the creation of reservoirs

Reservoir	Maximum downwarping (cm)	Period
Lake Mead, USA	20.1	1950–63
Kariba, Central Africa	12.7	1959–68
Koyna, India	8–14	1962–8
Bratsk, Siberia	5.6	1961–6
Krasnoyarsk, Siberia	3	1967–71
Plyava, Baltic	0.5	1965–70

Source: from data in Nikonov, 1977

produced by the melting of ground ice, permafrost. The development of thermokarst is due primarily to the disruption of the thermal equilibrium of the permafrost and an increase in the depth of the active layer. This is illustrated in figure 6.9. Following French (1976: 106), consider an undisturbed tundra soil with an active layer of 45 cm. Assume also that the soil beneath 45 cm is supersaturated permafrost and yields on a volume basis upon thawing 50 per cent excess water and 50 per cent saturated soil. If the top 15 cm were removed, the equilibrium thickness of the active layer, under the bare ground conditions, might increase to 60 cm. As only 30 cm of the original active layer remains, 60 cm of the permafrost must thaw before the active layer can thicken to 60 cm, since 30 cm of supernatant water will be released. Thus, the surface subsides 30 cm because of thermal melting associated with the degrading permafrost, to give an overall depression of 45 cm.

Thus the key process involved in thermokarst subsidence is the state of the active layer and its thermal relationships. When, for example, surface vegetation is cleared for agricultural or constructional purposes the depth of thaw will tend to increase. The movement of tracked vehicles has been particularly harmful to surface vegetation and deep channels may soon result from permafrost degradation. Similar effects may be produced by the siting of heated buildings on permafrost, and by the laying of oil, sewer and water pipes in or on the active layer (Ferrians et al., 1969; Lawson, 1986).

Thus subsidence is a diverse but significant aspect of the part humans play as geomorphological agents. The damage caused on a worldwide basis can be measured in billions of dollars each year (Coates, 1983), and among the effects are broken dams, cracked buildings, offset roads and railways, fractured well casings, de-

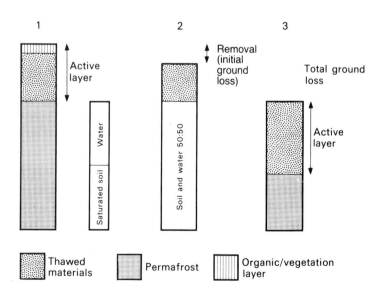

Figure 6.9
Diagram illustrating how the disturbance of high ice content terrain can lead to permanent ground subsidence. 1–3 indicate stages before, immediately after, and subsequent to, disturbance (after Mackay in French, 1976, figure 6.1)

Plate 6.11
In Alaska, and other areas
which are underlain by
permafrost, ground subsidence
caused by thermokarstic
processes presents
engineering problems.
Buildings sometimes have to
be abandoned

formed canals and ditches, bridges that need relevelling, saline encroachment and increased flood damage.

Arroyo trenching, gullies and peat haggs

In the south-western United States many broad valleys and plains became deeply incised with valley-bottom gullies (arroyos) over a short period between 1865 and 1915, with the 1880s being especially important (Cooke and Reeves, 1976). This cutting had a rapid and detrimental effect on the flat, fertile and easily irrigated valley floors, the most desirable sites for settlement and economic activity in a harsh environment.

Many students of this phenomenon have believed that thought-less human actions caused the entrenchment, and the apparent coincidence of white settlement and arroyo development tended to give credence to this viewpoint. The range of actions that could have been culpable is large: timber-felling, over-grazing, cutting grass for hay in valley bottoms, compaction along well-travelled routes, channelling of runoff from trails and railways, disruption of valley-bottom sods by animals' feet, and the invasion of grasslands by miscellaneous types of scrub.

On the other hand, study of the long-term history of the valley fills shows that there have been repeated phases of aggradation and incision and that some of these took place before the influence of humans could have been a significant factor. This has prompted debate as to whether the arrival of white communities was in fact responsible for this particularly severe phase of environmental degradation. Huntington (1914), for example, argued that valley filling would be a consequence of a climatic shift to more arid conditions. These, he believed, would cause a reduction in vegetation which in turn would promote rapid removal of soil from devegetated mountain slopes during storms, and overload streams with sediment. With a return to humid conditions vegetation would be re-established, sediment yields would be reduced and entrenchment of valley fills would take place. Bryan (1928) put forward a contradictory climatic explanation. He argued that a slight move towards drier conditions, by depleting vegetation cover and reducing soil infiltration capacity, would produce significant increases in storm runoff which would erode valleys. Another climatic interpretation was advanced by Leopold (1951), involving a change in rainfall intensity rather than quantity. He indicated that a reduced frequency of low-intensity rains would weaken the vegetation cover, while an increased frequency of heavy rains at the same time would increase the incidence of erosion. Support for this contention comes from the work of Balling and Wells (1990) in New Mexico. They attributed early twentieth-century arroyo trenching to a run of years with intense and erosive rainfall characteristics that succeeded a phase of drought conditions in which the productive ability of the vegetation had declined.

It is also possible, as Schumm et al. (1984) have pointed out, that arroyo incision could result from neither climatic change nor human influence. It could be the result of some natural geomorphological threshold being crossed.

It is therefore clear that the possible mechanisms that can lead to alternations of cut-and-fill of valley sediments are extremely complex, and that any attribution of all arroyos in all areas to human activities may be a serious oversimplification of the problem (figure 6.10). In addition, it is possible that natural environmental changes, such as changes in rainfall characteristics, have operated at the same time and in the same direction as human actions.

In the Mediterranean lands there have also been controversies surrounding the age and causes of alternating phases of aggradation and erosion in valley bottoms. Vita-Finzi (1969) has suggested that at some stage during historical times many of the streams in the Mediterranean area, which had hitherto been engaged primarily in downcutting, began to build up their beds. Renewed downcutting, still seemingly in operation today, has since incised the channels into the alluvial fill. He proposed that the reversal of the downcutting trend in the Middle Ages was both ubiquitous and confined in time, and that some universal and time-specific agency was required to explain it. He believed that devegetation by humans was not a medieval innovation and that some other mechanism was required. A solution he gave to account for the phenomenon was precipitation change during the climatic fluctuation known as the Little Ice Age (AD 1550–1850). This was not an interpretation which found favour with Butzer (1974). He reported that his investigations showed plenty of post-Classical and pre-1500 alluviation (which could not therefore be ascribed to the Little Ice Age), and he doubted whether Vita-Finzi's dating was precise enough to warrant a 1550–1850 date. Instead, he suggested that humans were

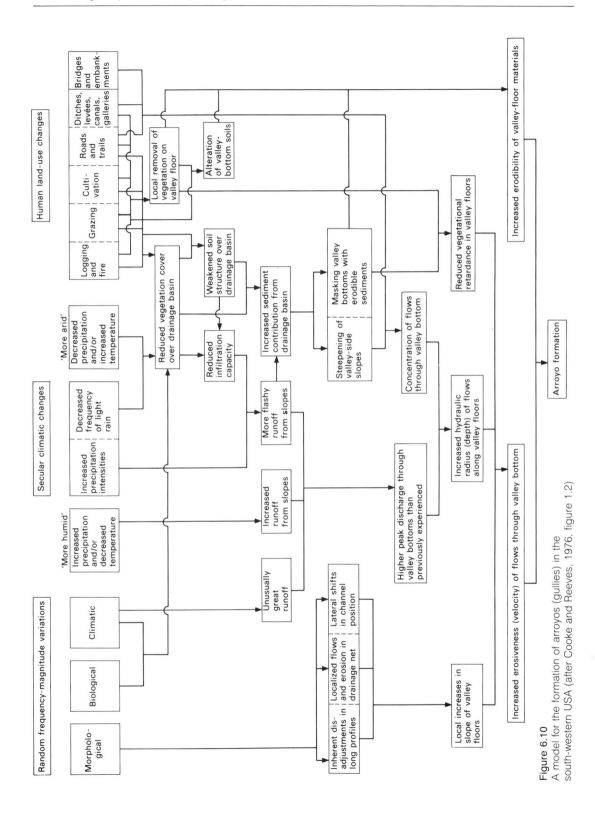

Figure 6.10
A model for the formation of arroyos (gullies) in the
south-western USA (after Cooke and Reeves, 1976, figure 1.2)

responsible for multiple phases of accelerated erosion from slopes, and accelerated sedimentation in valley bottoms, from as early as the middle of the first millennium BC.

Butzer's interpretation has found favour with van Andel et al. (1990) who have detected an intermittent and complex record of cut and fill episodes during the late Holocene in various parts of Greece (figure 6.11). They believe that this evidence is compatible with a model of the control of timing and intensity of landscape destabilization by local economic and political conditions. This is a view shared in the context of the Algarve in Portugal by Chester and James (1991).

Another example of drainage incision that demonstrates the problem of disentangling the human from the natural causes of erosion is provided by the eroding peat bogs of highland Britain. Over many areas, including the Pennines of northern England and the Brecon Beacons of Wales, blanket peats are being severely eroded to produce pool and hummock topography, areas of bare peat and incised gullies (haggs). Many rivers draining such areas are discoloured by the presence of eroded peat particles, and sediment yields of organic material are appreciable (Labadz et al., 1991).

Some of the observed peat erosion may be an essentially natural process, for the high water content and low cohesion of undrained peat masses make them inherently unstable. Moreover, the instability must normally become more pronounced as peat continues to accumulate, leading to bog slides and bursts round margins of expanded peat blankets. Conway (1954) suggested that an inevitable end-point of peat build-up on high altitude, flat or convex

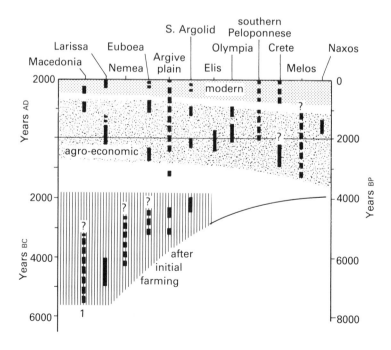

Figure 6.11
Chronology of Holocene alluviation in Greece and the Aegean. Broken bars are dated uncertainly or represent intermittent deposition (from various sources in Van Andel et al., 1990, figure 10)

surfaces is that a considerable depth of unconsolidated and poorly humidified peat overlies denser and well-humidified peat, so adding to the instability. Once a bog burst or slide occurs, this leads to the formation of drainage gullies which extend back into the peat mass, slumping-off marginal peat downslope, and leading to the drawing off of water from the pools of the hummock and hollow topography of the watershed.

Tallis (1985a) believes that there have been two main phases of erosion in the Pennines. The first, initiated 1000–1200 years ago, may have been caused by natural instability of the type outlined above. However, there has been a second stage of erosion, initiated 200–300 years ago, in which miscellaneous human activities appear to have been important. Radley (1962) suggested that among the pressures that had caused erosion were heavy sheep grazing, regular burning, peat cutting, the digging of boundary ditches, the incision of pack horse tracks, and military manoeuvres during the First World War. Other causes may include footpath erosion and severe air pollution (Tallis, 1965). In South Wales, there is some evidence that the blanket peats have degenerated as a result of contamination by particulate pollution (soot, etc) during the Industrial Revolution (Chambers et al., 1979). On the other hand, in Scotland lake core studies indicate that severe peat erosion was initiated between AD 1500 and 1700, prior to air pollution associated with industrial growth, and Stevenson et al. (1990) suggest that this erosion initiation may have been caused either by the adverse climatic conditions of the Little Ice Age or by an increasing intensity of burning as land use pressures increased.

Accelerated weathering and the tufa decline

Although fewer data are available and the effects are generally less immediately obvious, there is some evidence that human activities have produced changes in the nature and rate of weathering (Winkler, 1970). The prime cause of this is probably air pollution. It is clear that, as a result of increased emissions of sulphur dioxide through the burning of fossil fuels, there are higher levels of sulphuric acid in rain over many industrial areas. This in itself may react with stones and cause their decay. Chemical reactions involving sulphur dioxide can also generate salts such as calcium sulphate and magnesium sulphate, which may be effective in causing the physical breakdown of rock through the mechanism of salt weathering.

Similarly, atmospheric carbon-dioxide levels have been rising steadily because of the burning of fossil fuels, and deforestation. Carbon dioxide may combine with water, especially at lower temperatures, to produce weak carbonic acid which can dissolve limestone, marbles and dolomites. Weathering can also be accelerated by changes in ground-water levels resulting from irrigation. This can be illustrated by considering the Indus Plain in Pakistan

Plate 6.12
The ancient city of
Mohenjo-Daro in Pakistan was
excavated in the 1920s.
Irrigation has been introduced
into the area causing
ground-water levels to be
raised. This has brought salt
into the bricks of the ancient
city producing severe
disintegration

(Goudie, 1977), where irrigation has caused the water-table to be
raised by about 6 m since 1922. This has produced increased
evaporation and salinization. The salts that are precipitated by
evaporation above the capillary fringe include sodium sulphate, a
very effective cause of stone decay. Indeed buildings, such as the
great archaeological site of Mohenjo-Daro, are decaying at a
catastrophic rate.

In other cases accelerated weathering has been achieved by
moving stone from one environment to another. Cleopatra's Needle,
an Egyptian obelisk in New York City, is an example of rapid
weathering of stone in an inhospitable environment. Originally
erected on the Nile opposite Cairo about 1500 BC, it was toppled in
about 500 BC by Persian invaders, and lay partially buried in Nile
sediments until, in 1880, it was moved to New York. It immediately began to suffer from scaling and the inscriptions were largely
obliterated within ten years because of the penetration of moisture
which enabled the frost-wedging and hydration of salts to occur.

During the 1970s and 1980s an increasing body of isotopic dates
became available for deposits of tufa (secondary freshwater deposits
of limestone, also known as travertine). Some of these dates suggest
that over large parts of Europe, from Britain to the Mediterranean
basin and from Spain to Poland, rates of tufa formation were high
in the early and mid-Holocene, but declined markedly thereafter
(Weisrock, 1986: 165–7). Vaudour (1986) maintains that since
around 3000 BP 'the impact of man on the environment has
liberated their disappearance', but he gives no clear indication of
either the basis of this point of view or of the precise mechanism(s)
that might be involved. If the late Holocene reduction in tufa
deposition is a reality then it is necessary to consider a whole range
of possible mechanisms, both natural and anthropogenic (Nicod,
1986: 71–80; table 6.8). As yet the case for an anthropogenic role
is not proven.

Table 6.8 Some possible mechanisms to account for the alleged Holocene tufa decline

Climatic/natural	Anthropogenic
Discharge reduction following rainfall decline leading to less turbulence	Discharge reduction due to overpumping, diversions, etc.
Degassing leads to less deposition	Increased flood scour and runoff variability of channels due to deforestation, urbanization, ditching, etc.
Increased rainfall causing more flood scour	Channel shifting due to deforestation of floodplains leads to tufa erosion
Decreasing temperature leads to less evaporation and more CO_2 solubility	Reduced CO_2 flux in system after deforestation
Progressive Holocene peat development and soil podzol development through time leads to more acidic surface waters	Introduction of domestic stock causes breakdown of fragile tufa structures
	Deforestation = less fallen trees to act as foci for tufa barrages
	Increased stream turbidity following deforestation reduces algal productivity.

Accelerated mass movements

There are many examples of mass movements being triggered by human actions (Selby, 1979). For instance, landslides can be created either by undercutting or by overloading (figure 6.12). When a road is constructed, material derived from undercutting the upper hillside may be cast on to the lower hillslope as a relatively loose fill to widen the road bed. Storm water is then often diverted from the road on to the loose fill.

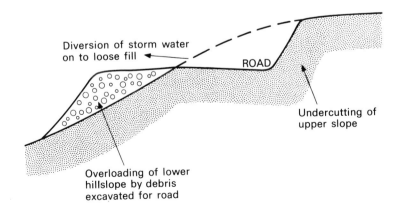

Diversion of storm water on to loose fill

ROAD

Undercutting of upper slope

Overloading of lower hillslope by debris excavated for road

Figure 6.12
Slope instability produced by road construction

Because of the hazards presented by both natural and accelerated mass movements, humans have developed a whole series of techniques to attempt to control them. Such methods, many of which are widely used by engineers, are listed in table 6.9. These techniques are increasingly necessary, as human capacity to change a hillside and to make it more prone to slope failure has been transformed by engineering development. Excavations are going deeper, buildings and other structures are larger, and many sites which are at best marginally suitable for engineering projects are now being used because of increasing pressure on land. This applies especially to some of the expanding urban areas in the humid parts of low latitudes – Hong Kong, Kuala Lumpur, Rio de Janeiro and many others. It is very seldom that human agency deliberately accelerates mass movements; most are accidentally caused, the exception possibly being the deliberate triggering of a threatening snow avalanche (Perla, 1978).

The forces producing slope instability and landsliding can usefully be divided into disturbing forces and resisting properties (Cooke and Doornkamp, 1990: 113–14). The factors leading to an increase in shear stress (disturbing forces) can be listed as follows (modified after Cooke and Doornkamp, 1990: 113):

1 *Removal of lateral or underlying support*
   Undercutting by water (for example, river, waves), or glacier ice
   Weathering of weaker strata at the toe of the slope
   Washing out of granular material by seepage erosion
   human cuts and excavations, drainage of lakes or reservoirs.
2 *Increased disturbing forces*
   Natural accumulations of water, snow, talus

Table 6.9 Examples of methods of controlling mass movements

Type of movement	Method of control
Falls	Flattening the slope Benching the slope Drainage Reinforcement of rock walls by grouting with cement, anchor bolts Covering of wall with steel mesh
Slides and flows	Grading or benching to flatten the slope Drainage of surface water with ditches Sealing surface cracks to prevent infiltration Sub-surface drainage Rock or earth buttresses at foot Retaining walls at foot Pilings through the potential slide mass

Source: after R. F. Baker and H. E. Marshall, in Dunne and Leopold, 1978, table 15.16

pressure caused by human activity (for example, stockpiles of ore, tip-heaps, rubbish dumps, or buildings).

3 *Transitory earth stresses*
  Earthquakes
  Continual passing of heavy traffic
4 *Increased internal pressure*
  Build up of pore-water pressures (for example, in joints and cracks, especially in the tension crack zone at the rear of the slide).

Some of the factors are natural, while others (italicized) are effected by humans. Factors leading to a decrease in the shearing resistance of materials making up a slope can also be summarized (also modified after Cooke and Doornkamp, 1990: 113):

1 *Materials*
  Beds which decrease in shear strength if water content increases (clays, shale, mica, schist, talc, serpentine) (for example, *when local water-table is artificially increased in height by reservoir construction*), or as a result of stress release (vertical and/or horizontal) following slope formation. Low internal cohesion (for example, consolidated clays, sands, porous organic matter)
  In bedrock: faults, bedding planes, joints, foliation in schists, cleavage, brecciated zones, and pre-existing shears.
2 *Weathering changes*
  Weathering reduces effective cohesion, and to a lesser extent the angle of shearing resistance
  Absorption of water leading to changes in the fabric of clays (for example, loss of bonds between particles or the formation of fissures).
3 *Pore-water pressure increase*
  High ground-water table as a result of increased precipitation *or as a result of human interference* (for example, *dam construction*) (see 1 above).

Once again the italics show that there are a variety of ways in which humans can play a role.

Some mass movements are created by humans piling up waste soil and rock into unstable accumulations that fail spontaneously. At Aberfan, in South Wales, a major disaster occurred when a 180 m high coal-waste tip began to move as an earth flow (plate 6.13). The tip had been constructed not only as a steep slope but also upon a spring line. This made an unstable configuration which eventually destroyed a school and claimed over 150 lives. Human-induced mass movements are also a problem in Los Angeles, California. Some result from the denudation of slope vegetation cover by fires, many of which are set by humans, while others are triggered by infiltration of water from cesspools and from irrigation water applied to lawns and gardens. In Hong Kong, where a large proportion of the population is forced to occupy steep slopes

Plate 6.13
The village of Aberfan in South Wales where in 1966 a massive debris flow caused severe loss of life when it destroyed a school and houses as it ran down from a steep coal-waste tip

developed on deeply weathered granites and other rocks, mass movements are a severe problem, and So (1971) has shown that many of the landslips and washouts (70 per cent of those in the great storm of June 1966, for example) were associated with road sections and slopes artificially modified through construction and cultivation.

One of the most serious mass movements partly caused by human activity was that which caused the Vaiont Dam disaster in Italy in 1963, in which 2600 people were killed (Kiersch, 1965). Heavy antecedent rainfall conditions and the presence of young, highly folded sedimentary rocks provided the necessary conditions for a slip to take place, but it was the construction of the Vaiont Dam itself which changed the local ground-water conditions sufficiently to affect the stability of a rock mass on the margins of the reservoir: 240 million m^3 of ground slipped with enormous speed into the reservoir, producing a sharp rise in water level which overtoppped the dam and caused flooding and loss of life downstream. Comparable slope instability resulted when the Franklin D. Roosevelt lake was impounded by the Columbia River in the USA (Coates, 1977), but the effects were, happily, less serious.

It is evident from what has been said about the predisposing causes of the slope failure *triggered* by the Vaiont Dam that human agency was only able to have such an impact because the natural conditions were broadly favourable. Exactly the same lesson can be learnt from the accelerated landsliding in southern Italy. Nossin (1972) has demonstrated how, in Calabria, road construction has triggered off (and been hindered by) landsliding, but he has also stressed that the area is fundamentally susceptible to such mass movement activity because of geological conditions. It is an area where recent rapid uplift has caused downcutting by rivers and the undercutting of slopes by erosion. It is also an area of incoherent metamorphic rocks, with frequent faulting. Further, water is often trapped by Tertiary clay layers, providing further stimulus to movement.

Although the examples of accelerated mass movements that have been given here are essentially associated with the effects of modern construction projects, more long-established activities, including deforestation and agriculture, are also highly important. For example, Innes (1983) has demonstrated, on the basis of lichenometric dating of debris-flow deposits in the Scottish Highlands, that most of the flows have developed in the last 250 years, and he suggests that intensive burning and grazing may be responsible.

However, as with so many environmental changes of that nature in the past, there are considerable difficulties in being certain about causation. This has been well expressed by Ballantyne (1991: 84):

Although there is growing evidence for Late Holocene erosion in upland Britain, the causes of this remain elusive. A few studies have presented evidence linking erosion to vegetation degradation and destruction due to human influence, but the validity of climatic deterioration as a cause of erosion remains unsubstantiated. This uncertainty stems from a tendency to link erosion with particular causes only through assumed coincidence in timing, a procedure fraught with difficulty because of imprecision in the dating of both putative causes and erosional effects. Indeed, in many reported instances it is impossible to refute the possibility that the timing of erosional events or episodes may be linked to high magnitude storms of random occurrence, and bears little relation to either of the causal hypotheses outlined above ...

Deliberate modification of channels

Both for purposes of navigation and flood control humans have deliberately straightened many river channels. Indeed, the elimination of meanders contributes to flood control in two ways. First, it eliminates some overbank floods on the outside of curves, against which the swiftest current is thrown and where the water surface rises highest. Second, and more imporantly, the resultant shortened course increases both the gradient and the velocity, and the floodwaters erode and deepen the channel, thereby increasing its flood capacity.

It was for this reason that a programme of channel cutoffs was initiated along the Mississippi in the early 1930s. By 1940 it had lowered flood stages by as much as 4 m at Arkansas City, Arkansas. By 1950 the length of the river between Memphis, Tennessee and Baton Rouge, Louisiana (600 km down the valley) had been reduced by no less than 270 km as a result of 16 cutouts.

The largest current scheme for channel modification is the Jonglei Canal in the Sudan (figure 6.13, plate 6.14). This scheme has involved the construction of a channel 360 km long that will bypass the Sudd swamps where more than half of the White Nile's flow is lost by evapotranspiration every year. This will provide an extra 4.75 billion m^3 of water each year for Sudan and Egypt, but its impact on the life of the local pastoral people and on the ecology of the floodplains is the cause of considerable misgivings (Charnock, 1983).

Some landscapes have become dominated by artificial channels, normally once again because of the need for flood alleviation and drainage. This is especially evident in an area like the English Fenlands where straight constructed channels contrast with the sinuous courses of original rivers such as the Great Ouse.

Non-deliberate river-channel changes

There are thus many examples of the intentional modification of river-channel geometry by humans, by the construction of embankments, by channelization, and by other such processes. However, major changes in the configuration of channels can be achieved accidentally (table 6.10), either because of human-induced changes in stream discharge or in their sediment load: both parameters affect channel capacity (Park, 1977).

Deliberate channel-straightening causes various types of sequential channel adjustment both within and downstream from straightened reaches, and the types of adjustment vary according to such influence as stream gradient and sediment characteristics. Brookes (1987) has recognized five types of change *within* the straightened reaches (types W1 to W5) and two types of change downstream (types D1 and D2). They are illustrated in figure 6.14.

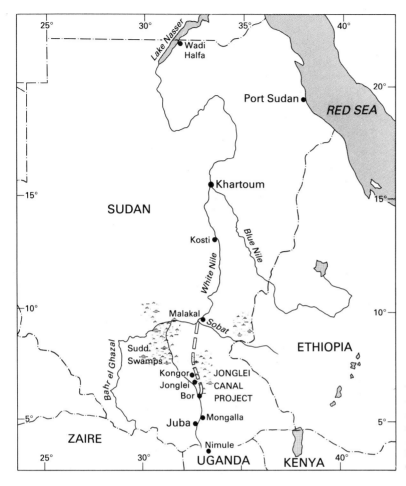

Figure 6.13
The diversion of the River Nile waters along the Jonglei Canal in Sudan. One of the major effects of this canal will be to reduce the degree of seasonal inundation of the Sudd and other floodplains

Table 6.10 Accidental channel changes

Phenomenon	Cause
Channel incision	'Clear-water erosion' below dams caused by sediment removal
Channel aggradation	Reduction in peak flows below dams Addition of sediment to streams by mining, agriculture, etc.
Channel enlargement	Increase in discharge level produced by urbanization
Channel diminution	Discharge decrease following water abstraction or flood control
Channel diminution	Trapping and stabilizing of sediment by artificially introduced plants

Plate 6.14
One of the most impressive
attempts to modify a major
world river is the construction
of the Jonglei Canal in the
Sudan to bypass the Sudd
swamps. Progress towards its
completion has been hindered
by political instability

Type W1   is degradation of the channel bed, which results
from the fact that straightening increases the slope
by providing a shorter channel path. This in turn
increases its sediment transport capability.

Type W2   is the development of an armoured layer on the
channel bed by the more efficient removal of fine
materials as a result of the increased sediment
transport capability referred to above.

Type W3   is the development of a sinuous thalweg in streams
which are not only straightened but which are also
widened beyond the width of the natural channel.

Type W4   is the recovery of sinuosity as a result of bank
erosion in channels with high slope gradients.

Type W5   is the development of a sinuous course by deposition in streams with a high sediment load and a
relatively low valley gradient.

Types D1 and D2   result from deposition downstream as the
stream tries to even out its gradient, the deposition
occurring as a general raising of the bed level, or as
a series of accentuated point bar deposits.

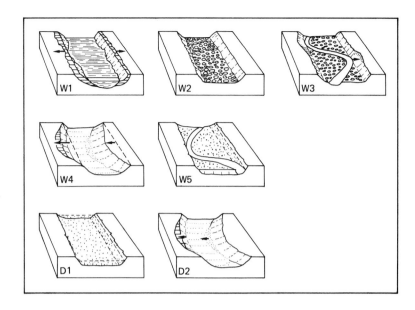

Figure 6.14
Principal types of adjustment in
straightened river channels
(after Brookes, 1987, figure 4).
For an explanation of the
different types see text

It is now widely recognized that the urbanization of a river basin
results in an increase in the peak flood flows in a river. It is also
recognized that the morphology of stream channels is related to
their discharge characteristics, and especially to the discharge at
which bank full flow occurs. As a result of urbanization, the
frequency of discharges which fill the channel will increase, with
the effect that the beds and banks of channels in erodible materials
will be eroded so as to enlarge the channel. This in turn will lead to
bank caving, possible undermining of structures, and increases in
turbidity (Hollis and Luckett, 1976).

Comparable changes in channel morphology result from discharge diminution produced by flood-control works and diversions
for irrigation. This can be shown for the North Platte and the South
Platte in America, where both peak discharge and mean annual
discharge have declined to 10–30 per cent of their pre-dam values.
The North Platte, 762–1219 m wide in 1890 near the Wyoming–
Nebraska border, has narrowed to about 60 m at present; while the
South Platte River was about 792 m wide, 89 km above its junction
with the North Platte in 1897, but had narrowed to about 60 m by
1959 (Schumm, 1977: 161). The tendency of both rivers has been
to form one narrow, well-defined channel in place of the previously
wide, braided channels, and, in addition, the new channel is
generally somewhat more sinuous than the old (figure 6.15).

Similarly, the building of dams can lead to channel aggradation
upstream from the reservoir and channel deepening downstream
because of the changes brought about in sediment loads (figure
6.16). Some data on observed rates of degradation below dams are
presented in table 6.11. They show that the average rate of
degradation has been of the order of a few metres over a few

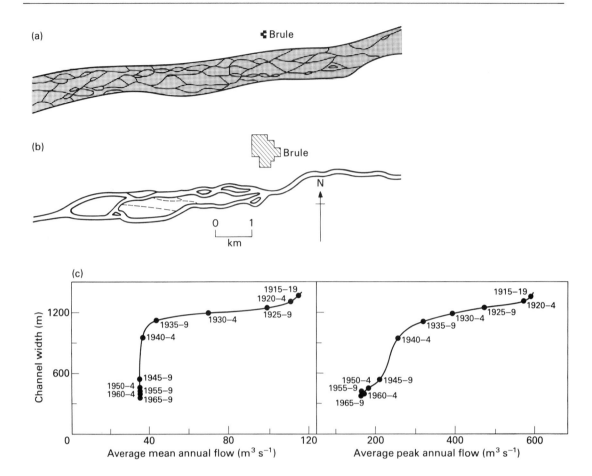

Figure 6.15
The configuration of the
channel of the South Platte
River at Brule in Nebraska,
USA: (a) in 1897 and (b) in
1959. Such changes in
channel form result from
discharge diminution (c)
caused by flood-control works
and diversions for irrigation
(after Schumm 1977, figure
5.32 and Williams, 1978)

decades following closure of the dams. However, over time the rate
of degradation seems to become less or to cease altogether, and
Leopold et al. (1964: 455) suggest that this can be brought about in
several ways. First, because degradation results in a flattening of the
channel slope in the vicinity of the dam – the slope may become so
flat that the necessary force to transport the available materials is
no longer provided by the flow. Second, the reduction of flood
peaks by the dam reduces the competence of the transporting
stream to carry some of the material on its bed. Thus if the bed
contains a mixture of particle size the river may be able to transport
the finer sizes but not the larger, and the gradual winnowing of the
fine particles will leave an armour of coarser material that prevents
further degradation.

The overall effect of the creation of a reservoir by the construction of a dam is to lead to a reduction in downstream channel
capacity (see Petts, 1979 for a review). This seems to amount to
between about 30 and 70 per cent (see table 6.12).

Equally far-reaching changes in channel form are produced by
land-use changes and the introduction of soil conservation mea-

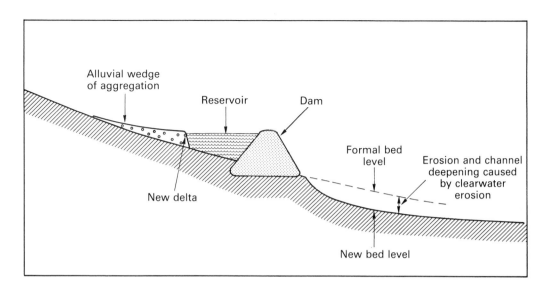

Table 6.11 Riverbed degradation below dams

River	Dam	Amount (m)	Length (km)	Time (years)
South Canadian (USA)	Conchos	3.1	30	10
Middle Loup (USA)	Milburn	2.3	8	11
Colorado (USA)	Hoover	7.1	111	14
Colorado (USA)	Davis	6.1	52	30
Red (USA)	Denison	2.0	2.8	3
Cheyenne (USA)	Angostura	1.5	8	16
Saalach (Austria)	Reichenhall	3.1	9	21
South Saskatchewan (Canada)	Diefenbaker	2.4	8	12
Yellow (China)	Samenxia	4.0	68	4

Source: from data in Galay, 1983

Figure 6.16
Diagrammatic long profile of a river showing the upstream aggradation and downstream erosion caused by dam and reservoir construction

Table 6.12 Channel capacity reduction below reservoirs

River	Dam	% channel capacity loss
Republican, USA	Harlan County	66
Arkansas, USA	John Martin	50
Rio Grande, USA	Elephant Buttre	50
Tone, UK	Clatworthy	54
Meavy, UK	Burrator	73
Nidd, UK	Angram	60
Burn, UK	Burn	34
Derwent, UK	Ladybower	40

Source: modified after Petts, 1979, table 1

sures. Figure 6.17 is an idealized representation of how the river basins of Georgia, USA have been modified through human agency between 1700 (the time of European settlement) and the present. Clearing of the land for cultivation (figure 6.17b) caused massive slope erosion which resulted in the transference of large quantities

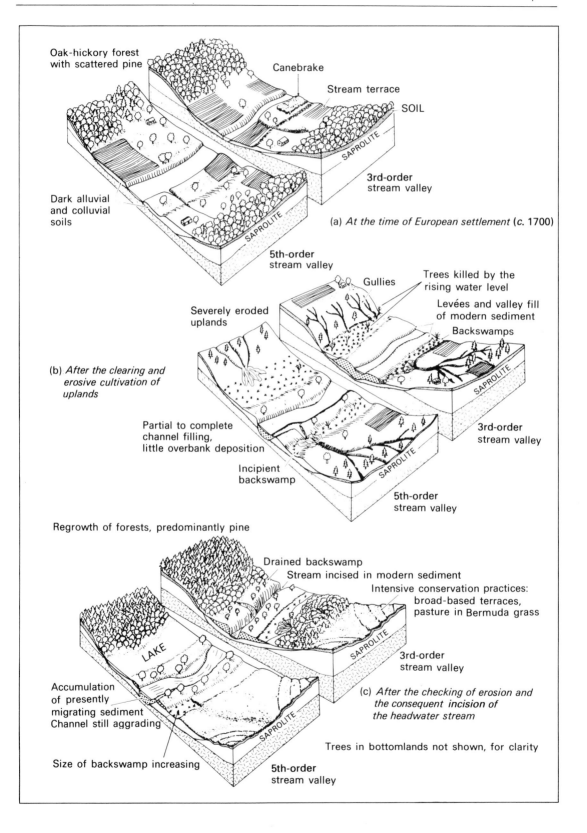

Oak-hickory forest
with scattered pine

Canebrake

Stream terrace

SOIL

SAPROLITE

3rd-order
stream valley

Dark alluvial
and colluvial
soils

SAPROLITE

(a) *At the time of European settlement* (c. 1700)

5th-order
stream valley

Gullies

Trees killed by the
rising water level

Severely eroded
uplands

Levées and valley fill
of modern sediment

Backswamps

(b) *After the clearing and
erosive cultivation of
uplands*

SAPROLITE

3rd-order
stream valley

Partial to complete
channel filling,
little overbank deposition

Incipient
backswamp

SAPROLITE

5th-order
stream valley

Regrowth of forests, predominantly pine

Drained backswamp

Stream incised in modern sediment

Intensive conservation practices:
broad-based terraces,
pasture in Bermuda grass

LAKE

SAPROLITE

3rd-order
stream valley

Accumulation
of presently
migrating sediment
Channel still aggrading

(c) *After the checking of erosion and
the consequent* incision *of
the headwater stream*

Trees in bottomlands not shown, for clarity

Size of backswamp increasing

SAPROLITE

5th-order
stream valley

of sediment into channels and floodplains. This phase of intense erosive land-use persisted and was particularly strong during the nineteenth century and the first decades of the twentieth century, but thereafter (figure 6.17c) conservation measures, reservoir construction and a reduction in the intensity of agricultural land use led to further channel changes (Trimble, 1974). Streams ceased to carry such a heavy sediment load, they became much less turbid, and incision took place into the floodplain sediments. By means of this active stream-bed erosion, streams incised themselves into the modern alluvium, lowering their beds by as much as 3–4 m.

In the Platte catchment of south-west Winsconsin a broadly comparable picture of channel change has been documented by Knox (1977). There, as in the Upper Mississippi valley (Knox 1987), it is possible to identify stages of channel modification associated with various stages of land use, culminating in decreased overbank sedimentation as a result of better land management in the last half century.

Other significant changes produced in channels include those prompted by accelerated sedimentation associated with changes in the vegetation communities growing along channels. The introduction of the salt cedar in the southern USA has caused significant floodplain aggradation. In the case of the Brazos River in Texas, for example, the plants encouraged sedimentation by their damming and ponding effect. They clogged channels by invading sand banks and sand bars, and so increased the area subject to flooding. Between 1941 and 1979 the channel width declined from 157 to 67 m, and the amount of aggradation was as much as 5.5 m (Blackburn et al., 1983). Another example of rapid aggradation is that produced by the introduction of mining waste into channels. In mid-Wales the Afon Ystwyth suffered 3 m of aggradation for this reason (Lewin et al., 1983).

Overall, the causes of observed cases of riverbed degradation are varied and complex and result from a variety of natural and human changes (table 6.13). A useful distinction can be drawn between degradation that proceeds downstream, and that which proceeds upstream, but in both cases the complexity of causes is evident.

Reactivation and stabilization of sand dunes

To George Perkins Marsh the reactivation and stabilization of sand dunes, especially coastal dunes, was a theme of the greatest importance in his analysis of human transformation of nature. He devoted fifty-four pages to it:

The preliminary steps, whereby wastes of loose, drifting, barren sands are transformed into wooded knolls and plains, and finally through the accumulation of vegetable mould, into arable ground, constitute a conquest over nature which precedes agriculture – a geographical revolution – and therefore, an account of the means by which the change has been effected belongs properly to the history of man's influence on the great features of physical geography. (1965: 393)

Figure 6.17 (opposite)
Changes in the evolution of the fluvial landscapes of the Piedmont of Georgia, USA, in response to land-use change between 1700 and 1970 (after Trimble, 1974, p. 117, in S. W. Trimble, *Man-induced soil erosion on the southern Piedmont*, Soil Conservation Society of America. © Soil Conservation Society of America)

Table 6.13 Causes of riverbed degradation

Type	Primary cause	Contributory cause
Downstream progressing	Decrease of bed-material discharge	Dam construction Excavation of bed material Diversion of bed material Change in land use Storage of bed material
	Increased water discharge	Diversion of flow Rare floods
	Decrease in bed-material size	
	Other	River emerging from lake Thawing of permafrost
Upstream progressing	Lower base level	Drop in lake level Drop in level of main river Excavation of bed material
	Decrease in river length	Cutoffs Channelization Stream capture Horizontal shift of base level
	Removal of control point	Natural erosion Removal of dam

Source: after Galay, 1983

He was fascinated by 'the warfare man wages with the sand hills' and asked (1965: 410) 'in what degree the naked condition of most dunes is to be ascribed to the improvidence and indiscretion of man'.

His analysis showed quite clearly that most of the coastal dunes of Europe and North America had been rendered mobile, and hence a threat to agriculture and settlement, through human action, especially because of grazing and clearing. In Britain the cropping of dune warrens by rabbits was a severe problem, and a most significant event in their long history was the myxomatosis outbreak of the 1950s, which severely reduced the rabbit population and led to dramatic changes in stability and vegetative cover.

Appreciation of the problem of dune reactivation on mid-latitude shorelines, and attempts to overcome it, go back a long way (Kittredge, 1948). For example, the menace of shifting sand following denudation is recognized in a decree of 1539 in Denmark which imposed a fine on those who destroyed certain species of sand plants on the coast of Jutland. The fixation of coastal sand dunes by planting vegetation was initiated in Japan in the seventeenth century, while attempts at the reafforestation of the spectacular Landes dunes in south-west France began as early as 1717, and came to fruition in the nineteenth century through the plans of the great Bremontier: 81 000 hectares of moving sand had been fixed in the Landes by 1865. In Britain possibly the most impressive example of sand control is provided by the reafforestation of the

Plate 6.15
The Culbin Sands of north-east
Scotland are coastal sand
dunes which were previously
unstable, as a result of
deforestation, grazing, rabbits
and severe storms. They have
now been stabilized by the
planting of pine forests

Culbin Sands in north-east Scotland with conifer plantations
(Edlin, 1976) (plate 6.15).

Human-induced dune instability is not, however, a problem that
is restricted to mid-latitude coasts. In inland areas of Europe,
clearing, fire and grazing have affected some of the late Pleistocene
dune fields that were created on the arid steppe margins of the great
ice sheets, and in eastern England the dunes of the Breckland
presented problems on many occasions. There are records of
carriages being halted by sand-blocked roads and of one village,
Downham, being overwhelmed altogether.

However, it is possibly on the margins of the great subtropical
and tropical deserts that some of the strongest fears are being
expressed about sand-dune reactivation. This is one of the facets of
the process of desertification and desertization. The increasing

population levels of both humans and their domestic animals, brought about by improvements in health and by the provision of boreholes, has led to an excessive pressure on the limited vegetation resources. As ground cover has been reduced, so dune instability has increased. The problem is not so much that dunes in the desert cores are relentlessly marching on to moister areas, but that fossil dunes, laid down during the more arid phase peaking around 18 000 years ago, have been reactivated *in situ*.

A wide range of methods is available to attempt to control drifting sand and moving dunes as follows:

1 *Drifting sand*
 Enhancement of deposition of sand by the creation of large ditches, vegetation belts and barriers and fences
 Enhancement of transport of sand by aerodynamic stream-lining of the surface, change of surface materials or panelling to direct flow
 Reduction of sand supply, by surface treatment, improved vegetation cover or erection of fences
 Deflection of moving sand, by fences, barriers, tree belts, etc.
2 *Moving dunes*
 Removal by mechanical excavation
 Destruction by reshaping, trenching through dune axis or surface stabilization of barchan arms
 Immobilization, by trimming, surface treatment and fences.

Yet in practice most solutions to the problem of dune instability and sand blowing have involved the establishment of a vegetation cover. This is not always easy. Species used to control sand dunes must be able to endure undermining of their roots, burning, abrasion and often severe deficiencies of soil moisture. Thus the species selected need to have the ability to recover after partial burying, to have deep and spreading roots, to have rapid height growth in the seedling stages, to promote rapid litter development, and to add nitrogen to the soil through root nodules. During the early stages of growth they may need to be protected by fences, sand traps and surface mulches. Growth can also be stimulated by the addition of synthetic fertilizers.

In the hearts of deserts sand dunes are naturally mobile because of the sparse vegetation cover. Even here, however, humans sometimes attempt to stabilize sand surfaces to protect settlements, pipelines, industrial plant and agricultural land. The use of relatively porous barriers to prevent or divert sand movement has proved comparatively successful, and palm fronds or chicken wire have made adequate stabilizers. Elsewhere surfaces have been strengthened by the application of high-gravity oil or by salt-saturated water (which promotes the development of wind-resistant surface crusts).

In temperate areas coastal dunes have been effectively stabilized by the use of various trees and other plants (Ranwell and Boar, 1986). In Japan *Pinus thumbergii* has been successful, while in the great Culbin Sands plantations of Scotland *P. nigra* and *P. laricio* have been used initially, followed by *P. sylvestris*. Of the smaller shrubs, *Hippophae* has proved highly efficient, sometimes too efficient, at spreading. Its clearance from areas where it is not welcome is difficult precisely because of some of the properties that make it such an efficient sand stabilizer: vigorous suckering growth and the rapid regrowth of cut stems (Boorman, 1977). Different types of grass have also been employed, especially in the early stages of stabilization. These include two grasses which are moderately tolerant of salt: *Agrophyron junceiforme*, now called *Elymus farctus* (sand twitch) and *Lye arenarius*, now called *Leymus arenarius* (lyme grass). Another grass which is much used, not least because of its rapid and favourable response to burial by drifting sand, is *Ammophila arenaria* (marram).

Further stabilization of coastal dunes has been achieved by setting up sand fences. These generally consist of slats about 1.0–1.5 m high, and have a porosity of 25–50 per cent. They have proved to be effective in building incipient dunes in most coastal areas. By installing new fences regularly, large dunes can be created with some rapidity (see figure 6.18). Alternative methods, such as using junk cars on the beaches at Galveston, Texas, have been attempted with little success.

Accelerated coastal erosion

Because of the high concentration of settlements, industries, transport facilities and recreational developments on coastlines, the pressures placed on coastal landforms are often acute, and the

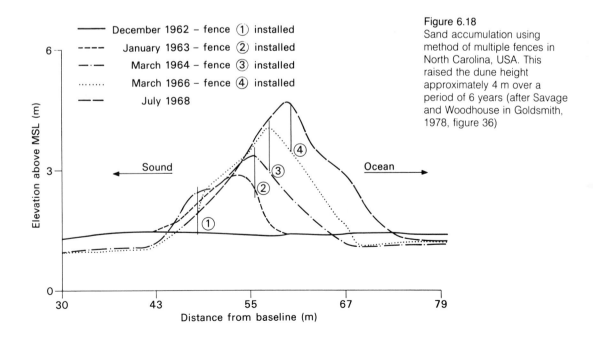

Figure 6.18
Sand accumulation using method of multiple fences in North Carolina, USA. This raised the dune height approximately 4 m over a period of 6 years (after Savage and Woodhouse in Goldsmith, 1978, figure 36)

consequences of excessive erosion serious. While most areas are subject to some degree of natural erosion and accretion, the balance can be upset by human activity in a whole range of different ways (table 6.14). However, humans seldom attempt to accelerate coastal erosion deliberately. More usually, it is an unexpected or unwelcome result of various economic projects. Frequently coast erosion has been accelerated as a result of human efforts to reduce it.

One of the best forms of coastal protection is a good beach. If material is removed from a beach accelerated cliff retreat may take place. Removal of beach materials may be necessary to secure valuable minerals, including heavy minerals, or to provide aggregates for construction. The classic example of the latter was the mining of 660 000 tonnes of shingle from the beach at Hallsands in Devon, England, in 1887 to provide material for the construction of dockyards at Plymouth. The shingle proved to be undergoing little or no natural replenishment and in consequence the shore level was reduced by about 4 m. The loss of the protective shingle soon resulted in cliff erosion to the extent of 6 m between 1907 and 1957. The village of Hallsands was cruelly attacked by waves and is now in ruins.

Table 6.14 Mechanisms of human-induced erosion in coastal zones

Human-induced erosion zones	Effects
1 Beach mining for placer deposits (heavy minerals) such as zircon, rutile, ilmenite and monazite	1 Loss of sand from frontal dunes and beach ridges
2 Construction of groynes, breakwaters, jetties and other structures	2 Downdrift erosion
3 Construction of offshore breakwaters	3 Reduction in littoral drift
4 Construction of retaining walls to maintain river entrances	4 Interruption of littoral drift resulting in downdrift erosion
5 Construction of sea-walls, revetments, etc	5 Wave reflection and accelerated sediment movement
6 Deforestation	6 Removal of sand by wind: sand drift
7 Fires	7 Migrating dunes and sand drift after destruction of vegetation
8 Grazing of sheep and cattle	8 Initiation of blow-outs and transgressive dunes: sand drift
9 Off-road recreational vehicles (dune buggies, trail bikes, etc.)	9 Triggering mechanism for sand drift attendant upon removal of vegetative cover
10 Reclamation schemes	10 Changes in coastal configuration and interruption of natural processes, often causing new patterns in sediment transport
11 Increased recreational needs	11 Accelerated deterioration, and destruction, of vegetation on dunal areas, promoting erosion by wind and wave action

Source: Hails, 1977, table 9.11, p. 348

Another common cause of beach and cliff erosion at one point is coast protection at another. As already stated, a broad beach serves to protect the cliffs behind, and beach formation is often encouraged by the construction of groynes. However, these structures sometimes merely displace the erosion (possibly in an even more marked form) further along the coast. This is illustrated in figure 6.19.

Piers or breakwaters can have similar effects to groynes. This has occurred at various places along the British coast: erosion at Seaford resulted from the Newhaven breakwater, while erosion at Lowestoft resulted from the pier at Gorleston. Figure 6.20 illustrates the changes in the location of beach erosion achieved by the building of jetties or breakwaters at various points. At Madras in south-east India, for example, a 1000 m long breakwater was constructed in 1875 to create a sheltered harbour on a notoriously inhospitable coast, dominated by sand transport from north to south. On the south side of the breakwater over 1 million m² of new land formed by 1912, but erosion occurred for 5 km north of the breakwater. At Ceara in Brazil, also in 1875, a detached breakwater was erected over a length of 430 m more or less parallel to the shore. It was believed that by using a detached structure

Plate 6.16
Near Brighton in southern England attempts have been made to reduce the rate of coast recession because of the threat to cliff-top housing. Defence mechanisms include a solid sea-wall and a large series of groynes. The reduction of erosion at this point may cause beach depletion elsewhere on the coast

Plate 6.17
The village of Hallsands in south Devon, England, displays the effects of sediment removal and dredging. The top photo shows the original state of the fishing village and its beach, while the lower one shows the ravages consequent upon interference

HALLSANDS IN 1894

Figure 6.20 (opposite)
Examples of the effects of shoreline installations on beach and shoreline morphology:
(a) Erosion of Bayocean Spit, Tillamook Bay, Oregon, after construction of a north jetty in 1914–17. The heavy dashed line shows the position of the new south jetty under construction
(b) The deposition–erosion pattern around the Santa Barbara breakwater in California
(c) Sand deposition in the protected lee of Santa Monica breakwater in California
(d) Madras Harbour, India, showing accretion on updrift side of the harbour and erosion on the downdrift side (Komar, *Beach processes and sedimentation,* p. 334, © 1976. Reprinted by permission of Prentice Hall, Inc.)

HALLSANDS IN 1904   After R. Hansford Worth

The two pictures are from the same view-point, and the state of the tide is the same in both. By 1904 the beach had fallen about 10 feet

littoral drift would be able to move along the coast uninterrupted by the presence of a conventional structure built across the surf zone. This, however, proved to be a fallacy (Komar, 1976), since the removal of the wave action which provided the energy for transporting the littoral sands resulted in their deposition within the protected area.

Plate 6.19 shows the evolution of the coast at West Bay in Dorset, southern England, following the construction of a jetty. As time goes on the beach in the foreground builds outwards, while the cliff behind the jetty retreats and needs to be protected with a sea-wall. Likewise, the construction of some sea-walls, erected to

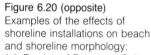

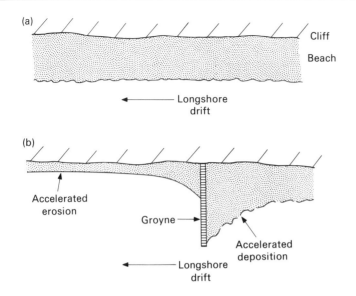

Figure 6.19
Diagrammatic illustration of the
effects of groyne construction
on sedimentation on a beach

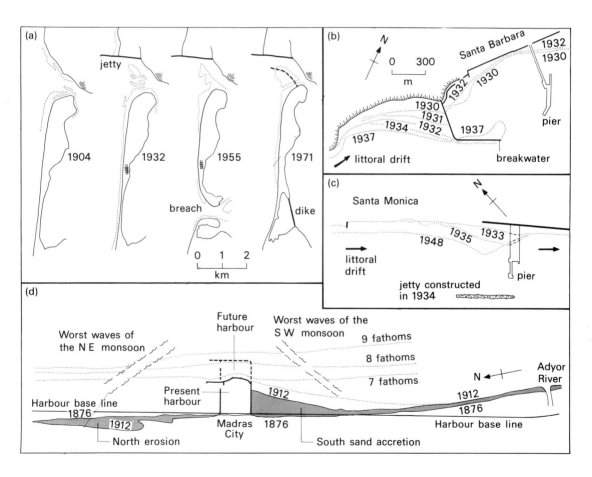

Plate 6.18
Folkestone Warren, Kent, England, has been subjected to a series of major landslides, where the chalk has slipped down over the impermeable Gault clays beneath. One cause of these events has been the construction of the breakwater for Folkestone Harbour. This has starved the protective beach at the seaward margin of the warren of its shingle supply

reduce coastal erosion and flooding, has had the opposite effect to the one intended (see figure 6.21). Given the extent to which artificial structures have spread along the coastlines of the world, this is a serious matter (Walker, 1988).

Problems of this type are exacerbated because there is now abundant evidence to suggest that much of the reservoir of sand and shingle that creates beaches is in some respects a relict feature. Much of it was deposited on the continental shelf during the maximum of the last glaciation (around 18 000 year BP), when sea-level was about 120–40 m below its present level. It was transported shoreward and incorporated in present-day beaches during the phase of rapidly rising post-Glacial sea-levels that characterized the Flandrian transgression until about 6000 years BP. Since that time, with the exception of minor oscillations of the order of a few metres, world sea-levels have been stable and much less material is, as a consequence, being added to beaches and shingle complexes. Therefore, according to Hails (1977: 322): 'in many areas, there is virtually no offshore supply to be moved onshore, except for small quantities resulting from seasonal changes.' It is because of these problems that many erosion prevention schemes now involve beach replenishment (by the artificial addition of appropriate sediments to build up the beach), or employ miscellaneous sand bypassing techniques (including pumping and dredging) whereby sediments are transferred from the accumulation side of an artificial barrier to the erosional side (King, 1974).

In some areas, however, sediment-laden rivers bring material into the coastal zone which becomes incorporated into beaches through the mechanism of longshore drift. Thus any change in the sediment

Plate 6.19
A jetty was built at West Bay to facilitate entry to the harbour.
Top: In 1860 it has had little effect on the coastline.
Centre: by 1900 sediment accumulation had taken place in the foreground but there was less sediment in front of the cliff behind the town.
Bottom: by 1976 the process had gone even further and the cliff had to be protected by a sea-wall. Even this has since been severely damaged by a winter storm

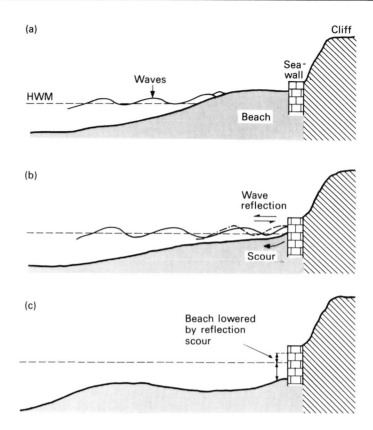

(a)

(b)

(c)

Figure 6.21
Sea-walls and erosion: a
broad, high beach prevents
storm waves breaking against
a sea-wall and will persist, or
erode only slowly; but where
the waves are reflected by the
wall (b) scour is accelerated,
and the beach is quickly
removed and lowered (c)
(modified after Bird, 1979,
figure 6.3)

load of such rivers may result in a change in the sediment budget of
neighbouring beaches. When accelerated soil erosion occurs in a
river basin the increased sediment load may cause coastal accretion
and siltation. But where the sediment load is reduced through
action such as the construction of large reservoirs, behind which
sediments accumulate, coastal erosion may result. This is believed
to be one of the less desirable consequences of the construction of
the Aswan Dam on the Nile: parts of its delta have shown recently
accelerated recession.

The Nile sediments, on reaching the sea, used to move eastward
with the general anticlockwise direction of water movement in that
part of the eastern Mediterranean, generating sand bars and dunes
which contributed to delta accretion. About a century ago an
inverse process was initiated and the delta began to retreat. For
example, the Rosetta mouth of the Nile has lost about 1.6 km of its
length from 1898–1954. The imbalance between sedimentation
and erosion appears to have started with the Delta Barrages (1861)
and then been continued by later works, including the Sennar Dam
(1925), Gebel Aulia Dam (1937), Khasm el Girba Dam (1966),
Roseires Dam (1966) and the High Dam itself. Much of the
Egyptian coast is now 'undernourished' with sediment and, as a
result of this overall erosion of the shoreline, the sand bars

bordering Lake Manzala and Lake Burullus on the seaward side are eroded and likely to collapse. If this were to happen, the lakes would be converted into marine bays, so that saline water would come into direct contact with low-lying cultivated land and fresh-water aquifers.

Likewise in Texas, where over the last century four times as much coastal land has been lost as has been gained, one of the main reasons for this change is believed to be the reduction in the suspended loads of some of the rivers discharging into the Gulf of Mexico (table 6.15). The four rivers listed carried, in 1961–70, on average only about one-fifth of what they carried in 1931–40. Comparably marked falls in sediment loadings occurred elsewhere in the eastern USA (figure 6.22). Likewise, in France the once mighty Rhône only carries about 5 per cent of the load it did in the nineteenth century; and in Asia, the Indus discharges less than 20 per cent of the load it did before construction of large barrages over the last half century (Milliman, 1990).

Table 6.15 Suspended loads of Texas rivers discharging into the Gulf of Mexico

| River | Suspended load (million tonnes) | | %* |
	1931–40	1961–70	
Brazos	350	120	30
San Bernard	1	1	100
Colorado	100	11	10
Rio Grande	180	6	3
TOTAL	631	138	20

*1961–70 loads as % 1931–40 loads
Source: modified from Hails, 1977, table 9.1 (after data from Stout et al. and Curtis et al.)

Construction of great levées on the Mississippi River since 1717 has also affected the Gulf of Mexico coast. The channelization of the river has increased its velocity, reduced overbank deposition of silt onto swamps, marshes and estuaries, and changed the salinity conditions of marshland plants (Cronin, 1967). As a result, the coastal marshes and islands have suffered from increased erosion or a reduced rate of development. This has been vividly described by Biglane and Lafleur (1967: 691):

Like a bullet through a rifle barrel, waters of the mighty Mississippi are thrust toward the Gulf between the confines of the flood control levées. Before the day of these man-made structures, these waters poured out over tremendous reaches of the coast . . . Freshwater marshes (salinities averaging 4–6%) were formed by deposited silts and vegetative covers of wire grass . . . As man erected his flood protection devices, these marshes ceased to form as extensively as before.

The changes between 1956 and 1978 are shown in figure 6.23. However, as with so many examples of environmental change, it is unlikely that just one factor, in this case channelization, is the sole

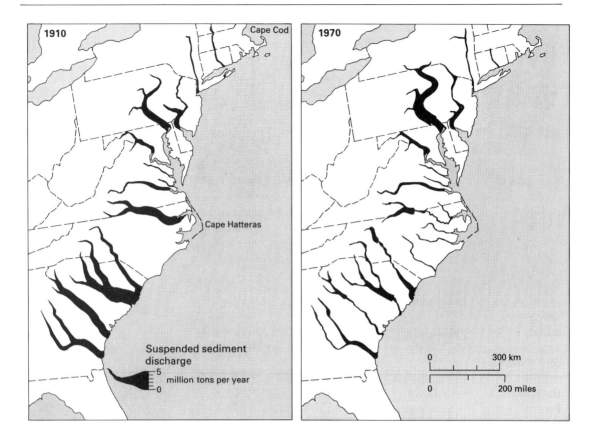

Figure 6.22
The decline in suspended
sediment discharge to the
eastern seaboard of the United
States between 1910 and 1970
as a result of soil conservation
measures, dam construction
and land-use changes (after
Meade and Trimble, 1974)

cause of the observed trend. In their study of erosion loss in the
Mississippi Delta and neighbouring parts of the Louisiana coast,
Walker et al. (1987) suggest that this loss is the result of a variety of
complex interactions among a number of physical, chemical,
biological and cultural processes. These processes include, in
addition to channelization, worldwide sea-level changes, subsidence resulting from sediment loading by the delta of the underlying crust, changes in the sites of deltaic sedimentation as the delta
evolves, catastrophic storm surges and subsidence resulting from
subsurface fluid withdrawal.

In some areas anthropogenic vegetation modification creates
increased erosion potential. This has been illustrated for the
hurricane-afflicted coast of Belize, Central America (Stoddart,
1971). He showed that natural, dense vegetation thickets on low,
sandy islands (cays) acted as a baffle against waves and served as a
massive sediment trap for coral blocks, shingle and sand transported during extreme storms. However, on many islands the
natural vegetation had been replaced by coconut plantations. These
had an open structure easily penetrated by sea water, they tended to
have little or no ground vegetation (thus exposing the cay surface to
stripping and channelling), and they had a dense but shallow root
net easily undermined by marginal sapping. Thus Stoddart found

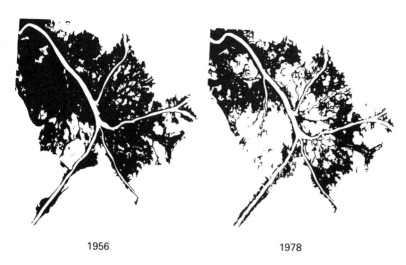

1956   1978

Figure 6.23
Comparison of the outlines of
the Mississippi Delta in 1956
and 1978 gives a clear
indication of the transformation
from marsh to open water.
Artificial controls up river have
decreased the amount of
sediment carried by the river;
artificial levées along much of
the lower course have kept
flood-borne sediment from
replenishing the wetlands; and
in the active delta itself rock
barriers installed across breaks
similarly confine the river. The
Gulf of Mexico is intruding as
the marshland sinks or is
washed away (*National
Geographic Magazine*, August,
1983)

(p. 191) that 'where the natural vegetation had been replaced by coconuts before the storm (Hurricane Hattie), erosion and beach retreat led to net vertical decreases in height of 3–7 ft; whereas where natural vegetation remained, banking of storm sediments against the vegetation hedge led to a net vertical increase in height of 1–5 ft.'

Other examples of markedly accelerated coastal erosion and flooding result from anthropogenic degradation of dune ridges. Frontal dunes are a natural defence against erosion, and coastal changes may be long-lasting once they are breached. Many of those areas in eastern England which most effectively resisted the great storm and surge of 1952 were those where humans had not intervened to weaken the costal dune belt.

Not all dune stabilization and creation schemes have proved desirable (Dolan et al., 1973). In North Carolina (see figure 6.24) the natural barrier–island system along the coastline met the challenge of periodic extreme storms, such as hurricanes, by placing no permanent obstruction in the path of the powerful waves. Under these natural conditions, most of the initial stress of such storms is sustained by relatively broad beaches (figures 6.24a). Since there is no resistance from impenetrable landforms, water can flow between the dunes (which do not form a continuous line) and across the islands, with the result that wave energy is rapidly exhausted. However, between 1936 and 1940, 1000 km of sand fencing was erected to create an artificial barrier dune along part of the Outer Banks, and 2.5 million trees and various grasses (especially *Ammophila breviligulata*) were planted to create large artificial dunes. The altered barrier islands (see figure 6.24b) not only have the artificial barrier–dune system; they also have beaches that are often only 30 m wide, compared with 140 m for the unaltered islands. This beach-narrowing process, combined with the presence of a permanent dune structure, has created a situation in which high wave energy is concentrated in an increasingly restricted run up area,

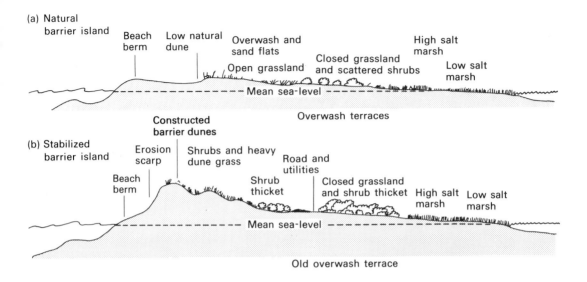

(a) Natural barrier island

Beach berm | Low natural dune | Overwash and sand flats | Closed grassland and scattered shrubs | High salt marsh | Low salt marsh

Open grassland

Mean sea-level

Overwash terraces

Constructed barrier dunes

(b) Stabilized barrier island

Erosion scarp | Shrubs and heavy dune grass | Road and utilities

Beach berm | Shrub thicket | Closed grassland and shrub thicket | High salt marsh | Low salt marsh

Mean sea-level

Old overwash terrace

Figure 6.24
Cross-sections of two barrier islands in North Carolina, USA. The upper diagram (a) is typical of the natural systems, and the lower (b) illustrates the stabilized systems (after Dolan et al., 1973, figure 4)

Plate 6.20 (opposite)
Low-lying coastal barrier systems, such as those that line much of the eastern seaboard of the USA, are especially prone to anthropogenic disturbance. Some 'beach stabilization' schemes have made them less resilient to hurricane attack

resulting in a steeper beach profile, increased turbulence and greater erosion. Another problem associated with artificial dune stabilization is the flooding that occurs when north-east storms pile the water of the lagoon, Pamlico Sound, up against the barrier island. In the past, these surge waters simply flowed out between the low, discontinuous dunes and over the beach to the sea, but with the altered dune chain the water cannot drain off readily and vast areas of land are at times submerged.

The human impact on seismicity and volcanoes

The seismic and tectonic forces which mould the relief of the earth and cause such hazards to human civilization are one of the fields in which efforts to control natural events have had least success and where least has been attempted. None the less, the fact that humans have been able, inadvertently, to trigger off small earthquakes by nuclear blasts, as in Nevada (Pakiser et al., 1969), by injecting water into deep wells, as in Colorado, by mining, by building reservoirs, and by fluid extraction, suggests that in due course it may be possible to 'defuse' earthquakes by relieving tectonic strains gradually in a series of non-destructive, low-intensity earthquakes. One problem, however, is that there is no assurance that an earthquake purposefully triggered by human action will be a small one, or that it will be restricted to a small area. The legal implications are immense.

The demonstration that increasing water pressures could initiate small-scale faulting and seismic activity was unintentionally demonstrated near Denver (Evans, 1966), where nerve-gas waste was being disposed of at great depth in a well in the hope of avoiding contamination of useful ground-water supplies. The waste

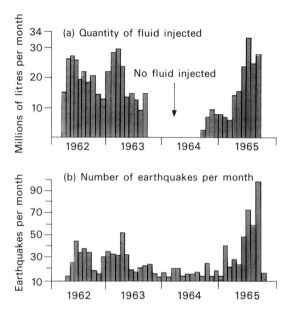

Figure 6.25
Correlation between quantity of
waste water pumped into a
deep well and the number of
earthquakes near Denver,
Colorado (after Birkeland and
Larson, 1978, p. 573)

was pumped in at high pressures and triggered off a series of
earthquakes (see figure 6.25), the timing of which corresponded
very closely to the timing of waste disposal in the well. It is also
now thought that the pumping of fluids into the Inglewood Oil
Field, Los Angeles, to raise the hydrostatic pressure and increase oil
recovery, may have been responsible for triggering the 1963
earthquake which fractured a wall of the Baldwin Hills Reservoir.
It appears that increased fluid pressure reduces the frictional force
across the contact surface of a fault, allows slippage to occur,
thereby causing an earthquake.

In general earthquake triggering has been related to fluid injection, but for reasons that are still obscure there may be some cases
where increased seismicity has resulted from fluid abstraction
(Segall, 1989).

The significance of these 'accidents' has now been verified
experimentally at an oilfield at Rangley in Colorado, USA, where
variations in seismicity have been produced by deliberately controlled variations in the fluid pressure in a zone that is seismically
active (Raleigh et al., 1976).

Perhaps the most important anthropogenically induced seismicity results from the creation of large reservoirs. Reservoirs impose
stresses of significant magnitude on crustal rocks at depths rarely
equalled by any other human construction. With the ever-
increasing number and size of reservoirs the threat rises. There are
at least six cases (Koyna, Kremasta, Hsinfengkiang, Kariba,
Hoover and Marathon) where earthquakes of a magnitude greater
than 5, accompanied by a long series of foreshocks and aftershocks,
have been related to reservoir impounding. However, as table 6.16
shows, there are many more locations where the filling of reservoirs

Table 6.16 List of dams for which associated seismic phenomena are reported

Dam	Year of completion	Year of largest earthquake since dam completion
Marathon, Greece	1929/30	1938
Oued Fodda, Algeria	1932	–
Boulder, USA	1936	1939
Pieve di Cadore, Italy	1949	–
Clark Hill, USA	1952	1974
Kariba, Zimbabwe/Zambia	1959	1963
Grandval, France	1959	1963
Hsingfengkiang, China	1959	1962
Kurobe, Japan	1960	1961
Camarillas, Spain	1960	1961
Cannelles, Spain	1960	1962
Vaiont, Italy	1961	1963
Monteynard, France	1962	1963
Koyna, India	1962	1967
Benmore, New Zealand	1965	1966
Kremasta, Greece	1965	1966
Piastra, Italy	1965	1966
Contra, Switzerland	1965	1965
Najina Bašta, Yugoslavia	1966	1967
Grančarevo, Yugoslavia	1967	–
Oroville, USA	1968	1975
Kastraki, Greece	1968	–
Nurek, Tajik Republic	1969	1972
Vouglans, France	1970	1971
Karnafusa, Japan	1970	–
Talbingo, Australia	1971	1972
Schlegeis, Austria	1971	–
H. Verwoerd, South Africa	1972	–
Jocasses, USA	1972	1975
Keban, Turkey	1973	1974
Manic 3, Canada	1975	1975

Source: from data in Judd, 1974 and Milne, 1976, processed by author

behind dams has led to appreciable, though less dramatic, levels of seismic activity. Detailed monitoring has shown that earthquake clusters occur in the vicinity of some dams after their reservoirs have been filled, whereas before construction activity was less clustered and less frequent. Similarly, there is evidence from Vaiont (Italy), Lake Mead (USA), Kariba (Central Africa), Koyna (India) and Kremasta (Greece) that there is a linear correlation between the storage level in the reservoir and the logarithm of the frequency of shocks. This is illustrated for Vaiont (figure 6.26a), Koyna (figure 6.26b) and Nurek (figure 6.26c). It is also apparent from Nurek that as the great reservoir has filled, so the depth of the more shallow-seated earthquakes appears to have increased (figure 6.27).

One reason why dams induce earthquakes involves the hydro-isostatic pressure exerted by the mass of the water impounded in the reservoir, together with changing water pressures across the contact surfaces of faults. Given that the deepest reservoirs provide

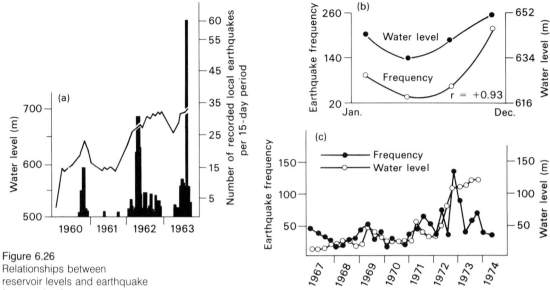

Figure 6.26
Relationships between
reservoir levels and earthquake
frequencies for:
(a) Vaiont Dam, Italy
(b) Koyna, India (these curves
    show the 3-monthly
    average of water level and
    the total numbers of
    earthquakes for the same
    months from 1964 to 1968)
(c) The Nurek Dam, Tajikistan
    (after Judd, 1974 and
    Tajikistan Academy of
    Sciences, 1975)

surface loads of only 20 bars or so, direct activation by the mass of
the impounded water seems an unlikely cause (Bell and Nur, 1978)
and the role of changing pore pressure assumes greater importance.
Paradoxically, there are some possible examples of reduced seismic
activity induced by reservoirs. (Milne, 1976). One possible explanation of this is the increased incidence of stable sliding (fault
creep) brought about by higher pore-water pressure in the vicinity
of the reservoir.

However, the ability to prove an absolutely concrete cause-and-
effect relationship between reservoir activity and earthquakes is
severely limited by our inability to measure stress below depths of
several kilometres, and some examples of induced seismicity may
have been built on the false assumption that because an earthquake
occurs in proximity to a reservoir it has to be induced by that
reservoir (Meade, 1991).

Miscellaneous other human activities appear to affect seismic
levels. In Johannesburg, South Africa, for example, gold-mining
and associated blasting activity have produced tens of thousands of
small tremors, and there is a notable reduction in the number that
occur on Sundays, a day of rest. In Staffordshire, England, coal-
mining has caused increased seismic activity. There are also cases
where seismicity and faulting can be attributed to fluid extraction,
for example, in the oilfields of Texas and California and the
gasfields of the Po Valley in Italy.

When looking at the human impact on volcanic activity human
impotence becomes apparent, though some success has been
achieved in the control of lava flows. Thus in 1937 and 1947 the US
Army attempted to divert lava from the city of Hilo, Hawaii, by

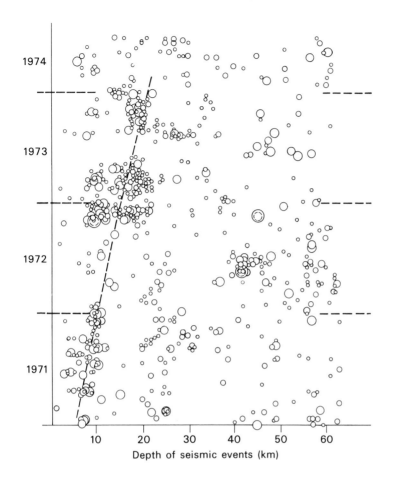

Depth of seismic events (km)

Figure 6.27
As the dashed line indicates there has been a progressive increase in the depth of the shallower seismic events in the vicinity of the Nurek Reservoir, Tajikistan since the dam was first used to pond up the lake that now lies behind it (after Tajikistan Academy of Sciences, 1975)

bombing threatening flows, while elsewhere, where lava rises in a crater, breaching of the crater wall to direct lava towards unin- habited ground may be possible. In 1973 an attempt was made to halt advance of lava with cold water during the Icelandic eruption of Kirkjufell. Using up to 4 million litres of pumped water per hour, the lava was cooled sufficiently to decrease its velocity at the flow front so that the chilled front acted as a dam to divert the still fluid lava behind (Williams and Moore, 1973).

SEVEN

The Human Impact on Climate and the Atmosphere

World climates

The climate of the world is now known to have fluctuated frequently and extensively in the three or so million years during which humans have inhabited the earth (Goudie, 1992). The bulk of these changes have nothing to do with human intervention. Climate has changed, and is currently changing, because of a wide range of different natural factors which operate over a variety of time scales (figure 7.1).

None the less, with the increasing human population and the rising level of technology, it is now apparent that over the last century human agency has probably become a significant factor in the variations in world climate which are taking place, but it remains extremely difficult to ascertain whether human influence or natural forces are responsible for observed trends. If human activity can be considered a factor then this is particularly because of inadvertent effects on atmospheric quality and on the albedo of land masses. Human influence on global climate can be summarized as follows:

> *Possible mechanisms*
> Gas emissions
> CO_2 – industrial and agricultural
>   methane
>   chlorofluorocarbons (CFCs)
>   nitrous oxide
>   krypton 85
>   water vapour
>   miscellaneous trace gases
> Aerosol generation
> Thermal pollution
> Albedo change
>   dust addition to ice caps
>   deforestation
>   over-grazing

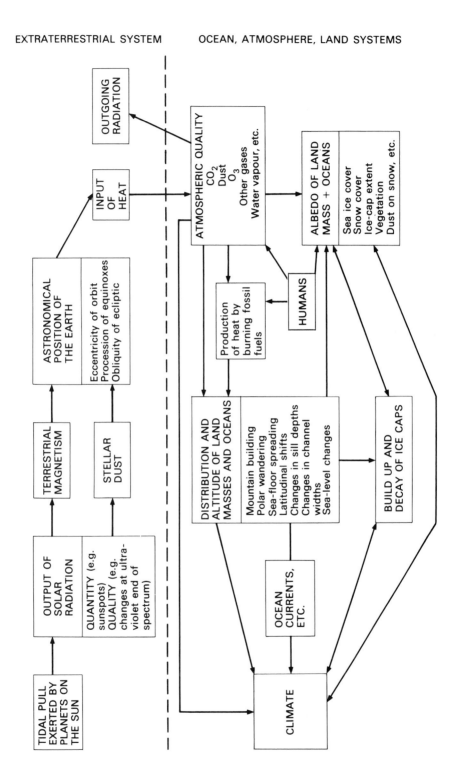

Figure 7.1
A schematic representation of some of the possible influences causing climatic change (after Goudie, 1992, figure 1)

Extension of irrigation
Alteration of ocean currents by constricting straits
Diversion of fresh waters into oceans

The CO_2 problem

Since the beginning of the Industrial Revolution humans have been taking stored carbon out of the earth in the form of coal, petroleum and natural gas, and burning it to make carbon dioxide (CO_2), heat, water vapour and smaller amounts of sulphur dioxide (SO_2) and other gases.

The pre-industrial level of carbon dioxide is a matter of some debate, but may have been as low as 260–70 ppm by volume (Wigley, 1983). The present level is over 350 ppmv, and the upward trend is evident in records from various parts of the world (figure 7.2). Forecasts suggest that the level by 2065 will be about 600 ppm (Carbon Dioxide Assessment Committee, 1983). If fossil-fuel consumption continues at present rates, a concentration of up to 2000 ppm of CO_2 might be reached in two or three centuries from now. However, predictions are not without their problems, depending as they do on two major imponderables: the future rate of fossil-fuel consumption, and the way in which CO_2 may be absorbed by the oceans.

Figure 7.2
Mean monthly concentrations of atmospheric CO_2 at Mauna Loa (an intermittently active volcano on a Hawaian Island). The yearly oscillation is explained mainly by the annual cycle of photosynthesis and respiration of plants in the northern hemisphere (based on data provided by the Scripps Institute of Oceanography)

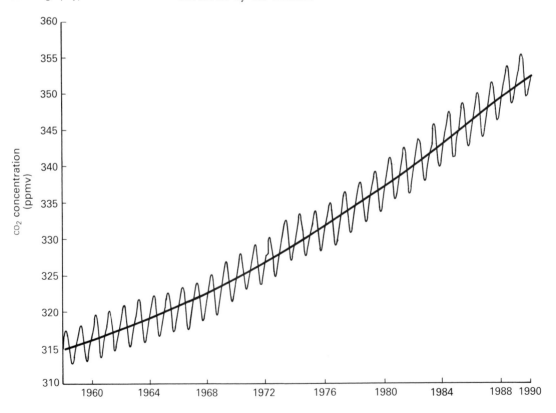

The prime cause of the release of CO_2 into the atmosphere is probably the burning of fossil fuels. Fossil-fuel burning throughout the world in 1981 released about 5.3 gigatons of carbon to the atmosphere as CO_2. The cumulative emissions from the mid-nineteenth century to 1981 totalled about 160 gigatons of carbon as CO_2. Fossil-fuel use for much of the post-Second World War period averaged an increase of around 4.5 per cent per year, but because of the changing nature of energy provision resulting from the massive increases in the cost of oil, and other factors such as the adoption of nuclear power, it is now widely believed that energy consumption using fossil fuels will rise relatively more slowly in the near future, probably averaging about 2.5 per cent per year.

Another factor that may contribute substantially to CO_2 levels in the atmosphere is the burning of forests and changes in the organic levels in soils that are subjected to deforestation and cultivation (Wilson, 1978; Wong, 1978). Woodwell (1978) estimated that this source may be currently more important than the burning of fossil

Plate 7.1
The combustion of fossil fuels, including coal, to generate electricity is a major source of carbon dioxide in the atmosphere

fuels, with deforestation releasing about 6 gigatons of carbon as CO_2 per year, and the decay of soil humus about 2 gigatons of carbon as carbon dioxide per year. However, there is considerable uncertainty about existing rates of deforestation, the proportion of deforestation that is caused by burning, the amount of carbon that may be kept stored as charcoal, and the rate at which regrowth of vegetation consumes carbon. Consequently, the rates initially proposed by Woodwell have now been reassessed, and a current consensus value for the amount of carbon dioxide released by land-use changes is of the order of 2 gigatons per annum (Clark, 1982). Attempts have also been made to assess longer-term release of carbon dioxide into the atmosphere by undertaking studies either of historical changes in land use, or by analysing $^{13}C/^{12}C$ ratios in tree rings. Both methods have given broadly similar results and indicate that since the middle of the last century human activities have released on average between 0.7 and 1.7 gigatons per year. From 1860 to the present, cumulative releases from changes in land use have been approximately equal to cumulative releases from the burning of fossil fuels.

The CO_2 levels in the atmosphere have an effect on the global heat balance, since CO_2 is virtually transparent to incoming solar radiation but absorbs outgoing terrestrial infra-red radiation; radiation that would otherwise escape to space and result in loss of heat from the lower atmosphere (figure 7.3). This is called 'the greenhouse effect'. Consequently one would expect increased CO_2 levels to lead to an increase in surface temperatures (Hansen et al., 1981). Indeed, there is some evidence of an overall global warming

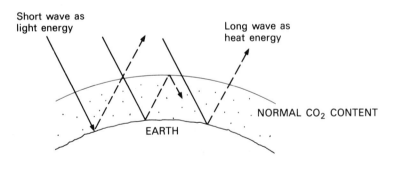

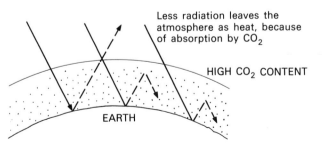

Figure 7.3
The greenhouse effect: short-wave radiation from the sun is absorbed by the earth's surface which, in turn, radiates heat at far longer wavelengths because of its temperature of around 280°K, compared with around 6000°K for the sun

trend of about 0.5°C in this century (see, for example, Jones et al., 1986) and the 1980s saw an exceptionally warm decade.

In reality the term 'greenhouse effect' is something of a misnomer. As Henderson-Sellers and Robinson (1986: 60) explain:

We now know that a greenhouse maintains its higher internal temperature largely because the shelter it offers reduces the turbulent transfers of energy away from the surface rather than because of any radiative considerations. Thus while the greenhouse effect remains valid, and vital, for the atmosphere it might be better to think of the physical processes in terms of the 'leaky bucket' analogy ... Here an increase in the amount of gas with absorption bands in the infrared part of the spectrum is represented by a decrease in the size of the hole in the bucket. The surface temperature, represented by the depth of the water in the bucket, rises as more absorbing gases enter the atmosphere.

Other gases

In addition to carbon dioxide, it is probable that other gases will contribute to the greenhouse effect. Individually their effects may be minor, but as a group they may be major (Ramanathan, 1988). Indeed, molecule for molecule some of them may be much more effective as greenhouse gases than CO_2, as the data in table 7.1 show.

Table 7.1 Radiative forcing relative to CO_2 per unit molecule change in the atmosphere

GAS	Relative radiative forcing
CO_2	1
CH_4 (methane)	21
N_2O (nitrous oxide)	206
CFC-11	12 400
CFC-12	15 800

Source: extracted from Houghton et al., 1990, table 2.3, p. 53

One of the more important of the trace gases is methane, which has a strong infrared absorption band at 7.66 μm. Ice core studies and recent direct observations (figure 7.4b) suggest that until the beginning of the eighteenth century background levels were stable at around 600 parts per billion by volume (ppbv). They rose steadily between AD 1700 and 1900, and then increased still more rapidly, attaining levels that averaged 1300 ppbv in the early 1950s and 1600 ppbv by the mid-1980s (Khalil and Rasmussen, 1987). This increase of 2.5 times over background levels results primarily from increased rice cultivation in waterlogged paddy fields, the enteric fermentation produced in the growing numbers of flatulent domestic cattle, and the burning of oil and natural gas (Crutzen et al., 1986) (table 7.2).

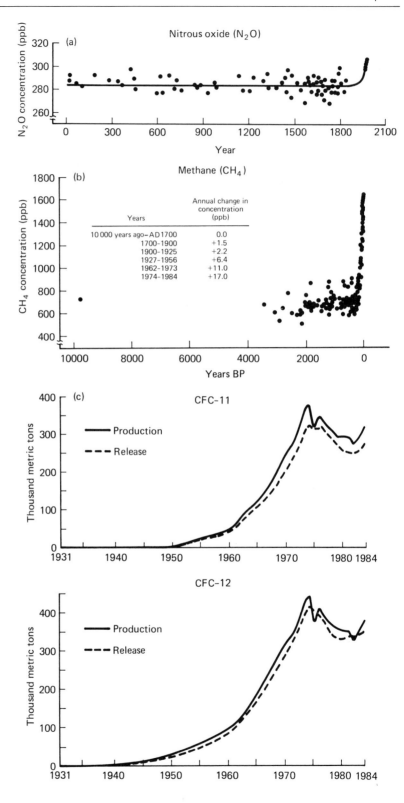

Figure 7.4
The changing concentrations
of accessory greenhouse
gases in the atmosphere:
(a) nitrous oxide (N₂O). Note
    that N₂O concentrations
    remained fairly constant
    between 3000 years ago
    and AD 1850 at
    approximately 285 parts
    per billion (after Khalil and
    Rasmussen, 1987)
(b) methane (Ch₄) (after Khalil
    and Rasmussen, 1987)
(c) the changing production
    and release of two CFC
    gases (CFC-11 and
    CFC-12) between 1931
    and 1984

Chlorofluorocarbons (CFCs), despite their relatively trace amounts in the atmosphere, have increased very markedly in terms of their emissions (figure 7.4c) and their concentrations in recent decades, and have a very strong greenhouse effect even in relatively small amounts. On the other hand, the ozone depletion they have caused in the stratosphere may counteract this effect, for stratospheric ozone depletion results in a decrease in radiative forcing (Houghton et al., 1992).

Nitrous oxide (N_2O) is also no laughing matter, for it can contribute to the greenhouse effect, primarily by absorption of infrared at the 7.8 and 17 μm bands. Combustion of hydrocarbon fuels, the use of ammonia-based fertilizers, deforestation and

Table 7.2 (a) Estimates of methane (CH_4) source strengths and sinks

Sources/sinks	Best estimate ($10^6 t\, a^{-1}$)	Range ($10^6 t\, a^{-1}$)
Sources		
Natural wetlands	115	100–200
Rice paddies	110	25–170
Enteric fermentation (animals)	80	65–100
Gas drilling, venting, transmission	45	25–50
Biomass burning	40	20–80
Termites	40	10–100
Landfills	40	20–70
Coal mining	35	19–50
Oceans	10	5–20
Fresh waters	5	1–25
CH_4 hydrate destabilization	5	0–100
Sinks		
Removal by soils	30	15–45
Reaction with OH	500	400–600
Atmospheric increase	44	40–48

(b) Estimates of nitrous oxide (N_2O) source strengths and sinks

Sources/sinks	Range ($10^6 t\, a^{-1}$)
Sources	
Oceans	1.4–2.6
Soils (tropical forests)	2.2–3.7
Soils (temperate forests)	0.7–1.5
Fossil fuel combustion	0.1–0.3
Biomass burning	0.02–0.2
Fertilizer (including ground water)	0.01–2.2
Sinks	
Removal by soils	Unknown
Photolysis in the stratosphere	7–13
Atmospheric increase	3–4.5

Source: data in UNEP, 1991

biomass burning are among the processes that could lead to increase in atmospheric N_2O levels (figure 7.4a and table 7.2b).

Warnings have been circulated (Boeck et al., 1975) about the possible role of a substance emitted from nuclear reactors, called Krypton 85. It is believed that its increasing presence will reduce the electrical resistance of the atmosphere between the oceans and the ionosphere. This in turn would affect the electrification of thunder-clouds and, through that, precipitation levels.

Supersonic aircraft, both civil and military, discharge some water vapour into the stratosphere as contrails. At present, the water content of the stratosphere is low, as is the exchange of air between the lower stratosphere and other regions. Consequently, comparat-ively modest amounts of water vapour discharged by aircraft could have a significant effect on the natural balance. It has been calculated that 400 supersonic aircraft, making four flights per day, would place 150 million kg of water into the lower stratosphere. This could, over a period of years, double the existing water content of the atmosphere, thereby leading to a small rise in temperature of perhaps 0.6°C (Sawyer, 1971).

However, statistics on the increase of cirrus levels as a result of stratospheric aircraft movements are still meagre (SCEP Report, in Matthews et al., 1971: 39) and their statistical significance is not always proven. Yet studies at Salt Lake City and Denver between 1949 and 1969 do show an upward trend in cirrus cloudiness which paralleled the increase in jet aircraft activity (figure 7.5). The expected rapid development of civil aviation in future decades may make this an important element of anthropogenic atmospheric modification.

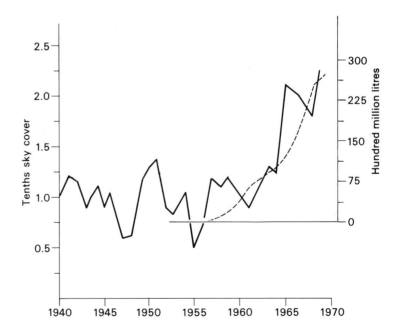

Figure 7.5
The history of the annual high cloudiness at Denver, Colorado. The dashed line shows the growth in jet-fuel consumption by domestic commercial jet aircraft. (After Matthews et al., 1971, figure 33.1, reprinted from *Man's impact on the climate*, ed. Matthews et al., by permission of The MIT Press, Cambridge, Massachusetts, © 1971 by The Massachusetts Institute of Technology)

Other trace gases that could play a greenhouse role include bromide compounds, carbon tetrafluoride, carbon tetrachloride and methyl chloride.

It is possible that the increase in global surface temperatures to be expected by 2050 as a result of these trace gases is 1.4–2.2°C, an increase that may be comparable to that anticipated from CO_2 itself (Macdonald, 1982). Indeed, Bolin et al. (1986: xxi) suggest:

The role of greenhouse gases other than CO_2 in changing the climate is already about as important as that of CO_2. If present trends continue, the combined concentrations of atmospheric CO_2 and other greenhouse gases would be radiatively equivalent to a doubling of CO_2 from pre-industrial levels possibly as early as the 2030s.

Aerosols

One consequence of the Industrial Revolution, which may be of equal significance to CO_2 emission, is the increased quantity of dust or smoke particles that are emitted to the lower atmosphere. A striking illustration of this is provided by the analysis of dust levels in glacier ice of known age in the southern CIS. Layers of ice dated AD 1800–1920 show a dust content of 10 mg/l, whereas by the 1950s, the figure had increased twenty times to 200 mg/l (Davitaya, 1969).

Figure 7.6 shows that the average turbidity factor for the atmosphere (*Linke turbidity*) has increased by 30 per cent in a

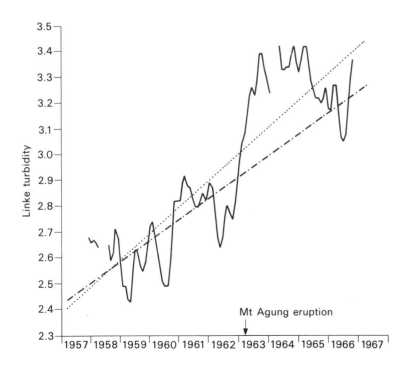

Figure 7.6
Increasing turbidity at Mauna Loa Observatory (after Peterson in Bryson and Kutzbach, 1968, figure 1); symbols are explained in the text

decade (the dot-and-dash line). It also shows the effect of a natural source of turbidity, the Mount Agung (Bali) eruption of 1963 (the single, continuous line). In the figure the dotted line represents the linear trend for the same period if the effects of the eruption are excluded from the computations.

However, not all investigations have demonstrated increases in atmospheric aerosol loadings on the regional or global scale. For example, studies by Hoyt and Fröhlich (1983) at Davos in Switzerland have provided no evidence for long-term changes in atmospheric transmission caused by this mechanism. They believe that trends in aerosol loadings from anthropogenic sources are local rather than regional. They suggest that changes in atmospheric transmission occur in urban and industrial areas, but believe that such changes are too localized to be the cause of hemispheric and global trends in temperature.

If an increase in atmospheric aerosols on the macro scale were to be proven, then they could be said to influence temperatures through their impact on the scattering and absorption of solar radiation. The exact effects, however, are still not clear, for whether added aerosols cause heating or cooling of the earth and atmosphere systems is a function not only of their intrinsic absorption and backscatter characteristics, but also of their location in the atmosphere with respect to such variables as cloud-cover, cloud reflectivity and underlying surface reflectivity (Weare et al., 1974). So, for example, over ice caps 'grey' aerosol particles would warm the atmosphere because they would be less reflective than the white snow surfaces beneath, while over a dark reflecting surface they would reflect a greater amount of radiation, leading to cooling (Peck, 1975). Thus precise quantitative assessment of the effects of increased aerosol content in the atmosphere is hazardous, but Rasool and Schneider (1971) have suggested that a rise by a factor of 4 or 5 in global aerosol concentrations could be sufficient to lower temperatures by as much as 4.5°C.

Similarly, Idso and Brazel (1978) and Brazel and Idso (1979) point to the two contrasting tendencies of dust: the backscattering effect producing cooling, and the thermal-blanketing effect causing warming. The second of these absorbs some of the earth's thermal radiation that would otherwise escape to space, and then re-radiates a portion of this back to the land surface, raising surface temperatures. They believe that natural dust from volcanic emissions tends to enter the stratosphere (where backscattering and cooling are the prime consequences), while anthropogenic dust more frequently occurs in the lower levels of the atmosphere causing thermal blanketing and warming.

Industrialization is not, however, the sole source of particles in the atmosphere, nor is a change in temperature the only possible consequence. Bryson and Barreis (1967), for example, argue that intensive agricultural exploitation of desert margins, such as in Rajasthan, India, would create a dust pall in the atmosphere by exposing larger areas of surface materials to deflation in dust

Plate 7.2
The removal of vegetation from desert surfaces as a result of over-cultivation and over-grazing may contribute to climatic change in a variety of ways: the changing of surface albedo and the liberation of dust particles into the atmosphere are two obvious mechanisms that deserve further investigation. Particularly important may be their role in accentuating drought, as in Ethiopia

storms. This dust pall, they believe, would so change atmospheric temperature that convection, and thus rainfall, would be reduced. Observations on dust levels over the Atlantic during the drought years of the late 1960s and early 1970s in the Sahel suggest that the degraded surfaces of that time led to a great (threefold) increase in atmospheric dust (Prospero and Nees, 1977). There is thus the possibility that human-induced desertization generates dust which could in turn increase the degree of desertization by its effect on rainfall levels.

Atmospheric aerosols can be an important source of cloud-condensation nuclei. Over the world's oceans a major source of such aerosols is dimethylsulphide (DMS). They are produced by planktonic algae in sea water and then oxidize in the atmosphere to form sulphate aerosols. Because the albedo of clouds (and thus the Earth's radiation budget) is sensitive to cloud-condensation nuclei density, any factor that controls planktonic algae may have an important impact on climate. The production of such plankton could be affected by water pollution in coastal seas or by global warming (Charlson et al., 1987). Charlson et al. (1992) believe that anthropogenically derived sulphate aerosols, through their direct scattering of short-wavelength solar radiation and their modification of the shortwave reflective properties of clouds, could significantly increase planetary albedo, thereby exerting a cooling influence on the planet. They maintain that the perturbation in global climate through this mechanism could be comparable in magnitude to current anthropogenic greenhouse gas forcing, but, of course, opposite in sign.

Any assessment of the likely consequences of human production of aerosols must bear in mind that natural production of aerosol materials is of a very much higher order. None the less the data suggest that on a global scale human activity is now producing 280 million tonnes per year of small particulate matter into the atmosphere, compared to a natural rate of 1250 million tonnes per year, or a ratio of about 4.5:1.

The most catastrophic effects of anthropogenic aerosols in the atmosphere could be those resulting from a nuclear exchange between the great powers. Explosion, fire and wind might generate a great pall of smoke and dust in the atmosphere which would make the world dark and cold. It has been estimated that if the exchange reached a level of several thousand megatons, a 'nuclear winter' would occur in which temperatures over much of the world would be as low as $-15°$ to $-25°C$ (Turco et al., 1983), though recent simulations by Schneider and Thompson (1988) suggest that some previous estimates may have been exaggerated. They suggest that in the northern hemisphere maximum average land surface summertime temperature depressions might be of the order of $5–15°C$.

Fears were also expressed that as a result of the severe smoke palls generated by the Gulf War in 1991 there might be severe climatic impacts. The situation is still not clear, but preliminary studies have suggested that because most of the smoke generated by the oil well fires will stay in the lower troposphere and only have a short residence time in the air, the effects will be local (some cooling) rather than global, and that the operation of the monsoon should not be affected to any significant degree (Browning et al., 1991; Bakan et al., 1991). Furthermore, in the event the emissions

Plate 7.3
The Gulf War of 1991 led to the deliberate release and burning of oil in Kuwait. Fears were expressed at the time that smoke palls might have regional and global climate effects. In general, subsequent research has suggested that such fears may have been exaggerated

of smoke particles were less than some forecasters had predicted, and they were also rather less black (Hobbs and Radke, 1992).

Thermal pollution

One consequence of the burning of fossil fuels is the production of heat. This is probably most important on the local scale where it can be identified as the 'urban heat island'. On a broader scale, the amount of energy used by humans has been negligible compared both to the resources of solar energy and to the energy of photosynthesis of plants. In global terms the total amount of heat released by all human activity is roughly 0.01 per cent of the solar energy absorbed at the surface (Kellogg, 1978: 215). Such a small fraction would have a negligible effect on the overall heat balance of the earth.

Vegetation and albedo

Incoming radiation of all wavelengths is partly absorbed and partly reflected. Albedo is the term used to describe the proportion of energy reflected and hence is a measure of the ability of the surface to reflect radiation.

   Land-use changes create differences in albedo which have important effects on the energy balance, and hence on the water balance, of an area. Tall rain forest may have an albedo as low as 9 per cent, while the albedo of a desert may be as high as 37 per cent (table 7.3).

   There has been growing interest recently in the possible consequences of deforestation on climate through the effect of albedo

Table 7.3 Albedo values for different land-use types

Surface type	Location	Albedo (%)
Tall rain forest	Kenya	9
Lake	Israel	11.3
Peat and moss	England and Wales	12
Pine forest	Israel	12.3
Heather morrland	England and Wales	15
Evergreen scrub (maquis)	Israel	15.9
Bamboo forest	Kenya	16
Conifer plantation	England and Wales	16
Citrus orchard	Israel	16.8
Towns	England and Wales	17
Open oak forest	Israel	17.6
Deciduous woodland	England and Wales	18
Tea bushes	Kenya	20
Rough grass hillside	Israel	20.3
Agricultural grassland	England and Wales	24
Desert	Israel	37.3

Source: from miscellaneous data in Pereira, 1973, collated by author

change. Ground deprived of a vegetation cover as a result of
deforestation and over-grazing (as in parts of the Sahel) has a very
much higher albedo than ground covered in plants. This could
affect temperature levels. Satellite imagery of the Sinai–Negev
region of the Middle East shows an enormous difference in image
between the relatively dark Negev and the very bright Sinai–Gaza
Strip area. This line coincides with the 1948–9 armistice line
between Israel and Egypt and results from different land-use and
population pressures. Otterman (1974) has suggested that the
albedo affected by land use has produced temperature changes of
the order of 5°C.

Charney and others (1975) have argued that the increase in
surface albedo, resulting from a decrease in plant cover, would lead
to a decrease in the net incoming radiation, and an increase in the
radiative cooling of the air. Consequently, they argue, the air would
sink to maintain thermal equilibrium by adiabatic compression,
and cumulus convection and its associated rainfall would be
suppressed. A positive feedback mechanism would appear at this
stage, for the lower rainfall would in turn adversely affect plants
and lead to a further decrease in plant cover. However, this view is
disputed by Ripley (1976) who suggests that Charney and his
colleagues, while considering the impact of vegetative changes on
albedo, have completely ignored the effect of vegetation on evapotranspiration. He points out that vegetated surfaces are usually
cooler than bare ground since much of the absorbed solar energy is
used to evaporate water, and conclude from this that protection
from over-grazing and deforestation might, in contrast to Charney's
views, be expected to lower surface temperatures and thereby
reduce, rather than increase, convection and precipitation.

Removal of humid tropical rain forests has also been seen as a
possible mechanism of anthropogenic climatic change through its
effect on albedo. Potter et al. (1975) have proposed the following
model for such change:

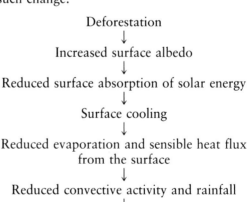

Deforestation
↓
Increased surface albedo
↓
Reduced surface absorption of solar energy
↓
Surface cooling
↓
Reduced evaporation and sensible heat flux
from the surface
↓
Reduced convective activity and rainfall
↓
Reduced release of latent heat, weakened Hadley circulation
and cooling in the mid and upper troposphere
↓

Increased tropical lapse rates
↓

Increased precipitation in the latitude bands
5–25°N and 5–25°S, and a decrease in the
equator–pole temperature gradient
↓

Reduced meridional transport of heat and
moisture out of equatorial regions
↓

Global cooling and a decrease in precipitation
between 45–85°N and 40–60°S

However, more recent studies (Potter et al., 1981) suggest that globally over the past few thousand years the climatic effects of albedo changes wrought by humans have been small and probably undetectable. Similarly, Henderson-Sellers and Gornitz (1984) have sought to model the possible future effects of albedo changes produced by humans and also predict that there will be but little alteration brought about by current levels of tropical deforestation.

On the other hand, Lean and Warrilow (1989) used a general circulation model (GCM) which showed greater changes than previous models and suggested that Amazon basin deforestation would, through the effects of changes in surface roughness and albedo, lead to reductions in both precipitation and evaporation. Likewise a UK Meteorological Office GCM indicates that the deforestation of both Amazonia and Zaire would by changing surface albedo cause a decrease in precipitation levels (Mylne and Rowntree, 1992).

Budyko (1974) believes that the present extension of irrigation to about 0.4 per cent of the earth's surface (1.3 per cent of the land surface) would decrease the albedo, possibly on average by 10 per cent. The corresponding change of the albedo of the entire earth–atmosphere system would amount to about 0.03 per cent; enough, according to Budyko, to maintain the global mean temperature at a level nearly 0.1°C higher than it would otherwise be.

Forests, irrigation and rain

The belief that forests can increase precipitation levels has a long history (Thornthwaite, 1956), and it has been the basis of action programmes in many lands. For example, the American Timber Culture Act of 1873 was passed in the belief that if settlers were induced to plant trees on the Great Plains and prairies, precipitation would be increased sufficiently to eliminate the climatic hazards to agriculture. On the other hand, at much the same time, the view was expressed that 'rain follows the plough'. Aughey, working in Nebraska, for example, believed that, after the soil is 'broken', rain as it falls is absorbed by the soil 'like a huge sponge', and that the soil gives this absorbed moisture slowly back to the atmosphere by evaporation (cited by Thornthwaite, 1956: 569), and so increases the rainfall.

These two early and contradictory views illustrate the confusion that still surrounds this question today. Forests undoubtedly influence rates of evapotranspiration, the flow of streams, the level of ground water and microclimates, but there is little reliable evidence to suggest that regional rainfall is either significantly increased by afforestation or decreased by deforestation. Certainly schemes to augment rainfall levels on desert margins by widespread planting of forest belts are likely to achieve relatively little, for the aridity of deserts and their margins is controlled dominantly by the gross features of the general circulation, especially the subsiding air associated with the big high-pressure cells of the subtropics.

Although forests may not necessarily have a proven effect on regional or continental rainfall levels, they are far more effective than other vegetation types at trapping other kinds of precipitation, especially cloud, fog and mist. Hence deforestation or afforestation can affect water budgets through the degree to which they intercept non-rainfall precipitation. For example, in Hawaii, Norfolk Island Pines (*Araucaria heterophylla*) in an area where the annual rainfall is 2600 mm, condensed an additional 760 mm from heavy cloud. In the San Francisco area, Parsons (1960) recorded about 250 mm of drip from a pine tree on the Berkeley Hills during each of four rainless summers. In Japan this phenomenon has been used to good effect: trees are planted along the coast to intercept the inland drift of sea fogs.

There is one other land-use change that may result in measurable changes in precipitation; namely large-scale crop irrigation in semi-arid regions. The High Plains of the USA are normally covered with sparse grasses and have dry soils throughout the summer; evapotranspiration is then very low. In the last four decades irrigation has been developed throughout large parts of the area, greatly increasing summer evapotranspiration levels. Barnston and Schickedanz (1984) have produced strong statistical evidence of warm-season rainfall enhancement through irrigation in two parts of this area: one extending through Kansas, Nebraska and Colorado, and a second in the Texas Panhandle. The largest absolute increase was in the latter area and, significantly, occurred in June, the wettest of the three heavily irrigated months. The effect appears to be especially important when stationary weather fronts occur, for this is a situation which allows for maximum interaction between the damp irrigated surface and the atmosphere. Hail storms and tornados are also significantly more prevalent than over non-irrigated regions (Nicholson, 1988).

The possible effects of water diversion schemes

The levels of the Aral and Caspian Seas in Central Asia are falling, as are water-tables all over the wide continental region. There are proposals to divert some major rivers to help overcome these problems. However, this raises difficult questions 'because it appears to touch a peculiarly sensitive spot in the existing climatic regime of the northern hemisphere' (Lamb, 1977: 671). The low-salinity water which forms a 100–200 m upper layer to the Arctic Ocean is in part caused by the input of fresh water from the large Russian and Siberian rivers. This low-salinity water is the medium in which the pack ice at present covering the polar ocean is formed. The tapping of any large proportion of this river flow might augment the area of salt water in the Arctic Ocean and thereby reduce the area of pack ice correspondingly. Temperatures over large areas might rise, which in turn might change the position and alignment of the main thermal gradients in the northern hemisphere and, with them, the jet-stream and the development and steering of cyclonic activity. However, assessment of this particular climatic impact is still very largely speculative, and some recent numerical models indicate that the climate of the Arctic will not be drastically affected by river diversions (Semtner, 1984).

Lakes

It has often been implied that the presence of a large body of inland water must modify the climate around its shores, and therefore that artificial lakes have a significant effect on local or regional climates. Climatic changes produced by the construction of a reservoir are the result of a variety of factors (Vendrov, 1965): the creation of a body of water with a large heat capacity that reduces the continentality of the climate; the substitution of a water surface for a land surface and the rise of the ground-water level in the littoral zone supplying moisture to the evaporating surface (leading to a rise in wind velocity above the lake and in the littoral zone). Schemes have been put forward for augmenting desert rainfall by flooding desert basins in the Sahara, Kalahari and Middle East (see, for example, Schwarz, 1923). However, whether evaporation from lake surfaces can raise local precipitation levels is open to question, for precipitation depends more on atmospheric instability than upon the humidity content of the air. Moreover, most lakes are too small to affect the atmosphere materially in depth, so that their influence falls heavily under the sway of the regional circulation. In addition, one needs to remember that some of the world's driest deserts occur along coastlines. Thus a relatively small artificial lake would be even more impotent in creating rainfall (see Crowe, 1971: 443–50).

However, the climatic effect of artificial lakes is evident in other ways, notably in terms of a local reduction in frost hazard. In the case of the Rybinsk reservoir (c.4500 km^2) in the CIS, it has been calculated that the climatic influence extends 10 km from the lake and that the frost-free season has been extended by 5–15 days on average (D'Yakanov and Reteyum, 1965).

Urban climates

It has been said that 'the city is the quintessence of man's capacity to inaugurate and control changes in his habitat' (Detwyler and Marcus, 1972). One way in which such control becomes evident is in a study of urban climates (Landsberg, 1981). Individual urban areas can at times, with respect to their weather, 'have similar impacts as a volcano, a desert, and as an irregular forest' (Changnon, 1973: 146). Some of the changes that can result are listed in table 7.4.

Compared with rural surfaces, city surfaces (table 7.4b) absorb significantly more solar radiation, because a higher proportion of the reflected radiation is retained by the high walls and dark-coloured roofs of the city streets. The concreted city surfaces have both great thermal capacity and conductivity, so that heat is stored during the day and released by night. By contrast the plant cover of the countryside acts like an insulating blanket, so that rural areas tend to experience relatively lower temperatures by day and night, an effect enhanced by the evaporation and transpiration taking place.

Table 7.4 (a) Average changes in climatic elements caused bi cities

Element	Parameter	Urban compared with rural (−, less; +, more)
Radiation	On horizontal surface	−15%
	Ultraviolet	−30% (winter); −5% (summer)
Temperature	Annual mean	+0.7°C
	Winter maximum	+1.5°C
	Length of freeze-free season	+2 to 3 weeks (possible)
Wind speed	Annual mean	−20 to −30%
	Extreme gusts	−10 to −20%
	Frequency of calms	+5 to 20%
Relative	Annual mean	−6%
humidity	Seasonal mean	−2% (winter); −8% (summer)
Cloudiness	Cloud frequency + amount	+5 to 10%
	Fogs	+100% (winter); −30% (summer)
Precipitation	Amounts	+5% to 10%
	Days	+10%
	Snow days	−14%

(b) Effect of city surfaces

Penomenon	Consequence
Heat production (the heat island)	Rainfall + Temperature +
Retention of reflected radiation by high walls and dark-coloured roofs	Temperature +
Surface roughness increase	Wind − Eddying +
Dust increase (the dust dome)	Fog + Rainfall + (?)

Source: H. Landsberg in Griffiths, 1976, p. 108

Another thermal change in cities, contributing to the development of the 'urban heat island', is the large amount of artificial heat produced by industrial, commercial and domestic users.

In general the highest temperature anomalies are associated with the densely built-up area near the city centre, and decrease markedly at the city perimeter. Observations in Hamilton, Ontario, and Montreal, Quebec, suggested temperature changes of 3.8 and 4.0°C respectively per kilometre (Oke, 1978). Temperature differences also tend to be highest during the night. The form of the urban temperature effect has often been likened to an 'island' protruding distinctly out of the cool 'sea' of the surrounding landscape. The rural–urban boundary exhibits a steep temperature gradient or 'cliff' to the urban heat island. Much of the rest of the urban area appears as a 'plateau' of warm air with a steady but weaker horizontal gradient of increasing temperature towards the city centre. The urban core may be a 'peak' where the urban maximum temperature is found. The difference between this value and the background rural temperature defines the *urban heat island intensity* ($\Delta T_{u-r}(max)$) (Oke, 1978: 225).

Table 7.5 lists the average annual urban–rural temperature differences for several large cities. Values range from 0.6 to 1.3°C. The relationship between city size and urban–rural difference, however, is not necessarily linear; sizeable nocturnal temperature contrasts have been measured even in relatively small cities. Factors such as building density are at least as important as city size, and high wind velocities will tend to eliminate the heat island effect.

None the less, Oke (1978: 257) has found that there is some relation between heat-island intensity and city size. Using population as a surrogate of city size $\Delta T_{u-r}(max)$ is found to be proportional to the log of the population. Other interesting results of this study include the tendency for quite small centres to have a heat island, the observation that the maximum thermal modification is about 12°C, and the recognition of a difference in slope between the North American and the European relationships (figure 7.7). The explanation for this last result is not clear, but it may be related to the fact that population is a surrogate index of the central building density.

Table 7.5 Annual mean urban–rural temperature differences of cities

City	Temperature differences (°C)
Chicago, USA	0.6
Washington DC, USA	0.6
Los Angeles, USA	0.7
Paris, France	0.7
Moscow, Russia	0.7
Phildelphia, USA	0.8
Berlin, Germany	1.0
New York, USA	1.1
London, UK	1.3

Source: from data in J. T. Peterson, table 11.2, p. 136 and other sources, in Detwyler, 1971

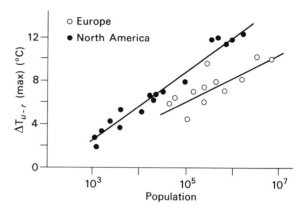

Figure 7.7
Relationship between
maximum observed heat-island
intensity ΔT_{u-r}(max) and
population of North American
and European settlements
(after Oke, 1978, figure 8.16)

In many older towns and cities in Western Europe and North America, the process of 'counter-urbanization' has in recent years led to a decline in population, and it is worth considering whether this is reflected in a decline in the intensity of urban heat islands. One attempt to do this, in the context of London (Lee, 1992), revealed the perplexing finding that the heat-island intensity has decreased by day, but increased by night. The explanation that has been tentatively advanced to explain this is that there has been a decrease in the receipt of daytime solar radiation as a result of vehicular atmosphere pollution, whereas at night the presence of such pollution absorbs and re-emits significant amounts of outgoing terrestrial radiation, maintaining higher urban nocturnal minimum temperatures.

The existence of the urban heat island has a number of implications; city plants bud and bloom earlier, some birds are attracted to the thermally more favourable urban habitat, humans find the added warmth stressful if the city is already situated in a warm area, less winter space-heating is required, but, conversely, more summer air-conditioning is necessary.

The urban-industrial effects on clouds, rain, snowfall and associated weather hazards such as hail and thunder are harder to measure and explain than the temperature changes (Darungo et al., 1978). The changes can be related to various influences (Changnon, 1973: 143):

> thermally induced upward movement of air
> increased vertical motions from mechanically-induced turbulence
> increased cloud and raindrop nuclei
> industrial increases in water vapour.

Table 7.6 illustrates the differences in summer rainfall, thunderstorms and hailstorms between various rural and urban areas in the USA. These data indicate that in cities rainfall increases ranged from 9 to 27 per cent, the incidence of the thunderstorms increased by 10 to 42 per cent, and hailstorms increased by 67 to 430 per cent.

Table 7.6 Areas of maximum increases (urban–rural difference) in summer rainfall and severe weather events for eight American cities

City	Rainfall		Thunderstorms		Hailstorms	
	%	Location*	%	Location*	%	Location*
St Louis	+15	B	+25	B	+276	C
Chicago	+17	C	+38	A,B,C	+246	C
Cleveland	+27	C	+42	A,B	+ 90	C
Indianapolis	0	–	0	–	0	–
Washington DC	+ 9	C	+36	A	+ 67	B
Houston	+ 9	A	+10	A,B	+430	B
New Orleans	+10	A	+27	A	+350	A,B
Tulsa	0	–	0	–	0	–

* A = within city perimeter
  B = 8–24 km downwind
  C = 24–64 km downwind
Source: after Changnon, 1973, figure 1.5, p. 144

An interesting example of the effects of major conurbations on precipitation levels is provided by the London area. In this case it seems that the mechanical effect of the city was dominant in creating localized maxima of precipitation both by being a mechanical obstacle to air flow, on the one hand, and by causing frictional convergence of flow, on the other (Atkinson, 1975). A long-term analysis of thunderstorm records for south-east England is highly suggestive – indicating the higher frequencies of thunderstorms over the conurbation compared to elsewhere (Atkinson, 1968). The similarity in the morphology of the thunderstorm isopleth and the urban area is striking (figure 7.8a and b). Moreover, Brimblecombe (1977) shows a steadily increasing thunderstorm frequency as the city has grown (figure 7.8c).

Similarly the detailed Metromex investigation of St Louis in the USA (Changnon, 1978) shows that in the summer the city affects precipitation and other variables within a distance of 40 km. Increases were found in various thunderstorm characteristics (about +10 to +115 per cent), hailstorm condition (+3 to +330 per cent), various heavy rainfall characteristics (+35 to +100 per cent) and strong gusts (+90 to +100 per cent).

Two main factors are involved in the effect that cities have on winds: the rougher surface they present in comparison with rural areas; and the frequently higher temperatures of the city fabric.

Buildings, especially those in cities with a highly differentiated skyline, exert a powerful frictional drag on air moving over and around them (Chandler, 1976). This creates turbulence, with characteristically rapid spatial and temporal changes in both direction and speed. The average speed of the winds is lower in built-up areas than over rural areas, but Chandler found that in London, when winds are light, speeds are greater in the inner city than outside, whereas the reverse relationship exists when winds are strong. The overall annual reduction of wind speed in Central London is about 6 per cent, but for the higher velocity winds (more than 1.5 m s^{-1}) the reduction is more than doubled.

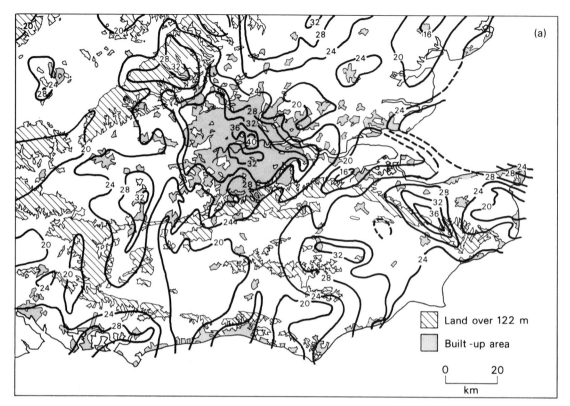

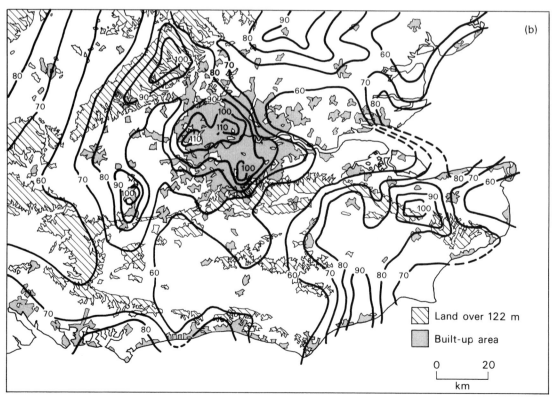

Studies in both Leicester and London, England, have shown that on calm, clear nights, when the urban heat-island effect is at its maximum, there is a surface inflow of cool air towards the zones of highest temperatures. These so-called 'country breezes' have low velocities and become quickly decelerated by intense surface friction in the suburban areas. A practical implication of these breezes is that they transport pollution from the outer parts of an urban area into the city centre, accentuating the pollution problem during smogs.

Smoke haze and photochemical smog

The emission of miscellaneous pollutants, especially into the atmospheres of industrial cities, increases the number of days with smoke-haze conditions as in eighteenth- and nineteenth-century London (figure 7.9a). This is also illustrated by the data for some Illinois cities (figure 7.9b). Chicago, the only major city in the group, was the first to show a sizeable increase, though the

Figure 7.8 ((a) and (b) opposite)
Thunder in south-east England:
(a) total thunder rain in south-east England, 1951–60, expressed in inches (after Atkinson, 1968, figure 6)
(b) number of days with thunder overhead in south-east England, 1951–60 (after Atkinson, 1968, figure 5)

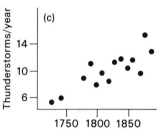

(c) thunderstorms per year in London (decadal means for whole year) (after Brimblecombe, 1977, figure 2)

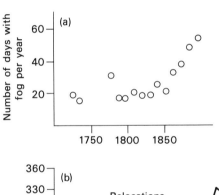

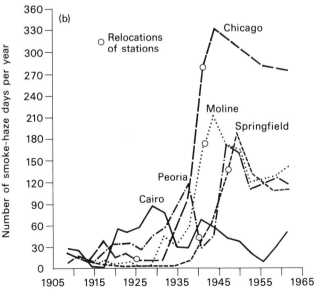

Figure 7.9
(a) Number of fogs per year in London based on decadal means (after Brimblecombe, 1977, figure 1)
(b) Annual number of days with smoke-haze conditions in Illinois cities based on a 3-year moving average (after Changnon, 1973, figure 1.5)

moderately industrialized cities of Moline, Springfield and Peoria also show considerable increases beginning in the 1930s. Cairo, a non-industrial city with minimal twentieth-century growth, has had no great temporal increase in smoke-haze day frequencies (Changnon, 1973). In general, Landsberg (1970) estimates that on average the number of particles present over urban areas is ten times greater than that over rural environs. The degree of pollution increases with city size (table 7.7). Many of these atmospheric particulates are hygroscopic and so tend to promote fog formation as water vapour readily condenses on them. However, although fog generally occurs more frequently in metropolitan areas, this is not true for very dense fog, for the heat-island effect of greater warmth in the inner city often prevents the thickest night fogs from reaching the densities reported in the suburbs and outlying districts (see data for London, in table 7.8).

Local dust plumes can create changes in precipitation. A notable example of this comes from a coal-fired power station near Boulder, Colorado where fly-ash aerosols inadvertently seeded

Table 7.7 Air pollution mechanisms

(a) Urban–rural air pollution gradients (concentrations in $\mu g/m^3$)

| | | Non-urban | | |
Pollution type	Urban	Proximate	Intermediate	Remote
Suspended particulates	102.0	45.0	40.0	21.0
Benzene soluble organics	6.7	2.5	2.2	1.1
Ammonium	0.9	1.22	0.28	0.15
Nitrate	2.4	1.40	0.85	0.46
Sulphate	10.1	10.0	5.29	2.51
Copper	0.16	0.16	0.08	0.06
Iron	1.41	0.56	0.27	0.15
Manganese	0.07	0.02	0.01	0.01
Nickel	0.02	0.01	0.00	0.00
Lead	1.11	0.21	0.10	0.00

(b) Air pollution related to city size

| | Concentration ($\mu g/m^3$) | | |
	Total suspended particulates	SO_2	NO_2
Population class			
Non-urban	25	10	33
< 10 000	57	35	116
10 000	81	18	64
25 000	87	14	63
50 000	118	29	127
100 000	95	26	114
400 000	100	28	127
700 000	101	29	146
1 000 000	134	69	163
3 000 000	120	85	153

Source: Berry, 1974, tables 2.3 and 2.4, modified by author

Table 7.8 Fog frequencies in London

| Location | Hours per year (based on four observations per day) with visibility less than | | | |
	40 m	200 m	400 m	1000 m
Kingsway (Central London)	19	126	230	940
Kew (inner suburbs)	79	213	365	633
London Airport (outer suburbs)	46	209	304	562
South-east England (mean of 7 stations outside London)	20	177	261	494

Source: after Chandler, 1965

supercooled fog and induced local snowfall. The shape of the plume and the area of snowfall were the same (Darungo et al., 1978).

Measures can be taken to control smoke haze, and Brimble-combe (1987) provides a detailed and scholarly history of atmospheric pollution and control in 'the big smoke' – London. Edward I ordered a man to be tortured for burning coal and fouling the air in the fourteenth century, while more recently the Clean Air Acts in Britain in 1956 and 1963 have led to a reduction in smog in major cities (see figure 7.10). In London, emissions of smoke had, by 1970, been reduced to one-tenth of what they were in 1956, and in Manchester fog and SO_2 levels decreased, while sunshine levels went up (see Wood et al., 1974).

Because of the presence of particulate matter, the duration of bright sunshine shows significant reductions over urban areas, and in large cities a progressive decline can be traced towards the city centre. Over the period 1921–50, for example, the mean daily sunshine duration outside London was 4.33 hours, whereas in the central area it was 3.60 hours (Chandler, 1965). Since that time, on account of clean-air legislation, the situation has improved. Improvements have also been made with respect to smog pollution in Los Angeles and other great cities. New legislation to reduce the adverse effects of exhausts from cars caused the average monthly oxidant concentration in central Los Angeles to fall from 0.27 ppm in 1965 to 0.17 ppm in 1974. Annual maximum hourly ozone concentrations have also fallen markedly and a Stage 3 Emergency Pollution Alert (see figure 7.11) has not been called in Los Angeles since 1973. Even so, pollution remains a problem of some gravity in the conurbation and it is commonplace for the 0.10 ppm threshold for adverse health effects to be exceeded on most days of the month between May and September (Elsom, 1987: 35).

The downward trend in fog frequency since the introduction of pollution control measures is also demonstrated clearly for Oxford, central England (Gomez and Smith, 1984), where since the early 1960s fog has only been about half as frequent as it was in the preceding decades (figure 7.12).

Figure 7.13 shows in a more general way how the number of days per year with fog over the period 1950–83 has changed for 28

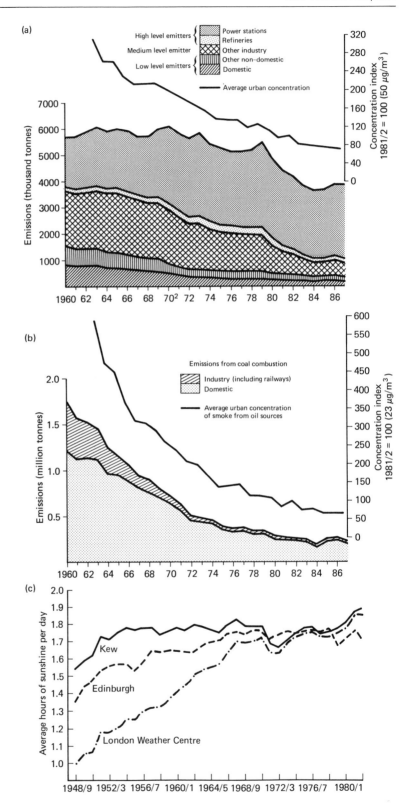

Figure 7.10
Trends in atmospheric quality in the United Kingdom:
(a) Sulphur dioxide emissions from fuel combustion and average urban concentrations (from DOE, 1989, figure 1.2)
(b) Smoke emissions from coal combustion and average urban concentrations of oil smoke (from DOE, 1989, figure 1.2)
(c) Increase in winter sunshine (10-year moving average) for London and Edinburgh city centres and for Kew, outer London (from DOE, 1984)

Plate 7.4
In December 1952 the city of London was affected by severe smog. Visibility was reduced and smog-masks had to be worn out-of-doors. Many people with weak chests died. Since then, because of legislation, the incidence of smog has declined markedly

stations in Great Britain. It demonstrates the high fog levels experienced in the inland industrial centres in comparison with coastal and rural situations, but, more impressively, illustrates the dramatic declines in fog frequency in those industrial/urban areas where Clean Air Act legislation has had a marked impact (Musk, 1991).

Smog produced by the burning of coal is often sulphurous. The sulphur dioxide (SO_2) generated can oxidize in air to produce sulphur trioxide (SO_3), which reacts with water vapour (H_2O) in the presence of catalysts to form sulphuric-acid mist (H_2SO_4). The precipitation of such acid droplets with sulphate particles creates the so-called 'acid rain' which is liable to contaminate rivers and lakes. The acidification of fresh waters in Scandinavia may be caused by emissions from the industrial heartland of Western Europe, including Britain.

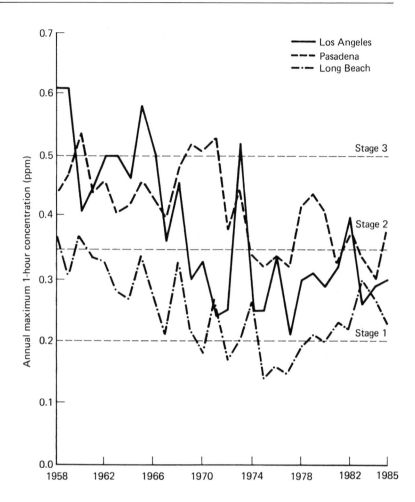

Figure 7.11
Annual maximum hourly ozone concentrations at selected sites in the Los Angeles basin, California, USA, 1958–85 (from Elsom, 1987, p. 37, figure 2.8)

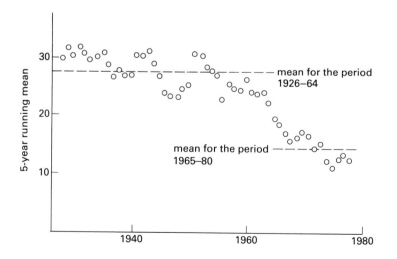

Figure 7.12
Annual fog frequency at 0900 GMT in Oxford, central England, 1926–80 (modified after Gomez and Smith, 1984, figure 3)

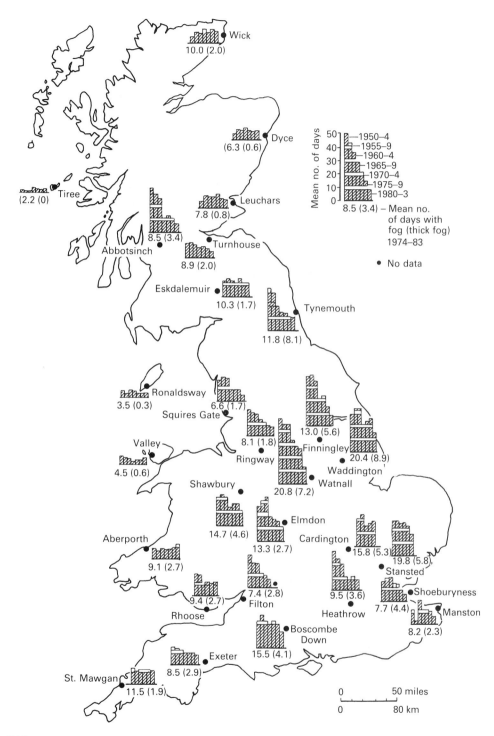

Figure 7.13
The spatial variation of fog over Great Britain, 1950–1983 (after Musk, 1991, figure 6.6)

While smog produced by burning coal is common, there is a second widespread type: photochemical smog. The name originates from the fact that most of the less desirable properties of such fog result from the products of chemical reactions induced by sunlight (Cadle, in Matthews et al., 1971: 340ff.).

Unburnt hydrocarbons play a major role in smog formation and result from evaporation of solvents and fuels, as well as incomplete combustion of fossil fuels. In the presence of oxides of nitrogen, strong sunlight and stable meteorological conditions, complex chemical reactions occur, forming a family of peroxyacyl nitrates (sometimes collectively abbreviated to PANs).

Photochemical smog appears 'cleaner' than other kinds of fog in the sense that it does not contain the very large particles of soot that are so characteristic of smog derived from coal-burning. However, the eye irritation and damage to plant leaves it causes make it unpleasant. Photochemical smog occurs particularly where there is large-scale combustion of petroleum products, as in car-dominated cities like Los Angeles. Its unpleasant properties include a high lead content; also a series of chemical reactions are triggered by sunlight (figure 7.14). For example, a photochemical decomposition of nitrogen dioxide into nitric oxide and atomic oxygen occurs, and the atomic oxygen can react with molecular oxygen to form ozone. Further ozone may be produced by the reaction of atomic oxygen with various hydrocarbons.

Photochemical smogs are not universal. Because sunlight is a crucial factor in their development they are most common in the tropics or during seasons of strong sunshine. Their especial notoriety in Los Angeles is due to a meteorological setting dominated at times by subtropical anti-cyclones with weak winds, clear skies and a subsidence inversion, combined with the general topographic situation and the high vehicle density (>1500 vehicles/km^2). However, photochemical ozone pollution can, on certain summer days, reach appreciable levels even in the UK (Colbeck, 1988), especially in large cities such as London (table 7.9).

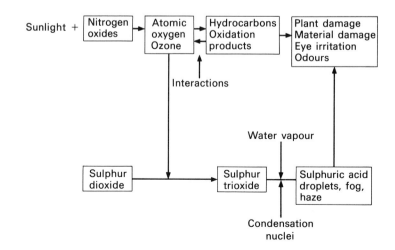

Figure 7.14
Possible reactions involving primary and secondary pollutants (after Haagen-Smit in Bryson and Kutzbach, 1968, figure 4. Reprinted by permission Association of American Geographers)

Table 7.9 Annual maximum hourly mean ozone concentrations at County Hall, Westminster in Greater London, 1975–85

Year	Peak value (ppm)	Date of peak	No of days ⩾ 0.080 ppm
1975	0.150	26 June	16
1976	0.212	27 June	26
1977	0.087	3 July	1
1978	0.103	29 July	2
1979	0.153	27 July	5
1980	0.116	27 August	2
1981	0.112	5 September	6
1982	0.091	3 August	1
1983	0.099	16 July	11
1984	0.088	6 July	3
1985	0.098	25 July	3

Source: Ball and Laxen, 1986

Rainmaking and other methods of deliberate climatic modification

The most fruitful human attempts to augment natural precipitation have been through cloud seeding. Rainmaking experiments of this type are based on three main assumptions (Chorley and More, 1967: 159).

1 Either the presence of ice crystals in a supercooled cloud is necessary to release snow and rain or the presence of comparatively large water droplets is necessary to initiate the coalescence process.
2 Some clouds precipitate inefficiently or not at all, because these components are naturally deficient.
3 The deficiency can be remedied by seeding the clouds artificially, either with solid carbon dioxide (dry ice) or silver iodide, to produce crystals, or by introducing water droplets or large hygroscopic nuclei.

These methods of seeding are not universally productive. Under conditions of orographic lift and in thunderstorm cells, when nuclei are insufficient to generate rain by natural means, some augmentation may be attained, especially if cloud temperatures are of the order of -10 to $-15°C$. The increase of precipitation gained under favourable conditions may be of the order of 10–20 per cent in any one storm. In lower latitudes, where cloud-top temperatures frequently remain above $0°C$, silver-iodide or dry-ice seeding is not applicable. Therefore alternative methods have been introduced whereby small water droplets of 50 mm diameter are sprayed into the lower layers of deep clouds, so that the growth of cloud particles will be stimulated by coalescence.

Other techniques of warm cloud seeding include the feeding of hygroscopic particles into the lower air layers near the updraught of a growing cumulus cloud (Breuer, 1980).

Plate 7.5
An example of the effects of cloud seeding using silver iodide crystals, undertaken at Socorro, New Mexico, USA, in June 1950. The upper plate shows the clouds in the area prior to seeding, while the lower plate shows the situation 73 minutes after seeding has taken place. Some large cumulus clouds have developed

The results of the many experiments now carried out on cloud seeding are still controversial, very largely because we have an imperfect understanding of the physical processes involved. This means that the evidence has to be evaluated on a statistical rather than a scientific basis so, although precipitation may occur after many seeding trials, it is difficult to decide to what extent artificial stimulation and augmentation is responsible. It also needs to be remembered that this form of planned weather modification applies

Plate 7.6
The damage caused by large hailstones can be considerable. These examples, the size of a baseball, fell near Neligh, Nebraska in June 1950. There is, therefore, a considerable incentive to explore ways of reducing their impact through climatic modification experiments

to small areas for short periods. As yet, no means exist to change precipitation appreciably over large areas on a sustained basis.

Other types of deliberate climatic modification have also been attempted (Hess, 1974). For example, it has been thought that the production of many more hailstone embryos by silver-iodide seeding will yield smaller hailstones which would both be less damaging and more likely to melt before reaching the ground. Some results in Russia have been encouraging, but one cannot exclude the possibility that seeding may sometimes even increase hail damage (Atlas, 1977). Similar experiments have been conducted in lightning suppression. The concept here is to produce in a thundercloud, again by silver-iodide seeding, an abnormal abundance of ice crystals that would act as added corona points and thus relieve the electrical potential gradient by corona discharge before a lightning strike could develop (Panel on Weather and Climate Modification, 1966: 4–8).

Hurricane modification is perhaps the most desirable aim of those seeking to suppress severe storms, because of the extremely favourable benefit-to-cost ratio of the work. The principle once again is that of introducing freezing nuclei into the ring of clouds around the hurricane centre to trigger the release of the latent heat of fusion in the eye-wall cloud system which, in turn, diminishes the maximum horizontal temperature gradients in the storm, causing a hydrostatic lowering of the surface temperature. This eventually should lead to a weakening of the damaging winds (Smith, 1975: 212). A 15 per cent reduction in maximum winds is theoretically possible, but as yet work is largely at an experimental stage. A 30 per cent reduction in maximum winds was claimed following seeding of Hurricane Debbie in 1969 but US attempts to seed hurricanes were discontinued in the 1970s following the development of new computer models of the effects of seeding on hurricanes. The models suggested that although maximum winds might be reduced by 10–15 per cent on average, seeding may increase the winds just outside the region of maximum winds by 10–15 per cent, may increase rainfall in the outer regions of the storms, may either increase or decrease the maximum storm surge, and may or may not affect the direction of the storm (Sorkin, 1982). Seeding of hurricanes is unlikely to recommence until such uncertainties can be resolved.

Much research has been devoted in Russia to the possibility of removing the Arctic Sea ice and to ascertain effects of such action on the climate of northern areas (see Lamb, 1977: 660). One proposal was to dam the Bering Strait, thereby blocking off water flow from the Pacific. The assumption was that more, and warmer, Atlantic water would be drawn into the central Arctic, improving temperature conditions in that area. Critics have pointed to the possibility of adverse changes in temperatures elsewhere, together with undesirable change in precipitation character and amount.

Fog dispersal, vital for airport operation, is another aim of weather modification. Seeding experiments have shown that fog consisting of supercooled droplets can be cleared by using liquid propane or dry ice. In very cold fogs this seeding method causes rapid transformation of water droplets into ice particles. Warm fogs with temperatures above freezing point occur more frequently than supercooled fogs in mid-latitudes and are more difficult to disperse. Some success has been achieved using sodium chloride and other hygroscopic particles as seeding agents but the most effective method is to evaporate the fog. The French have developed the 'turboclair' system in which jet engines are installed alongside the runway at major airports and the engines produce short bursts of heat to evaporate the fog and improve visibility as an aircraft approaches (Hess, 1974).

In regions of high temperature, dark soils then become over-heated, and the resultant high evapotranspiration rates lead to moisture deficiencies. Applications of white powders (India and Israel) or of aluminium foils (Hungary) increase the reflection from the soil surface and reduce the rate at which insulation is absorbed. Temperatures of the soil surface and sub-surface are lowered (by as much as 10°C), and soil moisture is conserved (by as much as 50 per cent).

The planting of windbreaks is an even more important attempt by humans to modify local climate deliberately. Shelter-belts have been in use for centuries in many windswept areas of the globe, both to protect soils from blowing and to protect the plants from the direct effects of high velocity winds. The size and effectiveness of the protection depends on their height, density, shape and frequency. However, the belts may have consequences for micro-climate beyond those for which they were planted. Evaporation rates are curtailed; snow is arrested, and its melting waters are available for the fields; but the temperatures may become more extreme in the stagnant space in the lee of the belt, creating an increase in frost danger.

Traditional farmers in many societies have been aware of the virtues of microclimatic management (Wilken, 1972). They manage shade by employing layered cropping systems or by covering the plant and soil with mulches; they may deliberately try to modify albedo conditions. Tibetan farmers, for example, reportedly throw dark rocks on to snow-covered fields to promote late spring melting; and in the Paris area of France some very dense stone walls were constructed to absorb and radiate heat.

Air pollution: some further effects

This chapter has already made much reference to the ways in which humans have changed the turbidity of the atmosphere and the gases within it. However, the consequences of air pollution go further than either their direct impact on human health or their impact on local, regional and global climates.

First of all, the atmosphere acts as a major channel for the transfer of pollutants from one place to another, so that some harmful substances have been transferred long distances from their sources of emission. DDT is one example; lead is another. Thus, since the start of the Industrial Revolution, the lead content of the Greenland ice cap, although far removed from the source of the pollutant (which is largely derived from either industrial or auto-mobile emissions) has risen very substantially (figure 7.15a). The same applies to its sulphate content (figure 7.15b). An analysis of pond sediments from a remote part of North America (Yosemite) indicates that lead levels have also been raised as a result of human activities, being more than twenty times the natural levels. The lead, which came in from atmospheric sources, shows a fivefold eleva-tion in the plants of the area and a fiftyfold elevation in the animals

Plate 7.7
Smoke, in its densest form, being emitted by coal-fired bottle ovens in England's 'Black Country' before the Clean Air Act of 1956

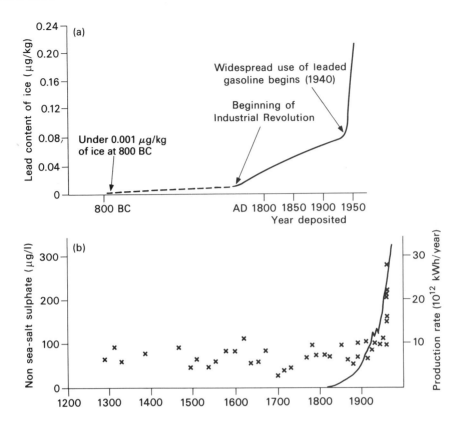

Figure 7.15
Trends in atmospheric quality:
(a) Lead content of the Greenland ice cap due to atmospheric fallout of the mineral on the snow surface. A dramatic upturn in worldwide atmospheric levels of lead occurred at the beginning of the Industrial Revolution of the nineteenth century and again after the more recent spread of the automobile (after Murozumi et al., 1969, p. 1247)
(b) The sulphate concentration on a sea-salt-free basis in north-west Greenland glacier ice samples as a function of year. The curve represents the world production of thermal energy from coal, lignite and crude oil (modified after Koide and Goldberg, 1971, figure 1)

compared with natural levels (Shirahata et al., 1980). Some estimates have also compared the total quantities of heavy metals that humans are releasing into the atmosphere with emissions from natural sources (Nriagu, 1979). The increase is eighteenfold for lead, ninefold for cadmium, sevenfold for zinc, and threefold for copper.

There are, it must be stated, signs that as a result of pollution control regulations some of these trends are now being reversed. Boutron et al. (1991), for example, have analysed ice and snow that has accumulated over Greenland in the last two decades and find that lead concentrations have decreased by a factor of 7.5 since 1970. They attribute this to a curbing of the use of lead additives in petrol. Over the same period cadmium and zinc concentrations have decreased by a factor of 2.5.

A second example of the possible widespread and ramifying ecological consequences of atmospheric pollution is provided by 'acid rain' (Likens and Bormann, 1974). Acid rain is rain which has a pH of less than 5.65, this being the pH which is produced by carbonic acid in equilibrium with atmospheric CO_2. In many parts of the world, rain may be markedly more acid than this normal, natural background level. Snow and rain in the north-east USA have been known to have pH values as low as 2.1, while in Scotland

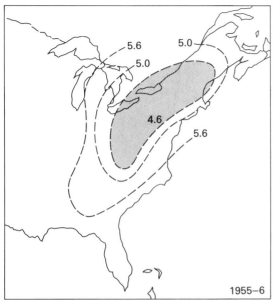

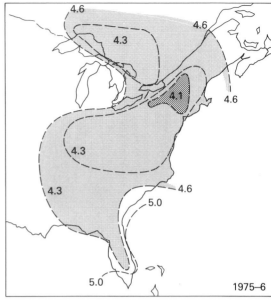

in one storm the rain was the acidic equivalent of vinegar (pH 2.4). In the eastern USA the average annual precipitation acidity values tend to be around pH 4 (see figure 7.16), and the degree of acidulation appears to have been increasing (see figure 7.17 for trends in Scandinavia).

It needs to be remembered that not all environmental acidification is caused by acid rain in the narrow sense. Acidity can reach the ground surface without the assistance of water droplets. This is as particulate matter and is termed 'dry deposition'. Furthermore, there are various types of 'wet precipitation' by mist, hail, sleet or snow, in addition to rain itself. Thus some people prefer the term 'acid deposition' to 'acid rain'. The acidity of the precipitation in

Figure 7.16
Isopleths showing annual average pH for precipitation in eastern North America (after Likens et al., 1979 with modification)

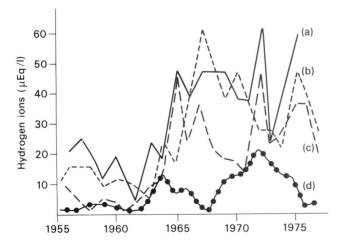

Figure 7.17
Acidity of rain and snow for four Scandinavian localities. The acidity is expressed in terms of the annual volume-weighted concentration of hydrogen ions in microequivalents per litre:
(a) Lista, Norway
(b) As, Norway
(c) Kise, Norway
(d) Kiruna, Sweden
(modified after Likens et al., 1974, p. 45, in *Science*, 184, 1176–9, © 1979 by the American Association for the Advancement of Science)

Plate 7.8
This statue – one of the emperors' heads in Oxford before their replacement in the 1970s – is made of limestone and has been attacked by urban pollution

turn leads to greater acidity in rivers and lakes. Studies by Beamish et al. (1975) demonstrated that between 1961 and 1975 pH had declined by 0.13 pH units per year in George Lake, Canada, and a comparable picture emerges from Sweden, where Almer et al. (1974) found that the pH in some lakes had decreased by as much as 1.8 pH units since the 1930s.

Very long-term records of lake acidification can provide dramatic evidence for a recent magnification of the acid deposition problem. These are obtained by extracting cores from lake floors and analysing their diatom assemblages at different levels. The diatom assemblages reflect water acidity levels at the time they were living. Two studies serve to illustrate the trend. In south-west Sweden (Renberg and Hellberg, 1982) for most of post-Glacial time (that is, the last 12 500 years) the pH of the lakes appears to have decreased gradually from around 7.0 to about 6.0 as a result of natural ageing processes. However, over the last few decades, and especially since the 1950s, a further, more marked decrease occurred to present-day values of about 4.5. In Britain, the work of

Battarbee and collaborators (1988a, b) shows that at sensitive sites pH values before around 1850 were close to 6.0 and that since then pH declines have varied between 0.5 and 1.5 units.

Fortunately there is some evidence that if industrially derived sulphate emissions decline lake acidity once again becomes reduced. Acidification of lakes thus appears to be a reversible process. This has been demonstrated through a study of lake water chemistry and diatom stratigraphy undertaken in Scotland by Battarbee et al. (1988b).

The causes of acid deposition are the increasing quantities of sulphur oxides (figure 7.18) and nitrogen oxides being emitted from fossil-fuel combustion. At present, roughly 60–70 per cent of the problem is due to sulphur-oxide emissions, and the balance to the nitrogen oxides. However, in some regions, such as the west coast of the United States and Japan, the nitric acid contribution may well be of greater relative importance. Moreover, with the emissions of oxides of nitrogen appearing to increase at a faster rate than sulphur-dioxide emissions, nitric acid will probably increase further in its relative importance.

Figure 7.19 shows how sulphate levels have increased in European precipitation. Two main factors contribute to the increasing seriousness of the problem. One is the replacement of coal by oil and natural gas. The second, paradoxically, is a result of the implementation of air pollution control measures (particularly increasing the height of smoke stacks and installing particle precipitators). These appear to have transformed a local 'soot problem' into a regional 'acid rain problem'. Coal-burning produced a great deal of sulphate, but was largely neutralized by high calcium contents in the relatively unfiltered coal smoke emissions. Natural gas burning creates less sulphate but that which is produced is not neutralized. The new higher chimneys pump the smoke so high that

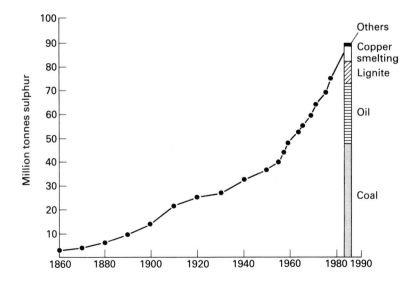

Figure 7.18
Global SO_2 emissions from anthropogenic sources including the burning of coal, lignite and oil, and copper smelting

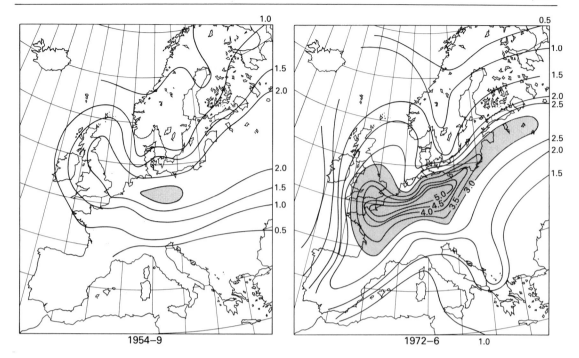

Figure 7.19
Annual mean concentration of
sulphate in precipitation in
Europe (mg S/l) (after Wallen in
Holdgate et al., 1982, figure
2.3)

it is dispersed over wide areas, whereas previously it returned to earth nearer the source. Some of the stacks are now enormous. One at the Sudbury copper-nickel smelter in Canada is 400 m tall. A possible subsidiary reason for some of the observed increases in acidity is that there may now be less calcium-rich dust in the atmosphere over North America since the passing of the Dust Bowl years (Stensland and Semonin, 1982).

The effects of acid rain (figure 7.20) are especially serious in areas underlain by highly siliceous types of bedrock (for example, granite, some gneisses, quartzite and quartz sandstone), such as the old shield areas of the Fenno-Scandian shield in Scandinavia and the Laurentide shield in Canada (Likens et al., 1979).

The ecological consequences of acid rain are still the subject of some debate. Krug and Frink (1983) have argued that acid rain only accelerates natural processes, and point out that the results of natural soil formation in humid climates include the leaching of nutrients, the release of aluminium ions and the acidification of soil and water. They also note that acidification by acid rain may be superimposed on longer-term acidification induced by changes in land use. Thus the regrowth of coniferous forests in what are now marginal agriculture areas, such as New England and highland Western Europe, can increase acidification of soils and water. Similarly, it is possible, though in general unproven (see Battarbee et al., 1985), that in areas like western Scotland, a decline in upland agriculture and the regeneration of heathland could play a role in increasing soil and water acidification. Likewise Johnston et al. (1982) have suggested that acid rain can cause either a decrease or

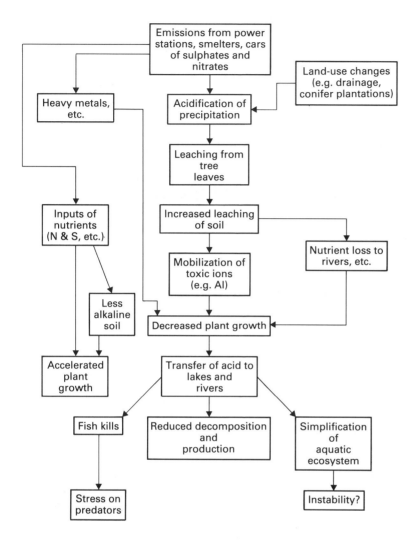

Figure 7.20
Pathways and effects of acid precipitation through different components of the ecosystem, showing some of the adverse and beneficial consequences

an increase in forest productivity, depending on local factors. For example, in soils where cation nutrients are abundant and sulphur or nitrogen are deficient, moderate inputs of acid rain are very likely to stimulate forest growth.

In general, however, it is the negative consequences of acid rain that have been stressed. One harmful effect is a change in soil character. The high concentration of hydrogen ions in acid rain causes accelerated leaching of essential nutrients, making them less available for plant use. Furthermore, the solubility of aluminium and heavy metal ions increases, and instead of being fixed in the soil's sorption complex, these toxic substances become available for plants or are transferred into lakes, where they become a major physiological stress for some aquatic organisms.

Fresh-water bodies with limited natural cations are poorly buffered and thus vulnerable to acid inputs. The acidification of

Plate 7.9
The acidification of fresh waters has some adverse effects on fish populations, both because fish are sensitive to acidity, but also because acidity may cause the release of certain toxic ions, including aluminium.

thousands of lakes and rivers in southern Norway and Sweden during the past two decades has been attributed to acid rain, and this increased acidity has resulted in the decline of various species of fish, particularly trout and salmon. But fish are not the only aquatic organisms that may be affected. Fungi and moss may proliferate, organic matter may start to decompose less rapidly, and the number of green algae may be reduced.

Forest growth can also be affected by acid rain, though the evidence is not necessarily proven. Acid rain can damage foliage, increase susceptibility to pathogens, affect germination and reduce nutrient availability. However, since acid precipitation is only one of many environmental stresses, its impact may enhance, be enhanced by, or be swamped by other factors. For example, Blank (1985), in considering the fact that an estimated one-half of the total forest area of the former West Germany was showing signs of damage, referred to the possible role of ozone or of a run of hot, dry summers on tree health and growth.

Further useful information on the contentious issue of the effects of acid rain is contained in Likens and Butler (1981), Hutchinson and Havas (1980), Hornbeck (1981) Park (1987) and Wellburn (1988).

A very rapidly developing area of concern in pollution studies is the current status of stratospheric ozone levels. The atmosphere has a layer of relatively high concentration of ozone (O_3) at a height of about 16 and 18 km in the polar latitudes and of about 25 km in equatorial regions. This ozone layer is important because it absorbs incoming solar ultraviolet radiation, thus warming the stratosphere and creating a steep inversion of temperature at heights between about 15 and 50 km. This in turn affects convective processes and atmospheric circulation, thereby influencing global weather and climate. However, the role of ozone in controlling receipts of ultraviolet radiation at the earth's surface has great ecological significance, because it modifies rates of photosynthesis. One class of organism that has been identified as being especially prone to the effects of increased ultraviolet radiation consequent upon ozone depletion are phytoplankton – aquatic plants that spend much of their time near the sea surface and are therefore exposed to such radiation. A reduction in their productivities would have potentially ramifying consequences, because these plants directly and indirectly provide the food for almost all fish.

Human activities appear to be causing ozone depletion in the stratosphere, most notably over the south polar regions, where an 'ozone hole' has been identified. Possible causes of ozone depletion are legion, and include various combustion products emitted from high-flying military and civil supersonic aircraft; nitrous oxide released from nitrogenous chemical fertilizers; and chlorofluorocarbons (CFCs) used in aerosol spray cans, refrigerant systems, and in the manufacture of foam fast-food containers.

However, in recent years the greatest attention has been focused on the role of CFCs, the production of which has been climbing in the last few decades (figure 7.4). These gases may diffuse upwards into the stratosphere where solar radiation causes them to become dissociated to yield chlorine atoms which react with and destroy the

Plate 7.10
In the 1980s ground observations and satellite monitoring of atmospheric ozone levels indicated that a 'hole' had developed in the stratospheric ozone layer above Antarctica. These Nimbus satellite images show the ozone concentrations (in Dobson Units) for the month of October between 1984 and 1988

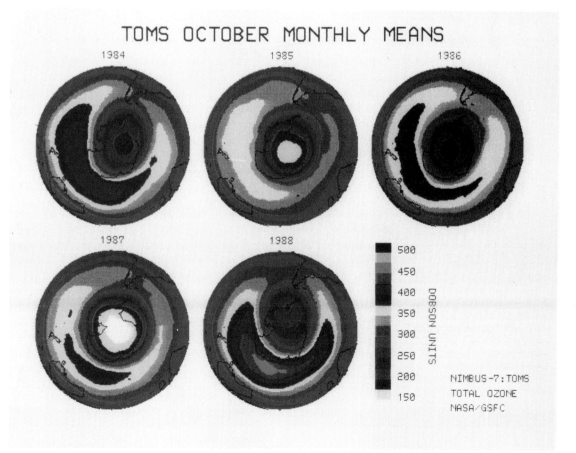

ozone. The process has been described thus by Titus and Seidel (1986: 4):

Because CFCs are very stable compounds, they do not break up in the lower atmosphere (known as the troposphere). Instead, they slowly migrate to the stratosphere, where ultraviolet radiation breaks them down, releasing chlorine.

Chlorine acts as a catalyst to destroy ozone; it promotes reactions that destroy ozone without being consumed. A chlorine (Cl) atom reacts with ozone (O_3) to form ClO and O_2. The ClO later reacts with another O_3 to form two molecules of O_2, which releases the Cl atoms. Thus two molecules of ozone are converted to three molecules of ordinary oxygen, and the chlorine is once again free to start the process. A single chlorine atom can destroy thousands of ozone molecules. Eventually, it returns to the troposphere, where it is rained out as hydrochloric acid.

The Antarctic ozone hole has been identified through satellite monitoring and by monitoring of atmospheric chemistry on the ground. The decrease in ozone levels at Halley Bay is shown in figure 7.21. The reasons why this zone of ozone depletion is so well developed over Antarctica include the very low temperatures of the polar winter, which seem to play a role in releasing chlorine atoms;

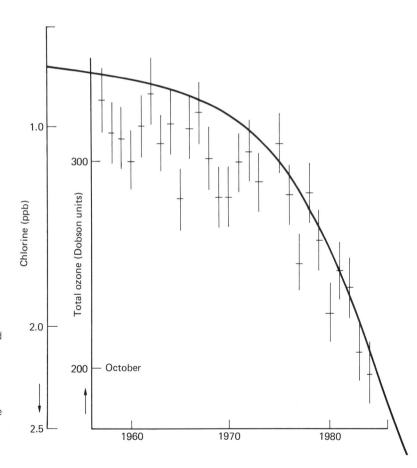

Figure 7.21
Decline in ozone concentrations at Halley Bay, Antarctica, between 1957 and 1984 for the month of October. This is shown by the elongated crosses, the vertical bar of which indicates the degree of uncertainty of the measurements. The smooth curve shows the total concentration of organochlorine (essentially CFCs in parts per billion ppb) at the same site) (Rowland, 1987)

the long sunlight hours of the polar summer, which promote photochemical processes; and the existence of a well-defined circulation vortex. This vortex is a region of very cold air surrounded by strong westerly winds, and air within this vortex is isolated from that at lower latitudes, permitting chemical reactions to be contained rather than more widely diffused. No such clearly defined vortex exists in the northern hemisphere, though ozone depletion does seem to have occurred in the Arctic as well (Proffitt et al., 1990). Furthermore, observations in the last few years indicate that the Antarctic ozone hole is spreading over wider areas and persisting longer into the Antarctic summer. It is also possible that the situation could be worsened by emissions of volcanic ash into the atmosphere (as from Mount Pinatubo), for these can also cause chemical reactions that lead to ozone depletion (Mintzer and Miller, 1992).

It is perhaps worth concluding this section by reiterating the fact that the variability of air pollution in both space and time, while undoubtedly owing much to human action, is also highly dependent on natural conditions, and in particular on meteorological conditions (Thompson, 1978). This is brought out in figure 7.22. Climatic factors control the efficiency with which effluents are diluted and dispersed away from primary sources, while local relief and microclimatic circumstances modify the interaction between the emission of pollutants and the atmosphere. Moreover, as we

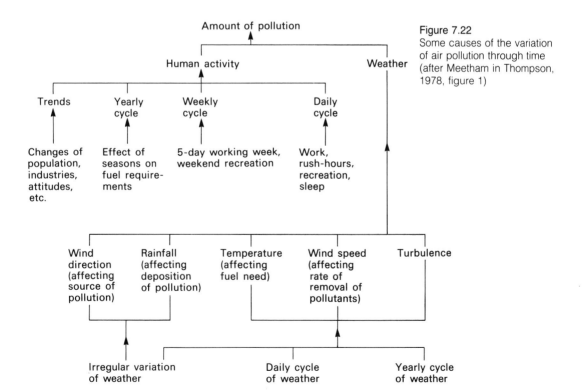

Figure 7.22
Some causes of the variation of air pollution through time (after Meetham in Thompson, 1978, figure 1)

have already stated, not all pollutants result from human endeavour: the vast majority of nitrogen oxides are emitted by bacteria rather than by car exhausts, most chlorides in rain water come from the oceans, and a considerable proportion of atmospheric dust is injected into the air by such mechanisms as volcanic eruptions, and by dust storms removing fine silt from glacial outwash fans or desert basins.

EIGHT

The Future

Much of this book has in one sense been historical, attempting to outline the various changes that have been wrought by humans on the face of the earth over the last few million years as levels of population and the nature of technological progress have changed. None the less, from time to time reference has been made to the future and to the uncertainties that surround it. Given the rapid expansion in human numbers and the increasing sophistication and pervasiveness of certain technological developments, the pace of change over the next decades is likely to become such that the human impact will be greatly magnified.

There is now an unprecedented interest in the environmental future, and few people are unaware of the threat and potential dangers posed by nuclear pollution, species extinction, acid rain, the carbon-dioxide problem and desertification. It is for this reason that various attempts have been made to formulate informed predictions of likely changes as a guide to governments and administrators. One of the most notable of these is the report prepared and first published in 1980 as *Global 2000*, an American compilation for the President of the USA (Council on Environmental Quality, 1982). Some of its findings on such matters as the extinctions of species that may result from deforestation in the moist tropics have already been referred to. The report is based on an enormous amount of informed research by a large number of often distinguished scientists. Although it is careful not to be excessively negative, and weighs up various alternative scenarios, its basic conclusions are fundamentally pessimistic (p. 1):

If present trends continue, the world in 2000 will be more crowded, more polluted, less stable ecologically, and more vulnerable to disruption than the world we live in now. Serious stresses involving population, resources and environment are clearly visible ahead. Despite greater material output, the world's people will be poorer in many ways than they are today. For hundreds of millions of the desperately poor, the outlook for food and other necessities of life will be no better. For many it will be worse. Barring revolutionary advances in technology, life for most people on earth will be more precarious in 2000 than it is now – unless the nations of the world act decisively to alter current trends.

In 1984, however, Simon and Kahn edited a volume entitled *The resourceful earth – a response to global 2000*, which took a very much more optimistic view, and which included contributions from a variety of distinguished scientists, including some eminent geographers. Note the alternative view of the future which they propose (pp. 1–2):

If present trends continue, the world in 2000 will be *less crowded* (though more populated), *less polluted, more stable ecologically* and *less vulnerable to resource-supply disruption* than the world we live in now. Stresses involving population, resources, and environment *will be less in the future than now* . . . The world's people will be *richer* in most ways than they are today . . . The outlook for food and other necessities of life will be *better* . . . life for most people on earth will be *less precarious* economically than it is now.

There are many possible reasons why there should be such a wide disparity in attitudes to the likely future of the earth. Amongst these are a paucity of background data on the present and past state of the earth, the problems of extrapolating data into the future by curve-fitting procedures, the crudeness and assumptions of so many models, the complexity of interactions referred to in the last section, uncertainty about the nature of technological developments and possible political constraints on the use of fossil fuels and nuclear power. Inevitably also differences in opinion may result from ideological and political differences between authors. However, whatever the standpoint of the investigator, it is unquestionable that our fundamental knowledge is still far too slim on which to build reliable prognoses of the future, and that geographers and practitioners in allied disciplines need to seek still more information.

The climatic future

Such is the uncertainty surrounding the climatic future that the *Global 2000* (Council on Environmental Quality, 1982) report considered three different scenarios feasible:

Case 1 (No change). Yearly rainfall and temperature statistics would be similar to those of the 1941–70 periods, with occasional droughts in areas like the Sahel, India and the USA. This scenario might result if the warming trends induced by the build up of CO_2 were counteracted by natural cooling tendencies.
Case 2 (Warming). Global temperatures would increase by 1°C, principally as a result of a warming trend caused by elevated CO_2 levels being sufficiently large to outweigh the effects of any natural cooling. Most of the warming would take place in polar regions and the higher middle latitudes, with only slight warming in the tropics. Annual precipitation would increase by 5–10 per cent on a global basis, but there would be more

likelihood of US drought conditons similar to those of the Dust Bowl years of the 1930s.

Case 3 (Cooling). As a result of changes in solar and volcanic activity, global temperatures would decrease by 0.5°C in the higher and middle latitudes. In general precipitation amounts would decline, and variability increase. Storm-tracks, and the precipitation they bring, would shift towards the equator, improving conditions in the upper latitudes of the great deserts and worsening them on their equatorial sides. Severe monsoon failures would be more frequent in India, with severe droughts in the Sahel.

Assessment of the future scale and trend of climatic change is thus full of imponderables. This has been well stated by Warrick (1988: 223–4):

Prediction of the rate at which atmospheric CO_2 will accumulate in the atmosphere depends largely on assumptions regarding future fossil fuel consumption and on knowledge of the global carbon cycle. Added to CO_2 are changes in the atmospheric concentrations of other 'greenhouse' gases – methane, chlorofluorocarbons, nitrous oxide, ozone – which in aggregate, are becoming just as important relatively speaking, as CO_2. Taking into account the projected concentrations of these gases (with the associated uncertainties), there could possibly be an 'equivalent' doubling of CO_2 by as early as the 2030s . . . But changes in climate will not follow step-by-step. Owing to the thermal inertia of the oceans, changes in global surface temperature will probably lag behind by several decades. Thus the question of 'when' is clouded.

The question of 'where' is even more problematic. Climatic changes will vary in direction and magnitude from one agricultural region to the next. Unfortunately, GCMs (General Circulation Models) cannot yet provide reliable predictions of climatic change at regional scales . . .

Future energy scenarios are notoriously variable, and hence there are huge uncertainties with respect to fossil fuel consumption. Major problems include the prediction of economic growth in different economies and the indeterminate nature of future energy choice (as, for example, between nuclear and conventional power generation) (Bach, 1988).

In addition, there are major limitations in our ability to predict what will happen to the CO_2 released by the burning of fossil fuels and by deforestation and soil humus destruction. Major uncertainties surround the role of oceans as sinks for CO_2, the role of oceanic biota in regulating global carbon dynamics (see Clark et al., 1982), and the role of different terrestrial biomes in extracting CO_2 from the atmosphere.

A major problem lies in the reliability of GCMs, many of which have been over simple and have many debatable assumptions. Dickinson (1986: 275–6) has urged caution:

The GCMs have tended to use submodels that are considerably oversimplified . . . Oversimplified ocean and sea-ice models, together with questionable descriptions for ice and snow albedos, suggest that ice-albedo radiative feedbacks in GCMs are of doubtful accuracy. Likewise, the models used for land surface water, snow budgets, and evapotranspiration have been too simple to allow quantitative evaluation of the effects of realistic changes in land surface processes on climate.

It is for these sorts of reasons that GCM scenarios for the future of climate have been so variable (Schlesinger and Mitchell, 1985).

The real world is more complex than the models that are designed to represent it, and the ocean–land–cryosphere–biosphere is governed by numerous feedback mechanisms which respond to changes in atmospheric composition and temperature. Macdonald (1982) lists the following for consideration:

1 *Temperature–infrared radiation feedback (negative)*
   As the temperature rises so does the outgoing infrared radiation.

2 *Water–greenhouse feedback (positive)*
   As temperature rises, the amount of water vapour increases since relative humidity is dependent on temperature. An increase in water vapour content increases the infrared blanket of the atmosphere and this causes a further temperature increase.

3 *Ice and snow cover albedo feedback (positive)*
   As temperature rises snow cover and ice will melt, decreasing the earth's surface albedo and thereby increasing the absorption of sunlight.

4 *Cloud cover feedback (positive)*
   As temperature rises, cloud cover becomes greater. This traps infrared radiation and so the temperature rises.

5 *Cloud cover feedback (negative)*
   As temperature rises producing more cloud, the atmospheric albedo increases, and this tends to reflect incoming solar radiation and so to lower the temperature (see Idso, 1982).

6 *Carbon dioxide biofeedback (negative)*
   Increasing the carbon dioxide concentration in the atmosphere will increase the growth of the biosphere, which will fix carbon via photosynthesis and so tend to reduce the atmospheric level.

7 *Iceberg formation feedback (negative and positive)*
   As temperature rises, ice sheets will break up, producing, at least initially, large numbers of icebergs which will increase the earth's albedo until all the ice melts. Eventual disappearance of the ice sheets and icebergs will lower the earth's surface albedo which will increase insolation absorption and increase surface temperature.

8 *Temperature–convection feedback (negative)*
   As temperature is increased at the ground, convection will increase, carrying sensible heat away from the ground and cooling it.

9 *Evaporation–precipitation feedback (negative and positive)*
   As temperature increases the intensity of the hydrological cycle will increase. Increased evaporation will carry off the

additional temperature rise in latent heat, while increased precipitation will be working to return that same energy.

10 *Sea-ice–ocean current feedback (positive)*

The bottom-water producing currents are driven in the Antarctic by the formation of sea-ice in the Weddell Sea. An increase in temperature will inhibit or stop sea-ice formation and remove the pump which drives these currents. As a result the cold bottom waters will not be replenished and an increase in ocean temperature will occur, leading to a loss of CO_2 from the oceans and a further increase in temperature.

11 *Biomass-albedo feedback (positive)*

Stimulation of new growth as a response to increased temperatures will cause a decrease in albedo which will stimulate a temperature rise.

A potential problem in forecasting future climates is that change may take place not solely in a gradual and progressive manner, but by stepwise changes across thresholds. Positive feedbacks may be involved. A notable example of this type of possibility is Flohn's (1982) contention that a global warming of 4°C will cause a sudden warming of central Arctic sea-ice. Because, in his view, the Antarctic ice will be less radically and less quickly affected, a globally asymmetric cryosphere will result, which will thereby cause a further profound change in the global general atmospheric circulation. Similarly, Bell (1982) has suggested that a marked warming in high latitudes could trigger the release of large quantities of methane, currently trapped in permafrost, subsea permafrost, and sea-ice, thereby adding significantly to the greenhouse effect (see also Nisbet, 1990).

A comparable position has been taken by Woodwell (1990: 5), who believes:

The speed of the warming may be determined less by human activities, unless rapid steps are taken to stop it, than by the warming itself, the climatically caused destruction of forests, and the oxidation of organic matter in soils and bogs. As the warming progresses and trees die at the warmer and drier margins of their distributions, organic matter from plants and soils will be released as carbon into the atmosphere, speeding the warming.

A second method of climate prediction is the climate analogue scenario, which is based on the use of a suitably defined ensemble of warm years from the recent instrumental record, which is then compared with either the long-term mean or a similarly defined cold-year ensemble (Webb and Wigley, 1985). This was the approach adopted by Wigley et al. (1980), who sought to predict the pattern of future temperature and precipitation changes by comparing the five warmest years in the period 1925–74 with the five coldest. As they recognized, however, they were dealing with a temperature shift of only 0.6°C, so the patterns obtained would not necessarily be representative of an even warmer climatic future.

Furthermore, as Webb and Wigley (1985: 242) point out, 'The use of past warm periods as potential analogs of a world with high CO_2 concentrations is based on the assumption that, given similar boundary conditions (as represented by the state of the oceans, the land surface and the cryosphere), the general circulation of the lower atmosphere will respond in a similar way, even with differing forcing mechanisms.' Such an assumption may not be valid.

A third approach to climatic prediction is the study of selected past warm periods. Flohn and Dansgard (1984), for example, examined the Early Medieval optimum, the Holocene optimum (at around 6000 years ago), the Last Interglacial (c.120 000 years ago), and the Late Tertiary. This method holds considerable promise, though there are major limitations imposed by data gaps and dating uncertainties for many parts of the world. Furthermore, boundary conditions may have been different in the Pliocene of the Late Tertiary, largely because the effect that the Himalayan uplift has had since that time on the location of the jet streams. Even the status of the Holocene 'altithermal' or 'hypsithermal' can be questioned, and Webb and Wigley (1985: 252) believe that

More data (and derived temperature estimates) are needed before it can be demonstrated that the global mean temperature was higher at 6000 B.P. than it is today. The current data suggest that the global mean temperature at 6000 B.P. was probably within 1°C of today's temperature.

Schneider (1984: 190) has also expressed doubts about using the altithermal as an analogue, stating that:

unless we can be sure that the Altithermal warm period was caused by an increase in CO_2 (or some equivalent annual, global scale external forcing) we cannot argue convincingly that a global warming induced by burning fossil fuels would result in similar regional climatic shifts – assuming, of course, we could know reliably what Altithermal seasonal and regional climatic patterns actually were.

The spatial patterning of change

Most models indicate that the increase in global surface temperature for a doubled CO_2 level will be around 1.5 to 4.5°C (Houghton et al., 1990).

The various GCMs indicate that warming will not occur in a regular and standard manner all over the world. There will be major differences between different regions. Models agree that there will be enhanced warming in higher latitudes in late autumn and winter. For example, the United Kingdom Meteorological Office GCM illustrated in figure 8.1 suggests that in the northern hemisphere winter there will be a great swathe of land across from the north-eastern parts of North America to central Europe, the Arctic Ocean and Siberia, where warming may amount to as much as 6–12°C. Likewise the south polar regions show a comparable pattern in the southern hemisphere winter (June, July, August). In

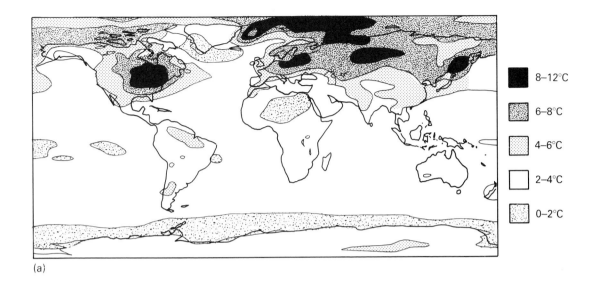

(a)

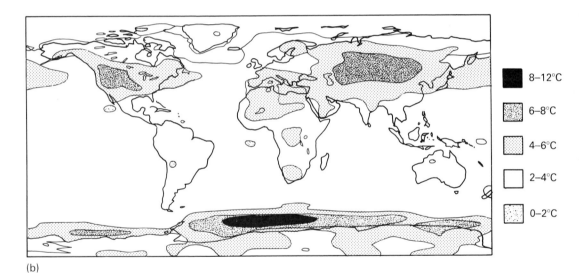

(b)

the same months parts of the western USA and central Asia will also show pronounced enhancements in their temperatures. The various GCMs also all tend to suggest that warming in the tropics will be smaller than the global mean and will vary little with season, being typically 2–3°C.

Simulations of the pattern of precipitation change are rather more complex and show marked differences between different GCMs. However, more recent models 'produce enhanced precipitation in high latitudes and the tropics throughout the year, and in mid-latitudes in winter' (Houghton et al., 1990: 145). Changes in the dry tropics are generally small, with both increases and

Figure 8.1
Changes in surface air temperature predicted for a doubling of carbon dioxide as simulated by the United Kingdom Meteorological Office GCM for (a) December, January and February and (b) June, July and August (modified after Houghton et al., 1990, figure 5.4)

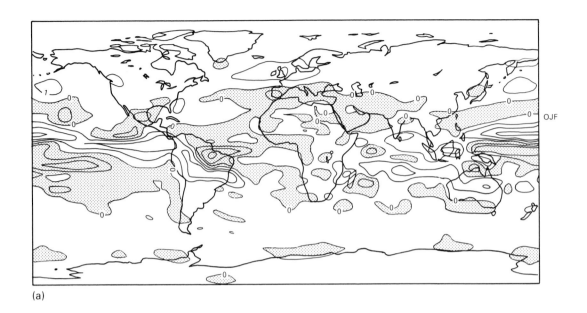

(a)

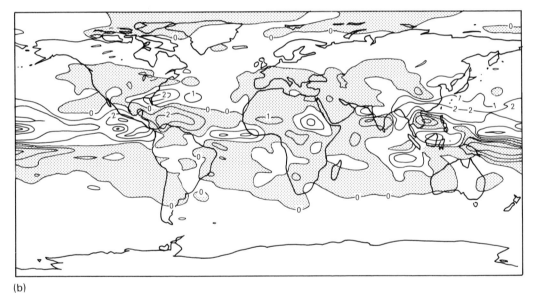

(b)

Figure 8.2
Changes in precipitation
predicted for a doubling of
carbon dioxide as simulated
by the United Kingdom
Meteorological Office GCM for
(a) December, January and
February and (b) June, July
and August. Contours are at ±
0, 1, 2, 5 mm per day. Areas of
decrease are stippled
(modified after Houghton et al.,
1990, figure 5.6)

decreases, though most models simulate an enhancement of precipitation associated with a strengthening of the south-west monsoon
in Asia. The pattern predicted by one GCM, the United Kingdom
Meteorological Office model, is shown in figure 8.2.

Predictions of precipitation change based on warm analogue
years (figure 8.3) and from the reconstruction of Holocene altithermal conditions (figure 8.4) show less complex patterns. However, both methods tend to suggest that global warming will cause
drier conditions in the huge agricultural region of the central United
States of America.

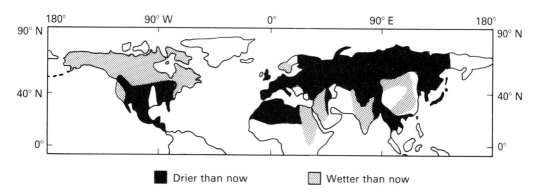

Figure 8.3
Mean annual precipitation
changes from cold to warm
years (after Wigley et al., 1980)

One of the few reasonably comprehensive attempts to predict future precipitation changes by combining the evidence from palaeoclimates, modern analogue data and GCMs is provided by Budyko and Izrael (1991). They argue on the basis of palaeoclimatic data that under conditions of warmth comparable to those that existed in the Pliocene (figure 8.5) precipitation would increase (by up to 30 cm per year) over most of Eurasia and the Sahara. On the other hand, under conditions comparable to those of the Holocene Atlantic Optimum, with temperatures up to 1°C warmer than the present, there is a large zone (between 50 and 30°N) where precipitation levels would decline (by up to 20 cm per year in central North America). There would, however, be improved moisture conditions in the subtropical regions (between 10 and 20°N) and in higher latitudes (more than 60°N). This pattern of changes is attributed to a northward shift both of the ITCZ and of westerly cyclone tracks. In sum, Budyko and Izrael believe that under conditions of marked warming both the mid-latitude and lower-latitude arid zones of the northern hemisphere will be wetter, whereas under conditions of less marked warming (i.e. by around 1°C) areas like the Sahara and Thar will become moister, but areas like the High Plains of the USA or the steppes of the CIS will become drier (figure 8.6).

Further consequences of a global warming

Tropical cyclones (hurricanes) are highly important geomorphological agents, in addition to being notable natural hazards, and are closely related in their places of origin to sea temperature conditions. They only develop where sea-surface temperatures are in excess of 26.5°C. Moreover, their frequency over the last century has changed in response to changes in temperature, and there is even evidence that their frequency was reduced during the Little Ice Age (see Spencer and Douglas, 1985). It is, therefore, likely that as the oceans warm up, so the geographical spread and frequency of hurricanes will increase. Furthermore, it is also likely that the intensity of these storms will be magnified. Emanuel (1987) has

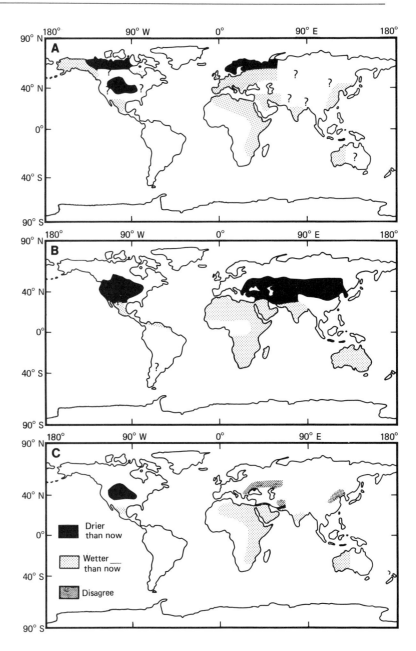

Figure 8.4
Estimates of surface moisture
conditions during the
Altithermal period c.8000 to
5000 BP:
(a) Kellogg, 1977
(b) Butzer, 1980
(c) Kellogg's (1982) attempt to
    compile regions of
    agreement and
    disagreement between (a)
    and (b)
(from Schneider, 1984, figures
3 and 4)

used a GCM which predicts that with a doubling of present
atmospheric concentrations of CO_2 there will be an increase of
40–50 per cent in the destructive potential of hurricanes.

An increase in hurricane intensity and frequency will have
numerous geomorphological consequences in low latitudes, including accentuated river flooding and coastal surges, severe coast
erosion, accelerated land erosion and siltation, and the killing of
corals (because of fresh-water and siltation effects) (De Sylva,
1986).

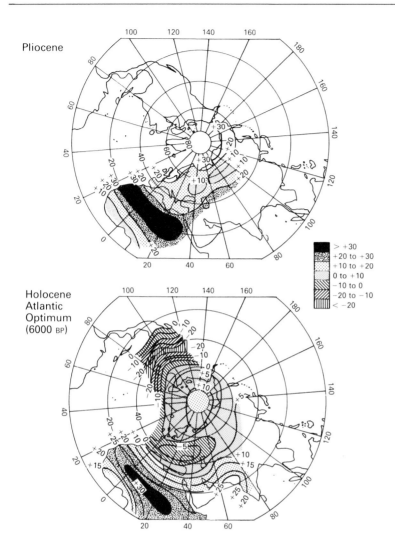

Pliocene

Holocene
Atlantic
Optimum
(6000 BP)

> +30
+20 to +30
+10 to +20
0 to +10
−10 to 0
−20 to −10
< −20

Figure 8.5
Deviations in annual
precipitation means (cm) for
two past warm phases (from
Budyko and Izrael, 1991)

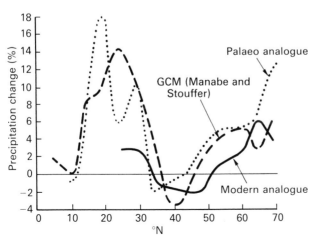

Figure 8.6
Relative changes in mean
latitudinal precipitation on the
continents of the northern
hemisphere with 1°C higher
mean surface air temperature
(after Budyko and Izrael, 1991,
figure 1.5)

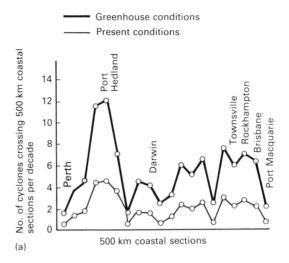

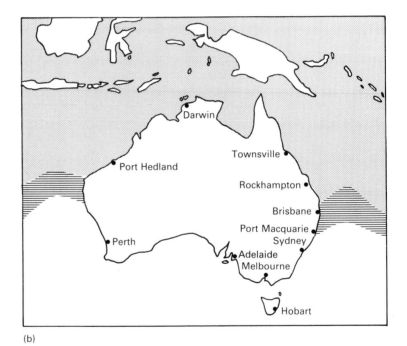

Figure 8.7
(a) The present frequency of tropical cyclones crossing 500 km long sections of the Australian coast, and an estimate of the frequency under conditions with a 2°C rise in temperature
(b) The area where February sea surface temperatures around Australia are currently greater than 27°C (stippled) and the additional area with such temperatures with a 2°C rise in temperature (hatched)
(modified after Henderson–Sellers and Blong, 1989, figures 4.12, 4.14)

Figure 8.7 indicates one scenario of the likely latitudinal change in the extent of warm, cyclone-generating sea water in the Australian region, using as a working threshold for cyclone genesis a summer (February) sea surface temperature of 27°C. Although cyclones do occur to the south of this line, they are considerably more frequent to the north of it. Under greenhouse conditions it is probable that on the margins of the Great Sandy Desert near Port Hedland the number of cyclones crossing the coast will approximately triple from around four per decade to twelve per decade (Henderson-Sellers and Blong, 1989).

However, it needs to be pointed out that the Intergovernmental Panel on Climate Change (Houghton et al., 1990, p. xxv) was somewhat equivocal on the question of the extent to which warming would stimulate cyclone activity, stating:

Although the area of sea having temperatures over this critical value [26.5°C] will increase as the globe warms, the critical temperature itself may increase in a warmer world. Although the theoretical maximum intensity is expected to increase with temperature, climate models give no consistent indication whether tropical storms will increase or decrease in frequency or intensity as climate changes; neither is there any evidence that this has occurred over the past few decades.

Of considerable importance in terms of both climatic feedbacks and the nature of future sea-level changes is the response of the Arctic and Antarctic ice masses to a global warming.

With respect to the Arctic ice, the situation is controversial and unclear. On the one hand there is the view of Budyko (1982: 233) who believes that 'If the mean global temperature rises to 2–3° higher than in preindustrial time, it will be enough for the many-year-ice in the Arctic to disappear.' By contrast, Flohn (1982) believes a temperature increase of 4–5°C would cause an ice-free Arctic Ocean, creating conditions broadly analogous to those of the Pliocene. These views are disputed by Parkinson and Kellogg (1979), who contend that while a 5°C atmosphere temperature rise would cause Arctic pack-ice to disappear in August and September, it would reappear during the winter months. Bentley (1984) argues that in all probability no less than a 10°C temperature increase would be required to remove Arctic sea-ice in summer.

Equal uncertainties surround the future of the West Antarctic ice-sheet. One extreme view has been postulated by Mercer (1978), who believes that predicted increases in temperature at 80°S would start a 'catastrophic' deglaciation of the area, leading to a sudden 5 m rise in sea-level. A less extreme view has been put forward by Thomas et al. (1979). They recognize that higher temperatures will weaken the ice-sheets by thinning them, enhancing lines of weakness, and promoting calving, but they contend that deglaciation would be rapid rather than catastrophic, the whole process taking 400 years or so. Robin (1986) takes a broadly intermediate position. He contends (p. 355) that:

A catastrophic collapse of the West Antarctic ice sheet is not imminent, but better oceanographic knowledge is required before we can assess whether a global temperature rise of 3.5°C might start such a collapse by the end of the next century. Even then it is likely to take at least 200 years to raise sea level by another 5 metres.

Predictions of the rate of ice sheet response to global warming are highly problematic because they involve the use of models with many assumptions or unknowns. Many factors are involved, including the rate at which accumulation occurs. This is crucial in the context of Antarctica, where warmer temperatures could cause more snow deposition in an environment which is currently extremely arid. Such increased accumulation might offset increased rates of ablation. Budd (1991) has calculated that increases of around 4°C in temperatures in the Antarctic would lead to a 30 per cent increase in net accumulation. Also important, however, are varying snow surface albedos, variability in rate of calving of outlet glaciers, the protective and buttressing role of ice shelves, and the geometry and slope of the ice-sheet (Huybrechts et al., 1990). The rapidity with which ice sheets and glaciers decay (and hence sea-level may rise) depends greatly on the terminal environment and certain topographic thresholds (Sugden, 1991).

Alpine glaciers are highly responsive to climatic change, as is made evident by their frequent and rapid fluctuations during the course of the Holocene (see Grove, 1988). Substantial neoglacial advances and retreats have been caused by quite limited fluctuations in temperature and/or precipitation. The historical picture of glacial response to such fluctuations is, however, complex, and this behoves us to exercise caution in predicting the future picture of response to just one control of glacial state, i.e. warming. Changes in precipitation (which control rate of accumulation) and cloudiness (which can control rate of ablation) will also be significant factors in determining glacier equilibrium, while topographic controls, the effects of size and varying propensity to surging will create local complexities of response.

Bearing such caveats in mind it is none the less highly likely that most alpine glaciers will show increasing rates of retreat in a warmer world, and given the rates of retreat experienced in many areas in response to the warming episode of the first decades of the twentieth century, it is probable that many glaciers will disappear altogether. For example, in New Zealand Chinn (1988) calculates that if temperatures rise by 3.6–6.3°C, snow lines will migrate vertically by 300–500 m, and around 1000 out of the country's 3000 glaciers will disappear.

As we have seen, under most models of climatic change, the degree of temperature increase will be greatest in high latitudes. With a doubling of atmospheric CO_2 concentration a high latitude temperature rise of perhaps even as much as 10°C might be expected (Bell, 1982). Given that the existence and distribution of permafrost is so closely related to temperature, it is probable that major changes will occur. Barry (1985) estimates that an average northward displacement of the southern permafrost boundary by 150 ± 50 km would be expected for each 1°C warming, so that a total *maximum* displacement of between 1000 and 2000 km is possible.

However, major uncertainty surrounds the rate at which permafrost degradation will occur. As it is probably a slow process, permafrost will continue to exist for an extended time in areas of *continous* permafrost. In areas of discontinuous or sporadic permafrost the rate will vary greatly depending on local material conductivity, snow cover and vegetation. Changes in vegetation type and in snow cover in a warmer world may modify the direct consequences of warmer surface temperatures (Boer et al., 1990).

Where the permafrost is ice-rich, or contains massive ground ice segregations, subsidence and settling due to thawing will occur, inducing a thermokarst relief which can alter drainage patterns and stream courses. Coastal retreat will also gather momentum as permafrost degrades in coastal lowlands, and large areas may be inundated as depressions are lowered to below current sea-levels because of thaw settlement. River banks and lake and reservoir shorelines might also become amenable to faster rates of erosion, while slopes could become less stable and the active zone become thicker.

Given that vegetation cover is a major control of the operation of geomorphological processes (Viles, 1988) it is inevitable that major changes in the nature and distribution of vegetation types will have a whole range of geomorphological consequences.

Some models suggest that wholesale changes in biomes will occur. Using the GCM developed by Manabe and Stouffer (1980) for a doubling of CO_2 levels, and mapping the present distribution of ecosystem types in relation to contemporary temperature conditions, Emmanuel et al. (1985) found *inter alia* that the following changes would take place: boreal forests would contract from their present position of comprising 23 per cent of total world forest cover to less than 15 per cent; grasslands would increase from 17.7 per cent of all world vegetation types to 28.9 per cent; deserts would increase from 20.6 to 23.8 per cent; and forests would decline from 58.4 to 47.4 per cent. More recently, Smith et al. (1992) have attempted to model the response of Holdridge's Life Zones to global warming and associated precipitation changes as predicted by a range of different GCMs. These are summarized for some major biomes in table 8.1. All the GCMs predict conditions that would lead to a very marked contraction in the tundra and desert biomes, and an increase in the areas of grassland and dry forests. There is, however, some disagreement as to what will happen to mesic forests as a whole, though within this broad class the humid tropical rainfall element will show an expansion.

Table 8.1 Changes in areal coverage (in $km^2 \times 10^3$) of major biomes as a result of the climatic conditions predicted for a $\times 2$ CO_2 world by various General Circulation Models (GCMs)

Biome	Current	OSU	GFDL	GISS	UKMO
Tundra	939	−302	−5.5	−314	−573
Desert	3699	−619	−630	−962	−980
Grassland	1923	380	969	694	810
Dry forest	1816	4	608	487	1296
Mesic forest	5172	561	−402	120	−519

OSU = Oregon State University
GFDL = Geophysical Fluid Dynamic Laboratory
GISS = Goddard Institute for Space Studies
UKMO = UK Meteorological Office

Source: from Smith et al., 1992, table 3

Theoretically a rise of 1°C in mean annual temperature could cause a poleward shift of vegetation zones of about 200 km (Ozenda and Borel, 1990). However, uncertainties surround the question of how fast plant species would be able to move to and to settle new habitats suitable to the changed climatic conditions. Postglacial vegetation migration rates appear to have been in the range of a few tens of kilometres per *century*. For a warming of 2–3°C forest bioclimates could shift northwards about 4–6° of latitude in a century, indicating the need for a migration rate of some tens of kilometres per *decade*. Furthermore, migration could be hampered because of natural barriers, zones of cultivation, etc.

Altitudinal changes in vegetation zones will be of considerable significance. In general, Peters (1988) believes that with a 3°C temperature change vegetation belts will move about 500 m in altitude. One consequence of this would be the probable elimination of the Douglas Fir (*Pseudotsuga taxifolia*) from the lowlands of California and Oregon, because rising temperatures would preclude the seasonal chilling this species requires for seed germination and shoot growth.

Vegetation will also probably be changed by variation in the role of certain extreme events that cause habitat disturbance, including fire (exacerbated by drought), wind storms, hurricanes and coastal flooding (Overpeck et al., 1990).

Other possible non-climatic effects of elevated CO_2 levels on vegetation include changes in photosynthesis, stomatal closure and carbon fertilization (Idso, 1983), though these are still matters of controversy (Solomon and West, 1986). However, especially at high altitudes, it is possible that elevated CO_2 levels would have potentially significant effects on tree growth, causing growth enhancement (La Marche et al., 1984).

Changes in temperatures, precipitation quantities, and the timing and form of precipitation would have highly important hydrological consequences. In areas affected by snowfall today, the changes may be especially marked. In a warmer world, there would be a tendency for a marked decrease to occur in the proportion of winter precipitation that falls as snow. Furthermore, there would be an earlier and shorter spring snowmelt. The first of these two effects would cause greater winter rainfall and hence winter runoff, since less overall precipitation would enter snowpacks to be held over until spring snowmelt. The second effect would intensify spring runoff, leading to additional adverse consequences for both summer runoff levels and for spring and summer soil moisture levels (Gleick, 1986). Table 8.2, taken from Gleick's work, indicates likely changes in runoff for two temperature and five precipitation scenarios in such regions, and demonstrates clearly the anticipated direction and seasonality of runoff changes.

On the other hand, in high-latitude tundra environments warmer winters may cause more snow to fall, thereby creating increased runoff levels in the summer months (Barry, 1985). Indeed, Budyko (1982: 242) predicts that because of increased precipitation

Table 8.2 The effects of hypothetical temperature and precipitation scenarios on runoff (% change) (after Gleick, 1986)

Temperature change	Precipitation change				
	−20%	−10%	0	+10%	+20%
(a) Winter (December, January, February)					
+2°C	−24	−9	+8	+25	+44
+4°C	−4	+14	+34	+54	+75
(b) Summer (June, July, August)					
+2°C	−42	−32	−22	−12	−1
+4°C	−73	−68	−62	−55	−49

(perhaps by as much as 500–600 mm y^{-1} in the tundra zone), runoff in Russia north of 58–60°N will increase by a factor of 2 to 3.

No less significant runoff changes may be anticipated for semi-arid environments, such as the south-west USA. For example, Revelle and Waggoner (1983) suggest that in the event of there being a 2°C rise in temperature and a 10 per cent reduction in precipitation, water supplies would be diminished by 76 per cent in the Rio Grande region and by 40 per cent in the Upper Colorado. Table 8.3 demonstrates that a 2°C rise in temperature would be most serious for water supplies and runoff in those regions where the mean annual precipitation is less than about 400 mm y^{-1}.

Projected summer dryness in areas such as the North American Great Plains (Manabe et al., 1981) may be accentuated by a positive feedback process involving changes in cloud cover. Manabe and Wetherald (1986: 627) suggest that:

When soil moisture is reduced, a large fraction of radiative energy absorbed by the continental surface is ventilated through the upward flux

Table 8.3 Approximate percentage decrease in runoff for a 2°C increase in temperature

Initial temperature (°C)	Precipitation (mm y^{-1})					
	200	300	400	500	600	700
−2	26	20	19	17	17	14
0	30	23	23	19	17	16
2	39	30	24	19	17	16
4	47	35	25	20	17	16
6	100	35	30	21	17	16
8		53	31	22	20	16
10		100	34	22	22	16
12			47	32	22	19
14			100	38	23	19

Source: data in Revelle and Waggoner, 1983

of sensible heat rather than through evaporation. Accordingly, the temperature of the continental surface and the overlying layer increases, resulting in the general reduction in relative humidity and precipitation in the lower troposphere of the model. Accompanying the reduction in relative humidity is a reduction in total cloud amount, causing an increase in solar energy reaching the continental surface. Thus the radiation energy absorbed by the continental surface also increases, raising the rate of potential evaporation.

Substantial hydrological changes may also occur in the catchments of the Great Lakes of North America because of changes in precipitation, snowmelt timing and evapotranspirational loss. Models produced by Croley (1990) and Hartmann (1990) suggest that for a doubling of CO_2 levels there may be a 23–51 per cent reduction in net basin supplies to all the Great Lakes, and that as a result lake levels will fall at rates ranging from 13 mm per decade (for Lake Superior) to 93 mm per decade (for Lake Ontario).

Projected sea-level rise

Among the most important geomorphological consequences of a global warming would be a worldwide rise in sea-level. This would occur as a result of two distinct processes: the thermal expansion of the upper layers of the oceans; and the melting of land ice (US Department of Energy, 1985; Titus, 1986).

The anticipated rise of sea-level over the next century is the subject of contention, largely because of uncertainties about the behaviour of Antarctic ice. In the 1980s there were expectations that sea-level could rise by over 3.5 m by 2100. Now, however, there is a tendency to view such values as excessive and the Intergovernmental Panel on Climate Change (Houghton et al., 1990: 279) concluded 'that a rise of more than 1 metre over the next century is unlikely'. None the less, this is a rate 3 to 6 times faster than that experienced over the last 100 years (1 to 2 mm yr). The Panel also recognized that there were still large uncertainties associated with the future contributions of Antarctica and Greenland to sea-level rise.

The degree of sea-level rise could be moderated by reservoir construction. Newman and Fairbridge (1986) have calculated that between 1957 and 1982 human intervention stored as much as 0.75 mm per year of sea-level rise potential in reservoirs and irrigation projects.

The degree of tidal inundation may show local variations. Goemans (1986), for example, has pointed out that the tidal system of the North Sea is rather complex, with several amphidromic points (points of zero tidal range). These points might move as a result of sea-level change and could therefore lead to a change in local tidal amplitudes and of the relative high-water level.

The effects of global sea-level rise will be compounded in those areas that suffer from local subsidence as a result of local tectonic movements, isostatic adjustments and fluid abstraction.

The results of sea-level rise in sensitive areas (e.g. low-lying coasts, deltas, marshes) will be marked. Broadus et al. (1986), for example, calculate that were the sea-level to rise by just 1 m in 100 years, 12–15 per cent of Egypt's arable land would be lost and 16 per cent of the population would have to be relocated. With a 3 m rise the figures would be a 20 per cent loss of arable land and a need to relocate 21 per cent of the population. In Bangladesh, a 1 m rise would inundate 11.5 per cent of the total land area of the state and affect 9 per cent of the population directly, while a 3 m rise in sea-level would inundate 29 per cent of the land area and affect 21 per cent of the population (figure 8.8). Many of the world's major conurbations might be flooded in whole or in part, and sewers and drains rendered inoperative (Kuo, 1986).

Salt marshes, including the mangrove swamps of the tropics, are potentially highly vulnerable in the face of sea-level rise, particularly in those circumstances where sea defences and other barriers prevent the landward migration of marshes as sea-level rises. However, salt marshes are highly dynamic features and in some situations may well be able to cope, even with quite rapid rises of sea-level.

Reed (1990) suggests that salt marshes in riverine settings may receive sufficient inputs of sediment that they are able to accrete

Plate 8.1
Bangladesh already suffers from severe flooding. This is a problem that may increase in coming decades should sea level rise or if tropical storms become more frequent and intense (as is predicted by certain General Circulation Models)

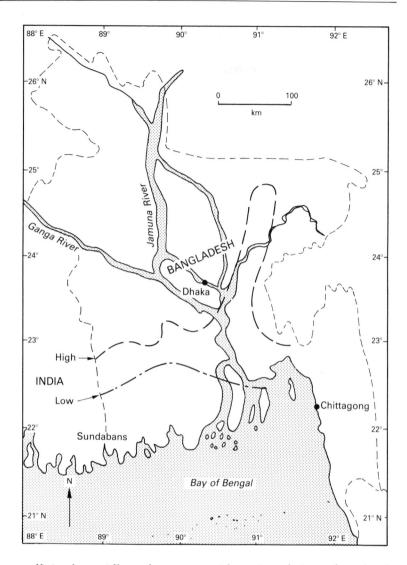

Figure 8.8
Projected areas of flooding as a result of sea-level change in Bangladesh, for two scenarios (low = 1 m and high = 3 m) (modified after Broadus et al., 1986, figure 7)

sufficiently rapidly to keep pace with projected rises of sea level. Areas of high tidal range, such as the marshes of the Severn Estuary in England/Wales, are also areas of high sediment transport potential and may thus be less vulnerable to sea-level rise. Likewise, some vegetation associations, e.g. *Spartina* swards, may be relatively more effective than others at encouraging accretion, and organic matter accumulation may itself be significant in promoting vertical build-up of some marsh surfaces. For marshes that are dependent upon inorganic sediment accretion, increased storm activity and beach erosion which might be associated with the greenhouse effect could conceivably mobilize sufficient sediments in coastal areas to increase their sediment supply.

Marsh areas that may be highly prone to sea-level rise include areas of deltaic sedimentation where, because of sediment move-

ment controls (e.g. reservoir construction) or because of cyclic changes in the locations of centres of deposition, rates of sediment supply are low. Such areas may also be areas with high rates of subsidence. A classic example of this is portions of the Mississippi delta. Park et al. (1986) undertook a survey of coastal wetlands in the USA and suggested that sea-level change could, by 2100, lead to a loss of between 22 and 56 per cent of the 1975 wetland area, according to the degree of sea-level rise that takes place.

One particular type of marsh that may be affected by anthropo-genically accelerated sea-level rise is the mangrove swamp. As with other types of marsh the exact response will depend on local setting, sources and rates of sediment supply, and the rate of sea-level rise itself. However, mangroves may respond rather differently to other marshes in that they are composed of relatively long-lived trees and shrubs which means that the speed of zonation change will be less (Woodroffe, 1990).

Another potentially susceptible environment is the coral reef (Stoddart, 1990). In the 1980s there were widespread fears that if rates of sea-level rise were high (perhaps 2 to 3 m or more by 2100) they would submerge whole atoll systems (e.g. the Maldives). However, with the reduced expectations for the degree of sea-level rise that may occur, there has arisen a belief that coral reefs may

Plate 8.2
Sonneratia and *Rhizophora* mangrove swamp in North Sulawesi, Indonesia. Such swamps, valuable wetland resources, are already under great human pressures in many areas, and may also be susceptible to the effects of any future sea-level rise

survive and even prosper with moderate rates of sea-level rise. Kinsey and Hopley (1991), for instance, believe that reef growth will be stimulated by the rising sea-levels of a warmer world and that reef production could double in the next 100 years from around 900 to 1800 million tonnes per year. On the other hand, Buddemeier and Smith (1988) took a rather pessimistic view of the response of reefs to a rate of sea-level rise of 15 mm per year over the next century, believing that sustained maxima of reef growth were generally less than 10 mm per year.

However, reef accretion is not the sole response of reefs to sea-level rises, for reef tops are frequently surmounted by small islands (cays and motus) composed of clastic debris. Such islands might be very susceptible to sea-level rise. On the other hand, were warmer seas to produce more storms, then the deposition of large amounts of very coarse debris could in some circumstances lead to their enhanced development.

Increased sea-surface temperatures could have deleterious consequences in corals which are near their thermal maximum, and increased temperatures in recent years have been identified as a cause of widespread coral bleaching (loss of symbiotic zooxanthellae). Those corals stressed by temperature or pollution might well find it more difficult to cope with rapidly rising sea-levels than would healthy corals. Moreover, it is possible that increased ultraviolet radiation due to ozone layer depletion could aggravate bleaching and mortality caused by global warming. Various studies suggest that coral bleaching was a widespread feature in the warm years of the 1980s (Brown, 1990).

Bird (1986: 84) has summarized the general effects of accelerated coastal submergence as follows:

- On cliffed coasts submergence is likely to accelerate coastline recession, except on outcrops of hard rock formations, where the high and low tide lines will simply move up the cliff face. Existing shore platforms and abrasion ramps will disappear beneath the sea.
- The shores of deltas and coastal plains will retreat, except where they are maintained by coastal sedimentation.
- Beaches will be narrowed, and beach erosion will become much more extensive and severe than it is now.
- Inlets, embayments, and estuaries will be enlarged and deepened, and increasing salinity penetration will cause a regression of coastal ecosystems: where possible, mangrove and salt marsh communities will move back into terrain presently occupied by freshwater vegetation.
- Coastal lagoons will also become larger and deeper, but the enclosing barriers may transgress landward on to them. If the barriers are submerged, or destroyed by erosion, the lagoons will become coastal inlets or embayments.
- Low-lying areas on coastal plains, such as sebkhas (saline depressions now subject to occasional marine flooding) on arid coasts, will be flooded to form permanent lagoons.

- Upward growth of coral and associated organisms will be stimulated on fringing biogenic reefs, keeping pace with the marine transgression or lagging somewhat behind it.
- Erosion, structural damage, and marine flooding caused by storm surges or tsunamis will intensify because of the greater heights of the waves arriving through deepening coastal waters.

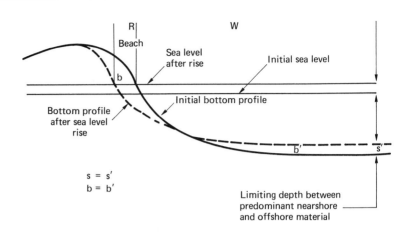

Figure 8.9
The so-called Bruun Rule, whereby a rise in sea-level causes beach erosion. If the sea-level(s) rises by a unit amount so will the offshore bottom (s'). The sand necessary to raise the bottom (area b') can be supplied by waves eroding the upper part of the beach (area b). R is the amount of shoreline recession and W is the 'active' portion of the profile participating in the adjustment (from Committee on Engineering Implications of Changes in Relative Mean Sea Level, 1987, p. 54, figure 5-3).

• Water tables will rise in coastal regions, and soil and water salinity will be augmented.

The relationship between sea-level rise and beach erosion is a notably important concern. In general, following the so-called Bruun Rule shown in figure 8.9, such erosion is to be expected, but the precise amounts are difficult to quantify. However, on sandy beaches exposed to ocean waves, the coastline may erode by the order of 1 m for every 1 cm rise in sea-level (Committee on Engineering Implications of Changes in Relative Mean Sea Level, 1987).

Environmental uncertainty

The debate about the relative power of human and natural influences is but one of many components of environmental uncertainty. In almost all attempts to predict the future major uncertainties arise. One can provide various explanations for this. First, environmental change is the result of very complex inter-actions between several closely coupled non-linear systems. The complexity creates problems for both modelling and comprehension, while the non-linearity means that the dimensions of a response are not by any means necessarily directly proportional to the size of the stimulus that promotes change. Secondly, prediction of environmental change depends on models and many models are imperfect because of the gross assumptions they employ (see the discussion of General Circulation Models for an exemplification of this problem). Thirdly, nature always has surprises in store (e.g. various types of extreme event or catastrophe) and these will not cease in the future. They may work to counteract or increase the consequences of various human actions. Fourthly, there are bound to be factors that we have ignored, which turn out to be highly important. For example, who several decades ago could have predicted the role of CFCs in creating the ozone hole? Fifthly, identification of future trends requires some knowledge of background or natural levels. Often we lack the necessary long-term data to enable us to ascertain whether observed trends have happened before and whether they are or are not cyclical. Sixthly, some key issues are less easy to predict than others and therefore preclude accurate prediction of phenomena that depend on them. For example, without a clear idea of precipitation patterns in a warmer world it is well nigh impossible to predict the response of rivers and biota. Seventh, we may often be able to predict that change will probably occur, but we find it much less easy to predict the speed of response. Ice caps and glaciers may well melt in a warmer world, but how fast? Finally, there are problems of definition. Without clear definitions of phenomena, measurement is difficult, 'results' can be meaningless and trends or changes difficult to identify. We find it difficult to define such phenomena as desertification and deforestation.

NINE

Conclusion

The power of non-industrial and pre-industrial civilization

Thus far we have looked at the impact that human societies have had on the different components of the physical environment: vegetation, animals, soil, water, landforms and climate. It has become apparent that Marsh was correct over a century ago to express his cogently argued views of the importance of human agency in environmental change. Since his time the impact that humans have had on the environment has increased, as has our awareness of this impact. However, it is worth making the point here that, although much of the concern expressed about the undesirable effects humans have tends to focus on the role played by sophisticated industrial societies, this should not blind us to the fact that many highly significant environmental changes were and are being achieved by non-industrial societies. In recent years it has become apparent that fire, in particular, enabled early societies to alter vegetation substantially, so that plant assemblages that were once thought to be natural climatic climaxes may in reality be in part anthropogenic fire climaxes. This applies to many areas of both savanna and mid-latitude grassland (see p. 38). Such alteration of natural vegetation has been shown to pre-date the arrival of European settlers in the Americas, New Zealand and elsewhere and the effects of fire may have been compounded by the use of the stone axe and by the grazing effects of domestic animals. In turn the removal and modification of vegetation would have led to adjustments in fauna. It is also apparent that soil erosion resulting from vegetation removal has a long history and that it was regarded as a threat by the classical authors. Recent studies (see p. 35) tend to suggest that some of the major environmental changes in highland Britain and similar parts of Western Europe that were once explained by climatic changes can be more effectively explained by the activities of Mesolithic and Neolithic peoples. This applies, for example, to the decline in the numbers of certain plants in the pollen record and to the development of peat bogs and podzolization (see p. 151). Even soil salinization started an an early date because of the adaptation of irrigation practices in arid areas, and its effects on crop yields were noted in Iraq more than 4000 years ago (see p. 145). Similarly (see p. 125) there is an increasing body of evidence that the hunting practices of early civilizations may have caused great changes in the world's mega-fauna as early as 11 000 years ago.

In spite of the increasing pace of world industrialization and urbanization, it is ploughing and pastoralism which are responsible for many of our most serious environmental problems and which are still causing some of our most widespread changes in the landscape. Thus soil erosion brought about by agriculture is, it can be argued, a more serious pollutant of the world's waters than is industry: many of the habitat changes which so affect wild animals are brought about through agricultural expansion (see p. 106); and soil salinization and desertification can be regarded as two of the most serious problems facing the human race. Land-use changes, such as the conversion of forests to fields, may be at least as effective in causing anthropogenic changes in climate as the more celebrated burning of fossil fuels and emission of industrial aerosols into the atmosphere. The liberation of CO_2 in the atmosphere through agricultural expansion, changes in surface albedo values and the production of dust, are all major ways in which agriculture may modify world climates.

The proliferation of impacts

A further point we can make is that, with developments in technology, the number of ways in which humans are affecting the environment is proliferating. It is these recent changes, because of the uncertainty which surrounds them and the limited amount of experience we have of their potential effects, which have caused greatest concern. Thus it is only since the Second World War, for example, that humans have had nuclear reactors for electricity generation, that they have used powerful pesticides such as DDT, and that they have sent supersonic aircraft into the stratosphere. Likewise, it is only since around the turn of the century that the world's oil resources have been extensively exploited, that chemical fertilizers have become widely used, and that the internal combustion engine has revolutionized the scale and speed of transport and communications.

Above all, however, the complexity, frequency and magnitude of impacts is increasing, partly because of steeply rising population levels and partly because of a general increase in *per capita* consumption. Thus some traditional methods of land use such as shifting agriculture (see p. 48) and nomadism, which have been thought to sustain some sort of environmental equilibrium, seem to break down and to cause environmental deterioration when population pressures exceed a particular threshold.

Figure 9.1
Land rotation and population
density. The relationship of soil
fertility to cycles of
slash-and-burn agriculture:
(a) fertility levels are
    maintained under the long
    cycles characteristic of
    low-density populations.
(b) fertility levels are declining
    under the shorter cycles
    characteristic of increasing
    population density. Notice
    that in both diagrams the
    curves of both depletion
    and recovery have the
    same slope (after Haggett,
    1979, figure 8.4)

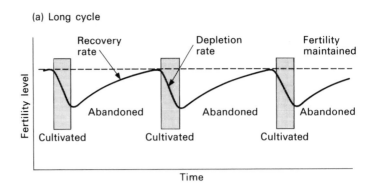

(a) Long cycle

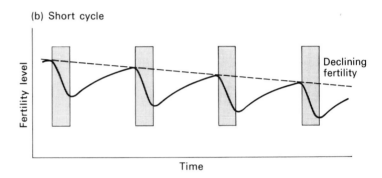

(b) Short cycle

This is illustrated for shifting agricultural systems by figure 9.1,
which shows the relationship of soil fertility levels to cycles of
slash-and-burn agriculture. Fertility can be maintained (figure 9.1a)
under the long cycles characteristic of low-density populations.
However, as population levels increase, the cycles necessarily
become shorter, and soil fertility levels are not maintained, thereby
imposing greater stresses on the land (figure 9.1b).

At the other end of the spectrum, increasing incomes, leisure and
ease of communication, have generated a stronger demand for
recreation and tourism in the developed nations. These have
created additional environmental problems (see p. 106), especially
in coastal and mountain areas (see, for example, Bayfield, 1973).
Some of the environmental consequences of recreation can be listed
as follows:

1 Desecration of cave formations by speleologists
2 Trampling by human feet leading to soil compaction
3 Nutrient additions at camp sites by people and their pets
4 Decrease in soil temperatures because of snow compaction
  by snowmobiles
5 Footpath erosion and off-road vehicle erosion
6 Dune reactivation by trampling
7 Vegetation change due to trampling and collecting
8 Creation of new habitats by cutting trails and clearing
  camp sites

9 Pollution of lakes and inland waterways by gasoline discharge from outboard motors and by human waste
10 Creation of game reserves and protection of ancient domestic breeds
11 Disturbance of wildlife by proximity of persons and by hunting, fishing and shooting
12 Conservation of woodland for pheasant shooting.

Plate 9.1
The impact of recreation pressures is well displayed at a prehistoric hill-fort, Badbury Rings, Dorset, England. Pedestrians and motorcyclists have caused severe erosion of the ramparts

Likewise, it is apparent when considering the range of possible impacts of one major type of industrial development that they are significant. As table 9.1 indicates, the exploitation of an oilfield and all the activities that it involves (for example, pipelines, new roads, refineries, drilling, etc.) have a wide range of likely effects on land, air, water and organisms.

A very substantial amount of change has been achieved in recent decades. Table 9.2, based on the work of Kates et al. (1991), attempts to make quantitative comparisons of the human impact on ten 'component indicators of the biosphere'. For each component they defined total net change clearly induced by humans to be 0 per cent for 10 000 years ago and 100 per cent for 1985. They then estimated dates by which each component had reached successive quartiles (i.e. 25, 50 and 75 per cent) of its 1985 total change. They believe that about half of the components have

Table 9.1 Qualitative environmental impacts of mineral industries with particular reference to an oilfield

| Facility | Direction of the impact and reaction of the environmental | | | |
	Land	Air	Water	Biocenosis
Well	Alienation of land surface Extraction of oil associated gas, ground water Pollution by crude oil, refined products, drilling mud Disturbance of internal structure of soil and subsoil, Destruction of soil	Pollution by associated gas and volatile hydrocarbons, products of combustion	Withdrawal of surface water and ground water Pollution by crude oil and refined products; salination of fresh water Disturbance of water balance of both subsurface and surface waters	Pollution by crude oil and refined products Disturbance and destruction over a limited surface area
Pipeline	Alienation of land Accidental oil spills Disturbance of landforms and internal structure of soil and subsoil	Pollution by volatile hydrocarbons	Disturbance and destruction over a limited surface area	
Motor roads	Alienation of land Pollution by oil products Disturbance of landforms and internal structure of soil and subsoil	Pollution by combustion products, volatile hydrocarbons, sulphur dioxide, nitrogen oxides	Pollution by combustion products Disturbances and destruction over limited surface area	
Collection point	Alienation of land Pollution by crude oil and refined products (spills) Disturbance of internal structure of soil and subsoil	Pollution by volatile hydrocarbons	Disturbance and destruction over limited surface area	

Source: Denisova, 1977, table 2, p. 650

Table 9.2 Chronologies of human-induced transformations

(a) Quartiles of change from 10,000 BC to mid 1980s

Form of transformation	Dates of quartiles		
	25%	50%	75%
Deforested area	1700	1850	1915
Terrestrial vertebrate diversity	1790	1880	1910
Water withdrawals	1925	1955	1975
Population size	1850	1950	1970
Carbon releases	1815	1920	1960
Sulphur releases	1940	1960	1970
Phosphorus releases	1955	1975	1980
Nitrogen releases	1970	1975	1980
Lead releases	1920	1950	1965
Carbon tetrachloride production	1950	1960	1970

(b) % change by the time of Marsh and Princeton symposium

Form of transformation	% change	
	1860	1950
Deforested area	50	90
Terrestrial vertebrate diversity	25–50	75–100
Water withdrawals	15	40
Population size	30	50
Carbon releases	30	65
Sulphur releases	5	40
Phosphorus releases	<1	20
Nitrogen releases	<1	5
Lead releases	5	50
Carbon tetrachloride production	0	25

Source: from Kates et al., 1991, table 1.3

changed more in the single generation since 1950 than in the whole of human history before that date.

Changes are reversible

It is evident that while humans have imposed many undesirable and often unexpected changes on the environment, they often have the capacity to modify the rate of such changes or to reverse them. There are cases where this is not possible: once soil has been eroded from an area it cannot be restored; once a plant or animal has become extinct it cannot be brought back; and once a laterite iron pan has become established it is difficult to destroy.

However, through the work of George Perkins Marsh and others, people became aware that many of the changes that had been set in train needed to be reversed or reduced in degree. Sometimes this has simply involved discontinuing a practice which has proved undesirable (such as the cavalier use of DDT), or replacing it with another which has been less detrimental in its effect. Often, however, specific measures have been taken which have involved deliberate decisions of management and conservation. Denson (1970) for example, outlines a sophisticated six-stage model for wildlife conservation:

1 Immediate physical protection from humans and from changes in the environment
2 Educational efforts to awaken the public to the need for protection and to gain acceptance of protective measures
3 Life-history studies of the species to determine their habitat requirements and the causes of their population decline
4 Dispersion of the stock to prevent loss of the species by disease or by a chance event such as a fire
5 Captive breeding of the species to assure higher survival of young, to aid research and to reduce the chances of catastrophic loss
6 Habitat restoration or rehabilitation when this is necessary before reintroducing the species.

Many conservation measures have been successful, while others have created as many problems as they were intended to solve. This applies, for example, to certain schemes for the reduction of coast erosion.

The concern with preservation and conservation has been long-standing, with many important landmarks. Interest has grown dramatically in recent years. Lowe (1983) has identified four stages in the history of British Nature Conservation:

1 The natural history/humanitarian period   (1830–90)
2 The preservationist period   (1870–1940)
3 The scientific period   (1910–70)
4 The popular/political period   (1960–present)

The first of these stages was rooted in a strong enthusiasm for natural history, and the crusade against cruelty to animals. Although many Victorian naturalists were avid collectors, numerous clubs were established to study nature and some of them sought to preserve species to make them available for observation. As we shall see, certain acts were introduced at this time to protect birds. During the preservationist period, there was the formation of a spate of societies devoted to preserving open land and its associated wildlife (for example, the National Trust, 1894; and the Council for the Preservation of Rural England, 1926). There was a growing sense of the vulnerability of wildlife and landscapes to urban and industrial expansion and geographers like Vaughan Cornish (see Goudie, 1972b) campaigned for the creation of national parks and the preservation of scenery, made possible through the National Parks and Access to Countryside Act of 1949. From the First World War onwards ecological research developed, and there arose an increasing understanding of ecological relationships. Scientists pressed for the regulation of habitats and species, and the Nature Conservancy was established in 1949. In the 1960s and the years that followed popular interest in conservation and widespread media attention first developed. This was partly generated by pollution incidents (such as the wrecks of the *Torrey Canyon* and *Amoco Cadiz*), and a gathering sense of impending environmental doom. Ecology became a political issue in various European nations, including the UK. In many countries major developments in land use, construction and industrialization now have to be preceded by the production of an Environmental Impact Assessment.

Thus in some countries, and in connection with particular species, conservation and protection have had a long and sometimes successful impact. In Britain, for example, the Wild Birds Protection Act dates back to 1880, and the Sea Birds Protection Act even further to 1869. The various acts have been modified and augmented over the years to outlaw egg-collecting, pole-trapping, plumage importation and the capture or possession of a range of species. The effectiveness of the different acts can be measured in real terms. Over the last 50 years no species have been lost as British breeding birds due to lack of protection – the only major loss has been the Kentish plover which was in any case on the edge of its range. Perhaps more importantly, several species have successfully recolonized Britain, the most celebrated being the avocet and the osprey. Today both are firmly established, together with other species lost in the nineteenth century: the black-tailed godwit, the goshawk and the bittern. Also as a result of protection the red kites of Wales have not only survived but increased in number, and the peregrine falcon maintains its largest numbers in Europe outside Spain.

One further ground which gives some basis for hope that humans may soon be reconciled with the environment is that there are some signs of a widespread shift in public attitudes to nature and the environment. These changing social values, combined with scientific facts, influence political action. This point of view, which acts as an antidote for some of the more pessimistic views of the world's future, has been elegantly presented by Ashby (1978). He contends that the rudiments of a healthy environmental ethic are developing, and explains (pp. 84–5):

Its premise is that respect for nature is more moral than lack of respect for nature. Its logic is to put the Teesdale Sandwort . . . into the same category of value as a piece of Ming porcelain, the Yosemite Valley in the same category as Chartres Cathedral; a Suffolk landscape in the same category as a painting of the landscape by Constable. Its justification for preserving these and similar things is that they are unique, or irreplaceable, or simply part of the fabric of nature, just as Chartres and the painting by Constable are part of the fabric of civilization; also that we do not understand how they have acquired their durability and what all the consequences would be if we destroyed them.

Although there may be considerable controversy surrounding the precise criteria that can be used to select and manage sites that are particularly worthy of conservation (Goldsmith, 1983), there are none the less many motives behind the increasing desire to protect species and landscapes. These can be listed under the following general headings:

1   *The ethical.* It is asserted that wild species have a right to coexist with us on our planet, and that we have no right to exterminate them. Nature, it is maintained, is not simply there for humans to transform and modify as they please for their own utilitarian ends.

2   *The scientific.* We know very little about our surrounding environments, including, for example, the rich insect faunas of the tropical rain forest; therefore such environments should be preserved for future scientific study.

3   *The aesthetic.* Plants and animals, together with landscapes, may be beautiful and so enrich the life of humans.

4   *The need to maintain genetic diversity.* By protecting species we maintain the species diversity upon which future plant- and animal-breeding work will depend. Once genes have been lost (see chapter 2, section on 'The change in genetic and species diversity') they cannot be replaced.

5   *Environmental stability.* It is argued that in general the more diverse an ecosystem is the more checks and balances there are to maintain stability. Thus environments that have been greatly simplified by humans may be inherently unstable, and prone to disease, etc.

6   *Recreational.* Preserved habitats and landscapes have enormous recreational value, and in the case of some game reserves and natural parks may have economic value as well (for example, the safari industry of East Africa).

7   *Economic.* Many of the species in the world are still little known, and there is the possibility that we have great storehouses of plants and animals which, when knowledge improves, may become useful economic resources.

8   *Future generations.* One of the prime arguments for conservation, whether of beautiful countryside, rare species, soil or mineral resources, is that future generations (and possibly ourselves later in life) will require them, and may think badly of a generation that has squandered them.

9   *Unintended impacts.* As we have seen so often in this book, profligate or unwise actions can lead to side-effects and consequences that may be disadvantageous to humans.

Some of these arguments are more utilitarian than others (for example, 4, 5, 6, 7 and 9), and some may be subject to doubt – it could, for example, be argued that future generations will have technology to use new resources and may not need some of those we regard as essential – but overall they provide a broadly based platform for the conservation ideal (Myers, 1979).

The susceptibility to change

Ecosystems respond in different ways to the human impact, and some are more vulnerable to human perturbation than others.

It has often been thought, for example, that complex ecosystems are more stable than simple ones. Thus in Clements' Theory of Succession the tendency toward community stabilization was ascribed in part to an increasing level of integration of community functions. As Goodman (1975: 238) has expressed it:

In general the predisposition to expect greater stability of complex systems was probably a combined legacy of eighteenth century theories of political economics, aesthetically and perhaps religiously motivated attraction to the belief that the wondrous variety of nature must have some purpose in an orderly world, and ageless folkwisdom regarding eggs and baskets.

Indeed, as Murdoch (1975) has pointed out, it makes good intuitive sense that a system with many links, or 'multiple fail-safes', is more stable than one with few links or feedback loops. As an example, if a type of herbivore is attacked by several predatory species, the loss of any one of these species will be less likely to allow the herbivore to erupt or explode in numbers than if only one predator species were present and that single predator type disappeared.

Various other arguments have been marshalled to support the idea that great diversity and complexity affords greater ability to minimize the magnitude, duration and irreversibility of changes brought about by some external perturbation such as humans (Noy-Meir, 1974). It has been stated that natural systems, which are generally more diverse than artificial systems such as crops or laboratory populations, are also more stable. Likewise, the tropical rain forest has been thought of as more diverse and more stable than less complex temperate communities, while simple Arctic ecosystems, exemplified by unstable lemming populations, are in turn held to be less stable than the more diverse temperate communities. In the same way the simple ecosystems of oceanic islands have always appeared highly vulnerable to disturbance brought about by anthropogenic plant and animal introductions (see p. 115).

However, considerable doubt has been expressed recently as to whether the classic concept of the causal linkage between diversity/complexity and stability is entirely valid (see, for example, Hurd et al., 1971). Murdoch (1975) indicates that there is no convincing field evidence that diverse natural communities are generally more stable than simple ones. He cites various papers which show that fluctuations of microtine rodents (lemmings, field voles, etc.) are as violent in relatively complex temperate zone ecosystems as they are in the less complex Arctic zone ecosystems. This is supported by Goodman (1975: 239) who writes:

As for the apparent stability of tropical biota, that could well be an illusion attributable to insufficient study of bewilderingly complex assemblages in which many species are so poorly represented in samples of feasible size that even considerable fluctuations might go undetected. Indeed, there are countervailing anecdotes regarding ecological instability in the tropics, such as recent reports of an insect virtually defoliating the wild Brazil-nut trees in Bolivia and of monkeys succumbing in large numbers to epidemics.

He goes on to add: 'There is growing awareness of the surprising susceptibility of the rain forest ecosystem to man-made perturbation.' This is a point of view supported by May (1979) and discussed by Hill (1975). Hill points out that a very high species diversity is frequently associated with areas which have relatively constant physical environmental conditions over the course of a year and a series of years. The rain forest may be construed to be such an environment, and one where this constancy has allowed the presence of many specialized species, each pursuing a narrow range of activities. It has been argued that because of the high degree of specialization, the indigenous species have a limited ability to recover from major stresses caused by human intervention.

Goodman (1975) has also queried the sufficiency of the argument in its reference to the apparent instability of island ecosystems, suggesting that islands, being evolutionary backwaters and dead-ends, may accumulate species that are especially susceptible to competitive or exploitative displacement. In this case, lack of diversity may not necessarily be the sole or prime cause of instability.

The apparent instability of agricultural compared to natural communities is also often attributed to lack of diversity (see p. 85), and indeed modern agriculture does involve significant ecosystem simplification. However, such instability as there is may not, once again, necessarily result from simplification. Other factors could promote instability: agricultural communities are disrupted, even destroyed, more frequently and more massively as part of the cultivation process than those natural systems we tend to think of as stable; the component species of natural systems are co-evolved (co-adapted), and this is not usually true of agricultural communities. As Murdoch (1975: 799) suggests, it may be that:

natural systems are more stable than crop systems because their interacting species have had a long shared evolutionary history . . . In contrast with these natural communities the dominant plant species of a crop system is thrust into an often alien landscape . . . the crops have undergone radical selection in breeding programmes, often losing their genetic defence mechanisms.

Thus the ideas that complex natural ecosystems will be less susceptible to human interference and that simple artificial ecosystems will inevitably be unstable are not necessarily tenable. None the less, it is apparent that there are differences in susceptibility between different ecosystem types, and these differences may result from factors other than the degree of diversity and complexity (Cairns and Dickson, 1977).

Some systems tend to be *vulnerable*. Lakes, for example, are natural traps and sumps and are thus more vulnerable to the effects of disadvantageous inputs than are rivers (which are continually receiving new inputs) or oceans (which are so much larger). Other systems display the property of *elasticity* – the ability to recover from damage. This may be because nearby epicentres exist to provide organisms to reinvade a damaged systesm. Small, isolated systems will often tend to possess low elasticity (see p. 132). Two of the most important properties, however, are *resilience* (being a measure of the number of times a system can recover after stress), and *inertia* (the ability to resist displacement of structural and functional characteristics).

Two systems which display resilience and inertia are deserts and estuaries. In both cases their indigenous organisms are highly accustomed to variable environmental conditions. Thus most desert fauna and flora evolved in an environment where the normal pattern is one of more or less random alternations of short favourable periods and long stress periods. They have pre-adapted resilience (Noy-Meir, 1974) so that they can tolerate extreme conditions, have the ability for rapid recovery, have various delay and trigger mechanisms (in the case of plants) and have flexible and opportunistic eating habits (in the case of beasts). Estuaries, on the other hand, though the subject of increasing human pressures, also display some resilience. The vigour of their water circulation continuously and endogenously renews the supply of water, food, larvae, etc.; this aids recovery. Also, many species have biological characteristics which provide special advantages in estuarine survival (Cronin, 1967). These characteristics usually protect the species against the natural violence of estuaries and are often helpful in resisting external forces such as humans.

Human influence or nature?

From many of the examples given in this book it is apparent that in many cases of environmental change it is impossible to state, without risk of contradiction, that people rather than nature are responsible. Most systems are complex and human agency is but one component of them, so that many human actions can lead to end-products which are intrinsically similar to those that may be produced by natural forces. It is a case of equifinality, whereby different processes can lead to basically similar results. Humans are not always responsible for some of the changes with which they are credited. This book has given many examples of this problem and a selection is presented in table 9.3. Deciphering the cause is often a ticklish problem, given the intricate interdependence of different components of ecosystems, the frequency and complexity of environmental changes, and the varying relaxation times that different ecosystem components may have when subject to a new impulse. This problem plainly does not apply to the same extent to changes which have been brought about deliberately and knowingly by

Table 9.3 Human influence or nature? Some examples with page references to this book

Change	Causes		Page reference
	Natural	Anthropogenic	
Late Pleistocene animal extinction	Climate	Hunting	125
Death of savanna trees	Soil salinization through climatically induced ground-water rise	Over-grazing	–
Desertification in semi-arid areas	Climatic change	Over-grazing, etc.	55
Holocene peat-bog development in highland Britain	Climatic change and progressive soil deterioration	Deforestation and ploughing	151
Holocene elm and linden decline	Climatic change	Feeding and stalling of animals	65
Tree encroachment into alpine pastures in USA	Temperature amelioration	Cessation of burning	–
Gully development	Climatic change	Land-use change	263
Early twentieth-century climatic warming	Changes in solar emission and volcanic activity	CO_2-generated greenhouse effect	304
Increasing coast recession	Rising sea-level	Disruption of sediment supply	286
Increasing coastal flood risk	Rising sea-level, natural subsidence	Pumping of aquifers creating subsidence	259
Increasing river flood intensity	Higher intensity rainfall	Creation of drainage ditches	254
Ground collapse	Karstic process	Dewatering by over-pumping	79
Forest decline	Drought	Air, soil and water pollution	–

humans, but it does apply to the many cases where humans may have initiated change inadvertently and unintentionally.

This fundamental difficulty means that environmental impact statements of any kind are extremely difficult to make. As we have seen, humans have been living on the earth and modifying it in different degrees for several millions of years, so that it is problematical to reconstruct any picture of the environment before human intervention. We seldom have any clear baseline against which to measure changes brought about by human society. Moreover, even without human interference, the environment would be in a perpetual state of flux on a great many different timescales (Goudie, 1992). In addition, as Wall and Wright (1977: 3) have demonstrated, there are spatial and temporal discontinuities between cause and effect. For example, erosion in one locality may lead to deposition in another, while destruction of key elements of an animal's habitat may lead to population declines throughout its range. Likewise, in a time context, a considerable interval may elapse before the full implications of an activity are apparent. Also, because of the complex interactions between different components of different environmental systems and subsystems it is almost impossible to measure total environmental impact. For example, changes in soil may lead to changes in vegetation, which in turn may trigger changes in water quality and in animal populations. Primary impacts give rise to a myriad of successive repercussions throughout ecosystems which may be impracticable to trace and monitor. Quantitative cause-and-effect relationships can seldom be established.

Into the unknown

During the 1980s and 1990s the full significance of possible future environmental changes has become apparent, and national governments and international institutions have begun to ponder whether the world is entering a spasm of unparalleled humanly induced modification. Our models and predictions are still highly inadequate, and there are great ranges in some of the values we give for such crucial changes as sea-level rise and global climatic warming, but the balance of scientific argument favours the view that change will occur and that change will be substantial. Some of the changes may be advantageous for humans or for particular ecosystems; others will be extremely disadvantageous. But all change, if it is rapid and of a great magnitude, is likely to create uncertainties and instabilities. The study of future events will not only become a focus for the environmental sciences, but will also become a major concern for economists, sociologists, lawyers and political scientists. George Perkins Marsh was a lawyer and politician, but it is only now, a century and a quarter since he wrote *Man and nature*, that the wisdom, perspicacity and prescience of his ideas have begun to be given the praise and attention they deserve.

Guide to Reading

Chapter One

Bell, M. and Walker, M. J. C., 1992, *Late Quaternary environmental change*. Harlow: Longman Scientific.
Berry, B. J. L., 1974, *Land use, urban form and environmental quality*. University of Chicago: Department of Geography Research Series.
Detwyler, T. R. (ed.), 1971, *Man's impact on environment*. New York: McGraw-Hill.
Douglas, I., 1983, *The urban environment*. London: Arnold.
Gregory, K. J. and Walling, D. E. (eds), 1986, *Man and environmental processes*. Chichester: Wiley.
Mannion, A. M., 1991, *Global environmental change*. Harlow: Longman Scientific and Technical.
Nesbit, E. G., 1991, *Leaving Eden: to protect and manage the Earth*. Cambridge: Cambridge University Press.
Simmonds, I., 1989, *Changing the earth: culture, environment and history*. Oxford: Basil Blackwell.

Chapter Two

Holzner, W., Werger, M. J. A., Werger, I. and Ikusima, I. (eds), 1983, *Man's impact on vegetation*. Hague: Junk.
Innes, J. L., 1987, Air pollution and forestry. *Forestry Commission Bulletin*, 70.
Kozlowski, T. T. and Ahlgren, C. C. (eds), 1974, *Fire and ecosystems*. New York: Academic Press.
Myers, N., 1984, *The primary source: tropical forests and our future*. New York: Norton.
Simmonds, I., 1979. *Biogeography: natural and cultural*. London: Arnold.
Wilson, E. O. (ed.), 1988, *Biodiversity*. Washington DC: National Academy Press.

Chapter Three

Elton, C. S., 1958, *The ecology of invasions by plants and animals*. London: Methuen.
Fitter, R., 1986, *Wildlife for man*. London: Collins.

Gilbert, O. L., 1989, *The ecology of urban habitats*. London: Chapman & Hall.

Holdgate, M. W., 1979, *A perspective on environmental pollution*. Cambridge: Cambridge University Press.

Jarvis, P. H., 1979, The ecology of plant and animal introductions. *Progress in Physical Geography*, 3, 187–214.

Martin, P. S. and Klein, R. G., 1984, *Pleistocene extinctions*. Tucson: University of Arizona Press.

Murton, R. K., 1971, *Man and birds*. London: Collins.

Chapter Four

Bidwell, O. W. and Hole, F. D., 1985, Man as a factor of soil formation. *Soil Science*, 99, 65–72.

Greenland, D. J. and Lal, R. 1977, *Soil conservation and management in the humid tropics*. Chichester: Wiley.

Hudson, N., 1971, *Soil conservation*. London: Batsford.

Morgan, R. P. C., 1986, *Soil erosion and conservation*. London: Longman.

Russell, J. S. and Isbell, R. F. (eds), 1986, *Australian soils: the human impact*. St Lucia: University of Queensland Press.

Chapter Five

Dunne, T. and Leopold, L. B., 1978, *Water in environmental planning*. San Francisco: Freeman.

Hollis, G. E. (ed.), 1979, *Man's impact on the hydrological cycle in the United Kingdom*. Norwich: Geobooks.

Petts, G. E., 1985, *Impounded rivers: perspectives for ecological management*. Chichester: Wiley.

Worthington, E. B. (ed.), 1977, *Arid land irrigation in developing countries: environmental problems and effects*. Oxford: Pergamon.

Chapter Six

Brokes, A., 1988, *Channelized rivers*. Chichester: Wiley.

Brown, E. H., 1970, Man shapes and earth. *Geographical Journal*, 136, 74–85.

Marsh, G. P., 1864. *Man and nature*. New York: Scribner.

Nir, D., 1983, *Man, a geomorphological agent: an introduction to anthropic geomorphology*. Jerusalem: Keter.

Chapter Seven

Bridgman, H., 1990, *Global air pollution*. London: Bellhaven Press.

Elsom, D., 1987, *Atmospheric pollution*. Oxford: Basil Blackwell.

Isaksen, I. S. A. (ed.), 1988, *Tropospheric ozone: regional and global scale interactions*. Dordrecht: Reidel.

Landsberg, H. E., 1981, *The urban climate*. New York: Academic Press.

Mason, B. J. (ed.), 1991, *The surface waters acidification programme*. Cambridge: Cambridge University Press.

Park, C. C., 1987, *Acid rain: rhetoric and reality*. London: Methuen.

Wellburn, A., 1988, *Air pollution and acid rain: the biological impact*. London: Longman.

Chapter Eight

Bolin, B., Doos, B. R., Jager, J. and Warrick, R. A. (eds), 1986, *The greenhouse effect, climatic change and ecosystems.* Chichester: Wiley.
Clark, W. G. (ed.), 1982, *Carbon dioxide review 1982.* Oxford: Oxford University Press.
Houghton, J. T., Jenkins, G. J. and Ephraums, J. J. (eds), 1990, *Climate change: the IPCC Scientific Assessment.* Cambridge: Cambridge University Press.
MacCracken, M. C. and Luther, F. M. (eds), 1985, *Detecting the climatic effects of increasing carbon dioxide levels.* Washington DC: US Department of Energy.
Macdonald, G. J., 1982, *The long-term impacts of increasing atmospheric carbon dioxide levels.* Cambridge, MA: Ballinger.
Schneider, S. H., 1989, The greenhouse effect: science and policy. *Science,* 243, 771–81.

Chapter Nine

Ashby, E., 1978, *Reconciling man with the environment.* London: Oxford University Press.
Clark, W. C. and Munn, R. E. (eds), 1986, *Sustainable development of the biosphere.* Cambridge: Cambridge University Press.
Simon, J. L. and Kahn, H., 1984, *The resourceful earth – a response to Global 2000.* Oxford: Basil Blackwell.
Turner, B. L., Clark, W. S., Kates, R. W., Richards, J. F., Matthews, J. T. and Meyer, W. B. (eds), 1990, *The earth as transformed by human action.* Cambridge: Cambridge University Press.
Warren, A. and Goldsmith, F. B. (eds), 1983, *Conservation in perspective.* Chichester: Wiley.

References

Abler, R., Adams, J. S. and Gould, P., 1971, *Spatial organisation: the geographer's view of the world.* Englewood Cliffs: Prentice Hall.

Abul-Atta, A. A., 1978, *Egypt and the Nile after the construction of the High Aswan Dam*, Cairo: Ministry of Irrigation and Land Reclamation.

Ackerman, W. C., Harmeston, R. H. and Sinclair, R. A., 1970, Some long-term trends in water quality of rivers and lakes. *Eos*, 51, 516–22.

Ahlgren, I. F., 1974, The effect of fire on soil organisms. In T. T. Kozlowski and C. C. Ahlgren (eds), *Fire and ecosystems.* New York: Academic Press, 47–72.

Alabaster, J. S., 1972, Suspended solids and fisheries. *Proceedings of the Royal Society*, 180B, 395–406.

Allchin, B., Goudie, A. S. and Hegde, K. T. M., 1977, *The prehistory and palaeogeography of the Great Indian Desert.* London: Academic Press.

Allen, J. C. and Barnes, D. F., 1985, The causes of deforestation in developing countries. *Annals of the Association of American Geographers*, 75, 163–84.

Allen, J., Golson, J. and Jones, R. (eds), 1977, *Sunda and Sahul. Prehistoric studies in southeast Asia, Melanesia and Australia.* London: Academic Press.

Almer, B., Dickson, W., Ekstrom, C., Hornstrom, E. and Miller, U., 1974, Effects of acidification on Finnish lakes. *Ambio*, 3, 30–6.

Andel, T. H. van, Zangger, E. and Demitrack, A., 1990, Land use and soil erosion in prehistoric and historical Greece. *Journal of Field Archaeology*, 17, 379–96.

Anderson, J. M. and Spencer, T., 1991, Carbon, nutrient and water balances of tropical rainforest ecosystem subject to disturbance. *MAB Digest*, 7.

Arber, M. A., 1946, Dust-storms in the Fenland around Ely. *Geography*, 31, 23–6.

Aron, W. I. and Smith, S. H., 1971, Ship canals and aquatic ecosystems. *Science*, 174, 13–20.

Ashby, E., 1978, *Reconciling man with the environment.* London: Oxford University Press.

Atkinson, B. W., 1968, A preliminary examination of the possible effect of London's urban area on the distribution of thunder rainfall, 1951–60. *Transactions of the Institute of British Geographers*, 44, 97–118.

——, 1975, The mechanical effect of an urban area on convective precipitation. Occasional paper 3, Department of Geography, Queen Mary College, University of London.

Atlas, P., 1977, The paradox of hail suppression. *Science*, 195, 139–45.

Aubertin, G. M. and Patric, J. H., 1974, Water quality after clearcutting a small watershed in West Virginia. *Journal of environmental quality*, 3, 243–9.

Aubréville, A., 1949, *Climats, forêts et desertification de L'Afrique tropicale*. Paris: Societé d'Edition Géographiques Maritimes et Coloniales.

Bach, W., 1979, Short-term climatic alterations caused by human activities: status and outlook. *Progress in physical geography*, 3, 55–83.

Bach, W., 1988, Modelling and climatic response to greenhouse gases. In S. Gregory (ed.), *Recent climatic change: a regional approach*. London: Bellhaven Press, 7–19.

Bakan, S., Chlono, A., Cubasch, U., Feichter, J., Graf, H., Grassl, H., Hasselman, K., Kirchner, I., Latif, M., Roeckner, E., Samsen, R., Schlese, U., Schrivener, D., Schult, I., Sielman, F. and Welks, W., 1991, Climate response to smoke from the burning oil wells in Kuwait. *Nature*, 351, 367–71.

Ball, D. F., 1975, discussion in J. G. Evans, S. Limbrey and H. Cleere (eds), The effect of man on the landscape: the Highland zone. *Council for British Archaeology Research Report*, 11, 26.

Ball, D. J. and Laxen, D. P. H., 1986, *Ambient ozone concentrations in Greater London 1975 to 1985*. London: London Scientific Services.

Ballantyne, C. K., 1991, Late Holocene erosion in upland Britain: climatic deterioration or human influence? *The Holocene*, 1, 81–5.

Balling, R. C. and Wells, S. G., 1990, Historical rainfall patterns and arroyo activity within the Zuni river drainage basin, New Mexico. *Annals of the Association of American Geographers*, 80, 603–17.

Banks, H. O. and Richter, R. C., 1953, Sea-water intrusion into groundwater basins bordering the California coast and inland bays. *Transactions of the American Geophysical Union*, 34, 575–82.

Barber, K. E., 1981, *Peat stratigraphy and climatic change*. Rotterdam: Balkema.

Bari, M. A. and Schofield, N. J., 1992, Lowering of a shallow, saline water table by extensive eucalypt reforestation. *Journal of Hydrology*, 133, 273–91.

Barnston, A. G. and Schickendanz, P. T., 1984, The effect of irrigation on warm season precipitation in the southern Great Plains. *Journal of climate and applied meteorology*, 23, 865–88.

Barrass, R., 1974, *Biology, food and people*. London: Hodder and Stoughton.

Barry, R. G., 1985, The cryosphere and climate change. In M. C. MacCracken and F. M. Luther (eds), *Detecting the climatic effects of increasing carbon dioxide*. Washington DC: US Dept of Energy, 111–48.

Bartlett, H. H., 1956, Fire, primitive agriculture, and grazing in the tropics. In W. L. Thomas (ed.), *Man's role in changing the face of the earth*. Chicago: University of Chicago Press, 692–720.

Barton, B. A. 1977, Short-term effect of highway construction on the limnology of a small stream in southern Ontario. *Freshwater biology*, 7, 99–108.

Bates, M. 1956, Man as an agent in the spread of organisms. In W. L. Thomas (ed.), *Man's role in changing the face of the earth*. Chicago: University of Chicago Press, 788–804.

Battarbee, R. W., 1977, Observations in the recent history of Lough Neagh and its drainage basin. *Philosophical transactions of the Royal Society of London*, 281B, 303–45.

Battarbee, R. W., Flower, R. J., Stevenson, A. C. and Rippey, B., 1985a, Lake acidification in Galloway: a palaeoecological test of competing hypotheses. *Nature*, 314, 350–2.

Battarbee, R. W., Appleby, P. G., Odel, K. and Flower, R. J., 1985b, ^{210}Pb dating of Scottish Lake sediments, afforestation and accelerated soil erosion. *Earth surface processes and landforms*, 10, 137–42.

Battarbee, R. W., Flower, R. J., Stevenson, A. C., Jones, V. J., Harriman, R. and Appleby, P. G., 1988a, Diatom and chemical evidence for reversibility of acidification of Scottish lochs. *Nature*, 332, 530–2.

Battarbee, R. W. and 15 collaborators, 1988b, *Lake acidification in the United Kingdom 1800–1986*. London: Ensis.

Battistini, R. and Verin, P., 1972, Man and the environment in Madagascar. *Monographiae biologicae*, 21, 331–7.

Baver, L. D., Gardner, W. H. and Gardner, W. R., 1972, *Soil physics* (4th edn). New York: Wiley.

Baxter, R. M., 1977, Environmental effects of dams and impoundments. *Annual Review of ecology and systematics*, 8, 255–83.

Bayfield, N. G., 1973, Human pressures on soils in mountain areas. *Welsh soils discussion group report*, 14, 36–49.

——, 1979, Recovery of four heath communities on Cairngorm, Scotland, from disturbance by trampling. *Biological conservation*, 15, 165–79.

Beamish, R. J., Lockhart, W. L., Van Loon, J. C. and Harvey, H. H., 1975, Long-term acidification of a lake and resulting effects on fishes. *Ambio*, 4, 98–102.

Beaumont, P., 1978, Man's impact on river systems: a world-wide view. *Area*, 10, 38–41.

Beckinsale, R. P., 1969, Human responses to river regimes. In R. J. Chorley (ed.), *Water, earth and man*. London: Methuen, 487–509.

——, 1972, The effect upon river channels of sudden changes in sediment load. *Acta geographica debrecina*, 10, 181–6.

Begon, M., Harper, J. L., and Townshend, C. R., 1986, *Ecology: individuals, populations and communities*. Oxford: Blackwell Scientific.

Behre, K.-E. (ed.), 1986, *Anthropogenic indicators in pollen diagrams*. Rotterdam: Balkema.

Bell, J. N. B. and Cox, R. A., 1975, Atmospheric ozone and plant damage in the United Kingdom. *Environmental pollution*, 8, 163–70.

Bell, M. L., 1982, The effect of land-use and climate on valley sedimentation. In A. F. Harding (ed.), *Climatic change in later prehistory*. Edinburgh: Edinburgh University Press, 127–42.

Bell, M. and Walker, M. J. C. 1992, *Late Quaternary environmental change*. Harlow: Longman Scientific and Technical.

Bell, M. L. and Nur, A., 1978, Strength changes due to reservoir-induced pore pressure and stresses and application to Lake Oroville. *Journal of geophysical research*, 83, 4469–83.

Bell, P. R., 1982, Methane hydrate and the carbon dioxide question. In W. G. Clark (ed.), *Carbon dioxide review 1982*. Oxford: Oxford University Press, 401–5.

Bendell, J. F., 1974, Effects of fire on birds and mammals. In T. T. Kozlowski and C. C. Ahlgren (eds), *Fire and ecosystems*. New York: Academic Press, 73–138.

Bender, B., 1975, *Farming in prehistory from hunter-gatherer to food-producer*. London: Baker.

Bennett, C. F., 1968, Human influences in the zoogeography of Panama. *Ibero-Americana*, 51.

Bennett, H. H., 1938, *Soil conservation*. New York: McGraw-Hill.

Bentley, C. R., 1984, Some aspects of the cryosphere and its role in climatic change. *Geophysical monograph*, 29, 207–20.

Bernabo, J. C. and Webb, T., III, 1977, Changing patterns in the Holocene pollen record of northeastern North America: mapped summary. *Quaternary research*, 8, 64–96.

Berry, B. J. L., 1974, *Land use, urban form and environmental quality*.

Chicago: Department of Geography Research Series.

Bidwell, O. W. and Hole, F. D., 1965, Man as a factor of soil formation. *Soil science*, 99, 65–72.

Biglane, K. E. and Lafleur, R. A., 1967, Notes on estuarine pollution with emphasis on the Louisiana Gulf Coast. *Publication 83, American association for the advancement of science*, 690–2.

Binford, M. W., Brenner, M., Whitmore, T. J., Higuera-Grundy, A., Deevey, E. S. and Leyden, B., 1987, Ecosystems, palaeoecology, and human disturbance in subtropical and tropical America. *Quaternary Science Review*, 6, 115–28.

Bird, E. C. F., 1979, Coastal processes. In K. J. Gregory and D. E. Walling (eds), *Man and environmental processes*, Folkestone: Dawson, 82–101.

——, 1985, *Coastline changes. A global view*. Chichester: Wiley.

——, 1986, Potential effects of sea level rise on the coasts of Australia, Africa and Asia. In J. G. Titus (ed.), *Effects of changes in stratospheric ozone and global climate*. Washington DC: UNEP/USEPA, 83–98.

Birkeland, P. W. and Larson, E. E., 1978, *Putnam's geology*. New York: Oxford University Press.

Birks, H. J. B., 1986, Late-Quaternary biotic changes in terrestrial and lacustrine environments, with particular reference to north-west Europe. In B. E. Berglund (ed.), *Handbook of Holocene palaeoecology and palaeohydrology*. Chichester: Wiley, 3–65.

——, 1988, Long-term ecological change in the British uplands. *British ecological society special publication*, 7, 37–56.

Blackburn, W. H. and Tueller, P. T., 1970, Pinjon and juniper invasion in Black sagebrush communities in east-central Nevada. *Ecology*, 51, 841–8.

Blackburn, W. H., Knight, R. W. and Schuster, J. L., 1983, Saltcedar influence on sedimentation in the Brazos River. *Journal of soil and water conservation*, 37, 298–301.

Blainey, G., 1975, *Triumph of the nomads. A history of ancient Australia*. Melbourne: Macmillan.

Blank, I. W., 1985, A new type of forest decline in Germany. *Nature*, 314, 311–14.

Blumer, M., 1972, Submarine seeps: are they a major source of open ocean oil pollution? *Science*, 176, 1257–8.

Böckh, A., 1973, Consequences of uncontrolled human activities in the Valencia lake basin. In M. T. Farvar and J. P. Milton (eds), *The careless technology*. London: Tom Stacey, 301–17.

Bockman, O. C., Kaarstad, O., Lie, O. H. and Richards, I., 1990, *Agriculture and fertilizers*. Oslo: Norsk Hydro.

Boeck, W. L., Shaw, D. T. and Vonnegut, B., 1975, Possible consequences of global dispersions of Krypton 85. *Bulletin of the American Meteorological Society*, 56, 527.

Bøer, E. M., Koster, E. and Lundberg, E., 1990, Greenhouse impact in Fennoscandia: preliminary findings of a European workshop on the effects of climatic change. *Ambio*, 19, 2–10.

Bolin, B., Doos, B. R., Jager, J. and Warrick, R. A. (eds), 1986, *The greenhouse effect, climatic change and ecosystems*. Chichester: Wiley.

Boorman, L. A., 1977, Sand-dunes. In R. S. K. Barnes (ed.), *The coastline*. London: Wiley, 161–97.

Booyson, P. de V. and Tainton, N. M. (eds), 1984, *Ecological effects of fire in South African ecosystems*. Berlin: Springer-Verlag.

Bormann, F. H., Likens, G. E., Fisher, D. W. and Pierce, R. S., 1968, Nutrient loss accelerated by clear cutting of a forest ecosystem. *Science*, 159, 882–4.

Bourne, W. R. P., 1970, Oil pollution and bird conservation. *Biological conservation*, 2, 300–2.

Boussingault, J. B., 1845, *Rural economy* (2nd edn). London: Baillière.

Boutron, C. F., Görlach, U., Candelone, J.-P., Bolshov, M. A. and Delmas, R. J., 1991, Decrease in anthropogenic lead, cadmium and zinc in Greenland snows since the late 1960s. *Nature*, 353, 153–6.

Brassington, F. C. and Rushton, K. R., 1987, A rising water table in central Liverpool. *Quarterly Journal of Engineering Geology*, 20, 151–8.

Brazel, A. J. and Idso, S. B., 1979, Thermal effects of dust on climate. *Annals of the Association of American Geographers*, 69, 432–7.

Breuer, G., 1980, *Weather modification: prospects and problems*. Cambridge: Cambridge University Press.

Bridges, E. M., 1978, Interaction of soil and mankind in Britain. *Journal of soil science*, 29, 125–39.

Brimblecombe, P., 1977, London air pollution 1500–1900. *Atmospheric environment*, 11, 1157–62.

——, 1987, *The big smoke*. London: Methuen.

Broadus, J., Milliman, J., Edwards, S., Aubrey, D. and Gable, F., 1986, Rising sea level and damming of rivers: possible effects in Egypt and Bangladesh. In J. G. Titus (ed.), *Effects of changes in stratospheric ozone and global climate*. Washington DC: UNEP/USEPA, 165–89.

Broecker, W. S., Takahashi, T., Simpson, H. J. and Peng, T. H., 1979, Fate of fossil fuel carbon dioxide and the global carbon budget. *Science*, 206, 409–18.

Brookes, A., 1985, River channelization: traditional engineering methods, physical consequences, and alternative practices. *Progress in physical geography*, 9, 44–73.

——, 1987, The distribution and management of channelized streams in Denmark. *Regulated rivers*, 1, 3–16.

Brown, A. A. and Davis, K. P., 1973, *Forest fire control and its use* (2nd edn). New York: McGraw-Hill.

Brown, B. E., 1990, Coral bleaching. *Coral Reefs*, 8, 153–232.

Brown, E. H., 1970, Man shapes the earth. *Geographical journal*, 136. 74–85.

Brown, J. H., 1989, Patterns, modes and extents of invasions by vertebrates. In J. A. Drake (ed.), *Biological invasions: a global perspective*. Chichester: Wiley, 85–109.

Brown, L. R., Flavin, C. and Kane, H. 1992, *Vital signs*. London: Earthscan.

Browning, K. A., Allah, R. J., Ballard, B. P., Barnes, R. T. H., Bennetts, D. A., Maryon, R. H., Mason, P. J., McKenna, D., Mitchell, J. F. B., Senior, C. A., Slingo, A. and Smith, F. B., 1991, Environmental effects from burning oil wells in Kuwait. *Nature*, 351, 363–7.

Bruijnzeel, L. A., 1990, *Hydrology of moist tropical forests and effects of conversion: a state of knowledge review*. Amsterdam: Free University for UNESCO International Hydrological Programme.

Brunhes, J., 1920, *Human geography*. London: Harrap.

Bryan, G. W., 1979, Bioaccumulation of marine pollutants. *Philosophical transactions of the Royal Society*, 286B, 483–505.

Bryan, K., 1928, Historic evidence of changes in the channel of Rio Puerco, a tributary of the Rio Grande in New Mexico. *Journal of geology*, 36, 265–82.

Bryson, R. A. and Barreis, D. A., 1967, Possibility of major climatic modifications and their implications: northwest India, a case for study. *Bulletin of the American Meteorological Society*, 48, 136–42.

Bryson, R. A. and Kutzbach, J. E., 1968, *Air pollution*. Commission on college geography resource paper 2. Washington DC: Association of American Geographers.

Budd, W. F., 1991, Antarctica and global change. *Climatic change*, 19, 271–99.

Buddemeier, R. W. and Smith, S. V., 1988, Coral reef growth in an era of rapidly rising sea level: predictions and suggestions for long-term research. *Coral Reefs*, 7, 51–6.

Budyko, M. I., 1974, *Climate and life*. New York: Academic Press.

——, 1982, *The earth's climate: past and future*. New York: Academic Press.

Budyko, M. I. and Izrael, Y. A., 1991, *Anthropogenic climatic change*. Tuscon: University of Arizona Press.

Bulmer, S., 1982, Human ecology and cultural variation in prehistoric New Guinea. *Monographiae biologicae* (New Guinea), 42, 169–206.

Burcham, L. T., 1970, Ecological significance of alien plants in California grasslands. *Proceedings of the Association of American Geographers*, 2, 36–9.

Burke, R., 1972, Stormwater runoff. In R. T. Oglesby, C. A. Carlson and J. A. McCann (eds), *River ecology and man*. New York: Academic Press, 727–33.

Burrin, P. J., 1985, Holocene alluviation in southeast England and some implications for palaeohydrological studies. *Earth surface processes and landforms*, 10, 257–71.

Burt, T. P., Donohoe, M. A. and Vann, A. R., 1983, The effect of forestry drainage operations on upland sediment yields: the results of a storm-based study. *Earth surface processes and landforms*, 8, 339–46.

Burt, T. P. and Haycock, N. E., 1992, Catchment planning and the nitrate issue: a UK perspective. *Progress in Physical Geography*, 16, 379–404.

Busack, S. D. and Bury, R. B., 1974, Some effects of off-road vehicles and sheep grazing on lizard population in the Mojave Desert. *Biological conservation*, 6, 179–83.

Bush, M. B., 1988, Early Mesolithic disturbance: a force on the landscape. *Journal of Archaeological Science*, 15, 453–62.

Bussing, C., 1972, The impact of feedlots. In D. D. MacPhail (ed.), *The High Plains: problem of semiarid environments*. Fort William: Colorado State University, 78–86.

Butzer, K. W., 1972, *Environment and archaeology – an ecological approach to prehistory*. London: Methuen.

——, 1974, Accelerated soil erosion: a problem of man–land relationships. In I. R. Manners and M. W. Mikesell (eds), *Perspectives on environments*. Washington DC: Association of American Geographers.

——, 1976, *Early hydraulic civilization in Egypt*. Chicago: University of Chicago Press.

——, 1977, Environment, culture and human evolution. *American scientist*, 65, 572–84.

——, 1980, Adaptation to global environmental change. *Professional geographer*, 32, 269–78.

Cadle, R. D., 1971, The chemistry of smog. In W. H. Matthews, W. W. Kellogg and G. D. Robinson (eds), *Man's impact on the climate*. Cambridge, Mass.: MIT Press, 339–59.

Cairns, J. and Dickson, D. K., 1977, Recovery of streams from spills of hazardous materials. In J. Cairns, K. L. Dickson and E. E. Herricks (eds), *Recovery and restoration of damaged ecosystems*. Charlottesville: University Press of Virginia, 24–42.

Carbon Dioxide Assessment Committee, 1983, *Changing climate*. Washington: National Academic Press.

Carlson, C. W., 1978, Research in ARS related to soil structure. In W. W. Emerson, R. D. Bond and A. R. Dexter (eds), *Modification of soil structure*. Chichester: Wiley, 279–84.

Carrara, P. E. and Carroll, T. R., 1979, The determination of erosion rates from exposed tree roots in the Piceance Basin, Colorado. *Earth surface processes*, 4, 407–17.

Carter, D. L., 1975, Problems of salinity in agriculture. *Ecological studies*, 15, 26–35.

Carter, L. J., 1977, Soil erosion: the problem persists despite the billions spent on it. *Science*, 196, 409–11.

Carter, R. W. G., 1988, *Coastal environments*. London: Academic Press.

Casey, H. and Clarke, R. T., 1979, Statistical analysis of nitrate concentrations from the River Frome (Dorset) for the period 1965–76. *Freshwater biology*, 9, 91–7.

Cathcart, R. B., 1983, Mediterranean Basin – Sahara reclamation. *Speculations in science and technology*, 6, 150–2.

Challinor, D., 1968, Alteration of surface soil characteristics by four tree species. *Ecology*, 49, 286–90.

Chambers, F. M., Dresser, P. Q. and Smith, A. G., 1979, Radiocarbon dating evidence on the impact of atmospheric pollution on upland peats. *Nature*, 282, 829–31.

Champion, T., Gramble, C., Shennan, S. and Whittle, A., 1984, *Prehistoric Europe*. London: Academic Press.

Chancellor, W. J., 1977, Compaction of soil by agricultural equipment. *Division of agricultural sciences, University of California bulletin*, 1881.

Chandler, T. J., 1965, *The climate of London*. London: Hutchinson.

——, 1976, The climate of towns. In T. J. Chandler and S. Gregory (eds), *The climate of the British Isles*. London: Longman, 307–29.

Changnon, S. A., 1968, The La Porte weather anomaly – fact or fiction? *Bulletin of the American Meteorological Society*, 49, 4–11.

——, 1973, Atmospheric alterations from man-made biospheric changes. In W. R. D. Sewell (ed.), *Modifying the weather: a social assessment*. Adelaide: University of Victoria, 135–84.

——, 1978, Urban effects on severe local storms at St Louis. *Journal of applied meteorology*, 17, 578–86.

Charlson, R. J., Lovelock, J. E., Andreae, M. O. and Warren, S. G., 1987, Oceanic phytoplankton, atmospheric sulphur, cloud albedo and climate. *Nature*, 326, 655–61.

Charlson, R. J., Schwartz, S. E., Hales, J. M., Cess, R. D., Coakley, J. A., Hansen, J. E. and Hoffmann, D. J., 1992, Climate forcing by anthropogenic aerosols. *Science*, 255, 423–30.

Charney, J., Stone, P. H. and Quirk, W. J., 1975, Drought in the Sahara: a bio-geophysical feedback mechanism. *Science*, 187, 434–5.

Charnock, A., 1983, A new course for the Nile. *New scientist*, October, 285–8.

Chester, D. K. and James, P. A., 1991, Holocene alluviation in the Algarve, southern Portugal: the case for an anthropogenic cause. *Journal of Archaeological Science*, 18, 73–87.

Chesters, G. and Konrad, J. F., 1971, Effects of pesticide usage on water quality. *Bio science*, 21, 565–9.

Chi, S. C. and Reilinger, R. E., 1984, Geodetic evidence for subsidence due to groundwater withdrawal in many parts of the United States of America. *Journal of Hydrology*, 67, 155–82.

Child, B. A., 1985, Bush encroachment and rangeland deterioration. Paper presented at a symposium on 'The deterioration of natural resources in Africa', School of Geography, University of Oxford, 1 July, 1985.

Childe, V. G., 1936, *Man makes himself*. London: Watts.

Chinn, T. J., 1988, Glaciers and snowlines. In Ministry for the Environment (ed.), *Climate Change: the New Zealand response*. Wellington: Ministry for the Environment, 238–40.

Chorley, R. J. and More, R. J., 1967, The interaction of precipitation and man. In R. J. Chorley (ed.), *Water, earth and man*. London: Methuen, 157–66.

Clark, G., 1962, *World prehistory*. Cambridge: Cambridge University Press.

——, 1977a, Domestication and social evolution. In J. Hutchinson, J. G. G. Clark, E. M. Jope and R. Riley (eds), *The early history of agriculture*. Oxford: Oxford University Press, 5–11.

——, 1977b, *World prehistory in new perspective*. Cambridge: Cambridge University Press.

Clark, J. R., 1977, *Coastal ecosystem management*. New York: Wiley.

Clark, R. R., 1963, *Grimes Graves*. London: Her Majesty's Stationery Office.

Clark, W. C. (ed.), 1982, *Carbon dioxide review 1982*. Oxford: Clarendon Press.

Clark, W. C. and 7 co-workers, 1982, The carbon dioxide question: perspectives for 1982. In W. C. Clark (ed.), *Carbon dioxide review 1982*. Oxford: Oxford University Press, 3–44.

Clarke, R. T. and McCulloch, J. S. G., 1979, The effect of land use on the hydrology of small upland catchments. In G. E. Hollis (ed.), *Man's impact on the hydrological cycle in the United Kingdom*. Norwich: Geo Abstracts, 71–8.

Coates, D. R. (ed.), 1976, *Geomorphology and engineering*. Stroudsburg: Dowden, Hutchinson and Ross.

——, 1977, Landslide perspective. *Reviews in engineering geology*, 3, 3–28.

——, 1983, Large-scale land subsidence. In R. Gardner and H. Scoging (eds), *Mega-geomorphology*. Oxford: Oxford University Press, 212–34.

Cochrane, R., 1977, The impact of man on the natural biota. In A. G. Anderson (ed.), *New Zealand in maps*. Section 14. London: Hodder & Stoughton.

Coe, M., 1981, Body size and the extinction of the Pleistocene megafauna. *Palaeoecology of Africa*, 13, 139–45.

——, 1982, The bigger they are . . . *Oryx*, 16, 225–8.

Coffey, M., 1978, The dust storms. *Natural history* (New York), 87, 72–83.

Colbeck, I., 1988, Photochemical ozone pollution in Britain. *Science progress*, 72, 207–26.

Cole, M. M., 1963, Vegetation and geomorphology in northern Rhodesia: an aspect of the distribution of the Savanna of Central Africa. *Geographical journal*, 129, 290–310.

Cole, M. M. and Smith, R. F., 1984, Vegetation as an indicator of environmental pollution. *Transactions of the Institute of British Geographers*, 9, 477–93.

Cole, S., 1970, *The Neolithic revolution* (5th edn). London: British Museum (Natural History).

Collier, C. R., 1964, Influences of strip mining on the hydrological environment of parts of Beaver Creek Basin, Kentucky, 1955–59. *United States geological survey professional paper*, 472–B.

Committee on the Atmosphere and the Biosphere, 1981, *Atmosphere–biosphere interactions: towards a better understanding of the ecological consequences of fossil fuel combustion*. Washington DC: National Academy Press.

Committee on Engineering Implications of Changes in Relative Mean Sea Level, 1987, *Responding to changes in sea level*. Washington DC: National Academy Press.

Conacher, A. J., 1979, Water quality and forests in Southwestern Australia: review and evaluation. *Australian geographer*, 14, 150–9.

Conway, G. R. and Pretty, J. N., 1991, *Unwelcome harvest: agriculture and pollution*. London: Earthscan.

Conway, V. M., 1954, Stratigraphy and pollen analysis of southern Pennine blanket peats. *Journal of Geology*, 42, 117–47.

Cooke, G. W., 1977, Waste of fertilizers. *Philosophical transactions of the Royal Society of London*, 281B, 231–41.

Cooke, R. U. and Doornkamp, J. C., 1990, *Geomorphology in environmental management* (2nd edn). Oxford: Clarendon Press.

Cooke, R. U. and Reeves, R. W., 1976, *Arroyos and environmental change in the American south-west*. Oxford: Clarendon Press.

Cooke, R. U., Brunsden, D., Doornkamp, J. C. and Jones, D. K. C., 1982, *Urban geomorphology in drylands*. Oxford: Oxford University Press.

Coones, P. and Patten, J. H. C., 1986, *The landscape of England and Wales*. Harmondsworth: Penguin Books.

Cooper, C. F., 1961, The ecology of fire. *Scientific American*, 204, 4, 150–60.

Costa, J. E., 1975, Effects of agriculture on erosion and sedimentation in the Piedmont province, Maryland. *Bulletin of the Geological Society of America*, 86, 1281–6.

Council on Environmental Quality and the Department of State, 1982, *The global 2000 report for the President*. Harmondsworth: Penguin Books.

Coupland, R. T. (ed.), 1979, *Grassland ecosystems of the world: analysis of grasslands and their uses*. Cambridge: Cambridge University Press.

Cowell, E. B., 1976, Oil pollution of the sea. In R. Johnson (ed.), *Marine pollution*, London: Academic Press, 353–401.

Crisp, D. T., 1977, Some physical and chemical effects of the Cow Green (Upper Teesdale) impoundment. *Freshwater biology*, 7, 109–20.

Croley, T. E., 1990, Laurentian Great Lakes double-CO_2 climate change hydrological impacts. *Climatic Change*, 17, 27–47.

Cronin, L. E., 1967, The role of man in estuarine processes. *Publication 83, American Association for the Advancement of Science*, 667–89.

Crowe, P. R., 1971, *Concepts in climatology*. London: Longman.

Crozier, M. J., Marx, S. L. and Grant, I. J., 1978, Impact of off-road recreational vehicles on soil and vegetation. *Proceedings of the 9th New Zealand Geography Conference*, Dunedin, 76–9.

Crutzen, P. J., Aselmann, I. and Sepler, W., 1986, Methane production by domestic animals, wild ruminants, other herbivores, fauna and humans. *Tellus*, 38B, 271–84.

Cumberland, K. B., 1961, Man in nature in New Zealand. *New Zealand geographer*, 17, 137–54.

Currey, D. T., 1977, The role of applied geomorphology in irrigation and groundwater studies. In J. R. Hails (ed.), *Applied geomorphology*, Amsterdam: Elsevier, 51–83.

Cwynar, L. C., 1978, Recent history of fire and vegetation from laminated sediment of Greenleaf Lake, Algonquin Park, Ontario. *Canadian journal of botany*, 56, 10–21.

Daniel, T. C., McGuire, P. E., Stoffel, D. and Millfe, B., 1979, Sediment and nutrient yield from residential construction sites. *Journal of environmental quality*, 8, 304–8.

Darby, H. C., 1956, The clearing of the woodland in Europe. In W. L. Thomas (ed.), *Man's role in changing the face of the Earth*. Chicago: University of Chicago Press, 183–216.

Darling, F. F., 1956, Man's ecological dominance through domesticated animals on wild lands. In W. L. Thomas (ed.), *Man's role in changing the face of the Earth*. Chicago: University of Chicago Press, 778–87.

Darungo, F. P., Allee, P. H. and Weickmann, H. K., 1978, Snowfall induced by a power plant plume. *Geophysical research letters*, 5, 515–17.

Daubenmire, R., 1968, Ecology of fire in grassland. *Advances in ecological research*, 5, 209–66.

Davis, B. N. K., 1976, Wildlife, urbanisation and industry. *Biological conservation*, 10, 249–91.

Davitaya, F. F., 1969, Atmospheric dust content as a factor affecting glaciation and climatic change. *Annals of the Association of American Geographers*, 59, 552–60.

Denisova, T. B., 1977, The environmental impact of mineral industries. *Soviet geography*, 18, 646–59.

Denson, E. P., 1970, The trumpeter swan, *Olor Buccinator*; a conservation success and its lessons. *Biological conservation*, 2, 251–6.

Department of Environment, 1984, *Digest of environmental pollution and water statistics for 1983*. London: HMSO.

——, 1989, *Digest of environmental pollution and water statistics for 1988*. London: HMSO.

De Sylva, D., 1986, Increased storms and estuarine salinity and other ecological impacts of the greenhouse effect. In J. G. Titus (ed.), *Effects of changes in stratospheric ozone and global climate*, vol. 4, *Sea level rise*. Washington DC: UNEP/USEPA, 153–64.

Detwyler, T. R. (ed.), 1971, *Man's impact on environment*. New York: McGraw-Hill.

Detwyler, T. R. and Marcus, M. G., 1972, *Urbanisation and environment: the physical geography of the city*. Belmont: Duxbury Press.

Di Castri, F., 1989, History of biological invasions with special emphasis on the old world. In J. A. Drake (ed.), *Biological invasions: a global perspective*. Chichester: Wiley, 1–30.

Dickinson, R. E., 1986, Impact of human activities on climate – a frame-work. In W. C. Clark and R. E. Munn (eds), *Sustainable development of the biosphere*, Cambridge: Cambridge University Press, 252–89.

Dicks, B., 1977, Changes in the vegetation of an oiled Southampton water salt marsh. In J. Cairns, K. L. Dickson and E. E. Herricks (eds), *Recovery and restoration of damaged ecosystems*. Charlottesville: University of Virginia, 72–101.

Dimbleby, G. W., 1974, The legacy of prehistoric man. In A. Warren and F. B. Goldsmith (eds), *Conservation in practice*. London: Wiley, 179–89.

Dodd, A. P., 1959, The biological control of the prickly pear in Australia. In A. Keast, R. L. Crocker and C. S. Christian (eds), *Biogeography and ecology in Australia*. The Hague: Junk, 565–77.

Doerr, A. and Guernsely, L., 1956, Man as a geomorphological agent: the example of coal mining. *Annals of the Association of American Geographers*, 46, 197–210.

Dolan, R., Godfrey, P. J. and Odum, W. E., 1973, Man's impact on the barrier islands of North Carolina. *American scientist*, 61, 152–62.

Donkin, R. A. 1979, Agricultural terracing in the Aboriginal New World. *Viking Fund publications in anthropology*, 561.

Doughty, R. W., 1974, The human predator: a survey. In I. R. Manners and M. V. Mikesell (eds), *Perspectives on environment*. Washington DC: Association of American Geographers, 152–80.

——, 1978, The English sparrow in the American landscape: a paradox in nineteenth century wildlife conservation. Research paper 19, School of Geography, University of Oxford.

Douglas, I., 1969, The efficiency of humid tropical denudation systems. *Transactions of the Institute of British Geographers*, 46, 1–6.

——, 1983, *The urban environment*. London: Arnold.

Down, C. G. and Stocks, J., 1977, *Environmental impact of mining*. London: Applied Science Publishers.

Dregne, H. E., 1986, Desertification of arid lands. In F. El-Baz and M. H.

A. Hassan (eds), *Physics of Desertification*. Dordrecht: Nijhoff, 4–34.

Dregne, H. E. and Tucker, C. J., 1988, Desert encroachment. *Desertification Control Bulletin*, 16, 16–19.

Drennan, D. S. H., 1979, Agricultural consequences of groundwater development in England. In G. E. Hollis (ed.), *Man's impact on the hydrological cycle in the United Kingdom*. Norwich: Geo Abstracts, 31–8.

Driscoll, R., 1983, The influence of vegetation on the swelling and shrinkage of clay soils in Britain. *Géotechnique*, 33, 293–326.

Dunne, T. and Leopold, L. B., 1978, *Water in environmental planning*. San Francsico: Freeman.

D'Yakanov, K. N. and Reteyum, A. Y., 1965, The local climate of the Rybinsk reservoir. *Soviet geography*, 6, 40–53.

Eden, M. J., 1974, Palaeoclimatic influences and the development of savanna in southern Venezuela. *Journal of biogeography*, 1, 95–109.

Edington, J. M. and Edington, M. A., 1977, *Ecology and environmental planning*. London: Chapman and Hall.

Edlin, H. L., 1976, The Culbin sands. In J. Leniham and W. W. Fletcher (eds), *Reclamation*. Glasgow: Blackie, 1–31.

Edmonson, W. T., 1975, Fresh water pollution. In W. W. Murdoch (ed.), *Environment*. Sunderland: Sinauer Associates, 251–71.

Edwards, A. M. C., 1975, Long term changes in the water quality for agricultural catchments. In R. D. Hey and T. D. Daniels (eds), *Science technology and environmental management*. Farnborough: Saxon House, 111–22.

Edwards, K. J., 1985, The anthropogenic factor in vegetational history. In K. J. Edwards and W. P. Warren (eds), *The Quaternary history of Ireland*. London: Academic Press, 187–200.

Ehrenfeld, D. W., 1972, *Conserving life on earth*. New York: Oxford University Press.

Ehrlich, P. R. and Ehrlich, A. H., 1970, *Population, resources, environment: issues in human ecology*. San Francisco: Freeman.

—— and ——, 1982, *Extinction*. London: Gollancz.

Ehrlich, P. R., Ehrlich, A. H. and Holdren, J. P., 1977, *Ecoscience: population, resources, environment*. San Francisco: Freeman.

Ellenberg, H., 1979, Man's influence on tropical mountain ecosystems in South America. *Journal of ecology*, 67, 401–16.

Ellis, J. B., 1975, Urban stormwater pollution. *Middlesex Polytechnic research report*, 1.

Elsom, D., 1987, *Atmospheric pollution* (revised reprint 1989). Oxford and Cambridge, Mass.: Basil Blackwell.

Elton, C. S., 1958, *The ecology of invasions by plants and animals*. London: Methuen.

Emanuel, K. A., 1987, The dependence of hurricane intensity on climate. *Nature*, 326, 483–5.

Emmanuel, W. R., Shugart, H. H. and Stevenson, M. P., 1985, Climatic change and the broad-scale distribution of terrestrial ecosystem complexes. *Climatic change*, 7, 29–43.

Engelhardt, F. R. (ed.), 1985, *Petroleum effects in the Arctic environment*. London: Elsevier Applied Science.

Evans, D. M., 1966, Man-made earthquakes in Denver. *Geotimes*, 10, 11–18.

Evans, J. G., Limbrey, S. and Cleere, H. (eds), 1975, *The effect of man on the landscape: the Highland zone*. Council for British Archaeology research report 11.

Evans, R. and Northcliffe, S. 1978, Soil erosion in North Norfolk. *Journal of agricultural science*, 90, 185–92.

Evenari, M., Shanan, L. and Tadmor, N. H., 1971, Runoff agriculture in the Negev Desert of Israel. In W. G. McGinnies, R. J. Goldman and P. Paylore (eds), *Food, fiber and the arid lands*. Tucson: University of Arizona Press, 312–22.

Feare, C. J., 1978, The decline of booby (Sulidae) population in the Western Indian Ocean. *Biological conservation*, 14, 295–305.

Ferrians, O. J., Kachadoorian, R. and Green, G. W., 1969, Permafrost and related engineering problems in Alaska. *United States geological survey professional paper*, 678.

Fillenham, L. F., 1963, Holme Fen Post. *Geographical journal*, 129, 502–3.

Fisher, J., Simon, N. and Vincent, J., 1969, *The red book – wildlife in danger*. London: Collins.

Flenley, J. R., 1979, *The equatorial rain forest: a geological history*. London: Butterworth.

Flenley, J. R., King, A. S. M., Jackson, J., Chew, C., Teller, J. and Prentice, M. E., 1991, The late Quaternary vegetational and climatic history of Easter Island. *Journal of Quaternary Science*, 6, 85–115.

Flohn, H., 1982, Climate change and an ice-free Arctic Ocean. In W. C. Clark (ed.), *Carbon dioxide review 1982*. Oxford: Oxford University Press, 145–79.

Flohn, H. and Dansgaard, W., 1984, Selected climates from the past and their relevance to possible future climate. In H. Flohn and R. Fantechi (eds), *The climate of Europe: past, present and future*. Dordrecht: Reidel, 198–268.

Food and Agricultural Organization of the United Nations, 1980, *1979 Production Yearbook*. Rome: FAO.

Foster, I. D. L., Dearing, J. A. and Appleby, R. G., 1986, Historical trends in catchment sediment yields: a case study in reconstruction from lake-sediment records in Warwickshire, UK. *Hydrological science journal*, 31, 427–43.

Foster, T., 1976, *Bushfire*. Sydney: Reed.

Fox, H. L., 1976, The urbanizing river: a case study in the Maryland Piedmont. In D. R. Coates (ed.), *Geomorphology and engineering*, Stroudsburg: Dowden, Hutchinson and Ross, 245–71.

Frankel, O. H., 1984, Genetic diversity, ecosystem conservation and evolutionary responsibility. In F. D. Castri, F. W. G. Baker and M. Hadley (eds), *Ecology in practice*, vol. I. Dublin: Tycooly, 4315–27.

Freeland, W. J., 1990, Large herbivorous mammals: exotic species in northern Australia. *Journal of Biogeography*, 17, 445–9.

French, H. M., 1976, *The periglacial environment*. London: Longman.

Frenkel, R. E., 1970, Ruderal vegetation along some California roadsides. *University of California publications in geography*, 20.

Fuller, R., Hill, D. and Tucker, G., 1991, Feeding the birds down on the farm: perspectives from Britain. *Ambio*, 20(6), 232–7.

Gade, D. W., 1976, Naturalization of plant aliens: the volunteer orange in Paraguay. *Journal of Biogeography*, 3, 269–79.

Galay, V. J., 1983, Causes of river bed degradation. *Water resources research*, 19, 5, 1057–90.

Gameson, A. L. H. and Wheeler, A., 1977, Restoration and recovery of the Thames Estuary. In J. Cairns, K. L. Dickson and E. E. Herricks (eds), *Recovery and restoration of damaged ecosystems*. Charlottesville: University Press of Virginia, 92–101.

Geertz, C., 1963, *Agricultural involution: the process of ecological change in Indonesia*. Berkeley: University of California Press.

Gerasimov, I. P., 1976, Problems of natural environment transformation in Soviet constructive geography. *Progress in geography*, 9, 75–99.

Gerasimov, I. P., Armand, D. L. and Yefron, K. M., 1971, *Natural resources of the Soviet Union: their use and renewal*. San Francisco: Freeman.

GESAMP (IMO/FAO/UNESCO/WMO/IAEA/UN/UNEP Joint Group of Experts on the Scientific Aspects of Marine Pollution), 1990, *The state of the marine environment*. UNEP Regional Seas Reports and Studies, 115. Nairobi.

Gifford, G. F. and Hawkins, R. H., 1978, Hydrological impact of grazing on infiltration: a critical review. *Water resources research*, 14, 305–13.

Gilbert, G. K., 1971, Hydraulic mining debris in the Sierra Nevada. *United States geological survey professional paper*, 105.

Gilbert, O. L., 1970, Further studies on the effect of sulphur dioxide on lichens and bryophytes. *New phytologist*, 69, 605–27.

——, 1975, Effects of air pollution on landscape and land-use around Norwegian aluminium smelters. *Environmental pollution*, 8, 113–21.

Gillespie, R., Horton, D. R., Ladd, P., Macumber, P. G., Rich, T. H., Thorne, R. and Wright, R. V. S., 1978, Lancefield Swamp and the extinction of the Australian megafauna. *Science*, 200, 1044–8.

Gillon, D., 1983, The fire problem in tropical savannas. In F. Bourlière (ed.), *Tropical savannas*. Oxford: Elsevier Scientific, 617–41.

Gimingham, C. H., 1981, Conservation: European Heathlands. In R. L. Spect (ed.), *Heathlands and related shrublands*. Amsterdam: Elsevier Scientific, 249–59.

Gimingham, C. H. and de Smidt, I. T., 1983, Heaths and natural and semi-natural vegetation. In W. Holzner, M. J. A. Werger and I. Ikusima (eds), *Man's impact on vegetation*, Hague: Junk, 185–99.

Glacken, C. J., 1956, Changing ideas of the habitable world. In W. L. Thomas (ed.), *Man's role in changing the face of the earth*. Chicago: University of Chicago Press, 70–92.

——, 1963, The growing second world within the world of nature. In F. R. Fosberg (ed.), *Man's place in the island ecosystem*. Honolulu: Bishop Museum Press, 75–100.

——, 1967, *Traces on the Rhodian shore: nature and culture in western thought from ancient times to the end of the eighteenth century*. Berkeley: University of California Press.

Gleick, P. H., 1986, Regional water resources and global climatic change. In J. G. Titus (ed.), *Effects of changes in stratospheric ozone and global climate*, vol. 3, *Climatic change*. Washington DC: UNEP/USEPA, 217–49.

Godbole, N. N., 1972, Theories on the origin of salt lakes in Rajasthan, India. *Proceedings of the 24th International Geological Congress*, Section 10, 354–7.

Goemans, T., 1986, The sea also rises: the ongoing dialogue of the Dutch with the sea. In J. G. Titus (ed.), *Effects of changes in stratospheric ozone and global climate*, vol. 4, *Sea level rise*. Washington DC: UNEP/USEPA, 47–56.

Goldberg, E. D., Hodge, V., Koide, M., Griffin, J., Gamble, E., Bicker, O. P., Metisoff, G., Holdren, G. R. and Brown, R. 1978, A pollution history of Chesapeake Bay. *Geochimica et cosmochimica acta*, 42, 1413–25.

Goldsmith, F. B., 1983, Evaluating nature. In A. Warren and F. B. Goldsmith (eds), *Conservation in perspective*. Chichester: Wiley, 233–46.

Goldsmith, V., 1979, Coastal dunes. In R. A. Davis (ed.), *Coastal sedimentary environments*. New York: Springer-Verlag.

Gomez, B. and Smith, C. G., 1984, Atmospheric pollution and fog frequency in Oxford, 1926–1980. *Weather*, 39, 379–84.

Gong, Zi-Tong, 1983, Pedogenesis of paddy soil and its significance in soil classification. *Soil science*, 135, 5–10.

Goodman, D., 1975, The theory of diversity–stability relationships in ecology. *Quarterly review of biology*, 50, 237–66.

Gorman, M., 1979, *Island ecology*. London: Chapman & Hall.

Gottschalk, L. C., 1945, Effects of soil erosion on navigation in Upper Chesapeake Bay. *Geographical review*, 35, 219–38.

Goudie, A. S., 1972a, The concept of post-glacial progressive desiccation. Research paper 4, School of Geography, University of Oxford.

——, 1972b, Vaughan Cornish: geographer. *Transactions of the Institute of British Geographers*, 55, 1–16.

——, 1973, *Duricrusts of tropical and subtropical landscapes*. Oxford: Clarendon Press.

——, 1977, Sodium sulphate weathering and the disintegration of Mohenjo-Daro, Pakistan. *Earth surface processes*, 2, 75–86.

——, 1983, Dust storms in space and time. *Progress in physical geography*, 7, 502–30.

—— (ed.), 1990, *Techniques for desert reclamation*. Chichester: Wiley.

——, 1992, *Environmental change*, 3rd edn. Oxford: Clarendon Press.

Goudie, A. S. and Middleton, N. J., 1992, The changing frequency of dust storms through time. *Climatic Change*, 20, 197–225.

Goudie, A. S. and Wilkinson, J. C., 1977, *The warm desert environment*. Cambridge: Cambridge University Press.

Gourou, P., 1961, *The tropical world* (3rd edn). London: Longman.

Gowlett, J. A. J., Harris, J. W. K., Walton, D. and Wood, B. A., 1981, Early archaeological sites, hominid remains and traces of fire from Chesowanja, Kenya. *Nature*, 284, 125–9.

Graf, W. K., 1977, Network characteristics in surburbanizing streams. *Water resources research*, 13, 459–63.

Graf, W. L., 1988, *Fluvial processes in dryland rivers*. Berlin: Springer-Verlag.

Grainger, A., 1990, *The threatening desert*: controlling desertification. London: Earthscan

——, 1992, *Controlling tropical deforestation*. London: Earthscan.

Grayson, D. K., 1977, Pleistocene avifaunas and the overkill hypothesis. *Science*, 195, 691–3.

——, 1988, Perspectives on the archaeology of the first Americans. In R. C. Carlisle (ed.), *Americans before Columbus: Ice Age origins*. Pittsburgh: University of Pittsburgh, 107–23.

Green, F. H. W., 1976, Recent changes in land use and treatment. *Geographical journal*, 142, 12–26.

——, 1978, Field drainage in Europe. *Geographical journal*, 144, 171–4.

Green, H. S. et al., 1981, Pontnewydd cave in Wales – a new Middle Pleistocene hominid site. *Nature*, 294, 707–13.

Green, R. C., 1975, Adaptation and change in Maori culture. *Monographiae biologicae* (New Guinea), 27, 591–661.

Greenland, D. J., 1977, Soil drainage by intensive arable cultivation: temporary or permanent. *Philosophical transactions of the Royal Society of London*, 281B, 193–208.

Greenland, D. J. and Lal, R., 1977, *Soil conservation and management in the humid tropics*. Chichester: Wiley.

Gregory, K. J., 1985a, The impact of river channelization. *Geographical journal*, 151, 53–74.

——, 1985b, *The nature of physical geography*. London: Arnold.

Gregory, K. J. and Walling, D., 1973, *Drainage basin form and process: a geomorphological approach*. London: Arnold.

Gregory, K. J. and Walling D., 1979, *Man and environmental processes*. Folkstone: Dawson.

Grieve, A. M., 1987, Salinity and waterlogging in the Murray–Darling basin. *Search*, 18, 72–4.

Griffiths, J. F., 1976, *Applied climatology, an introduction* (2nd edn). Oxford: Oxford University Press.

Grigg, D., 1970, *The harsh lands*. London: Macmillan.

Gross, M. G., 1972, Geological aspects of waste solids and marine waste deposits, New York Metropolitan region. *Bulletin of the Geological Society of America*, 83, 3163–76.

Grove, J. M., 1988, *The Little Ice Age*. London: Routledge.

Grove, R. H., 1983, *The future for forestry*. Cambridge: British Association of Nature Conservationists.

——, 1990, The origins of environmentalism. *Nature*, 345, 11–14.

Grover, H. D. and Musick, H. B., 1990, Shrubland encroachment in southern New Mexico, USA: an analysis of desertification processes in the American southwest. *Climatic Change*, 17, 305–30.

Guidon, N. and Delibrias, G., 1986, Carbon 14 dates point to man in the Americas 32 000 years ago. *Nature*, 321, 769–71.

Guilday, J. E., 1967, Differential extinction during late Pleistocene and recent time. In P. S. Martin and H. E. Wright (eds), *Pleistocene extinctions*. New Haven: Yale University Press, 121–40.

Haggett, P., 1979, *Geography: a modern synthesis* (3rd edn). London: Prentice Hall.

Haigh, M. J., 1978, Evolution of slopes on artificial landforms – Blaenavon, UK. Research paper 183, Department of Geography, University of Chicago.

Hails, J. R. (ed.), 1977, *Applied geomorphology*. Amsterdam: Elsevier.

Hanes, T. L., 1971, Succession after fire in the chaparral of Southern California. *Ecological monographs*, 41, 27–52.

Hansen, J., Johnson, D., Lacis, A., Lededeff, S., Lee, P., Rims, D. and Russell, G., 1981, Climatic impact of increasing atmospheric carbon dioxide. *Science*, 213, 957–66.

Hansen, J., Fung, I., Lacis, A., Rind, D., Lebedeff, S., Ruedy, R. and Russell, G., 1988, Global climate changes as forecast by Goddard Institute for Space Studies three dimensional model. *Journal of geophysical research*, 93, 9341–64.

Happ, S. C., 1944, Effect of sedimentation on floods in the Kickapoo Valley, Wisonsin. *Journal of geology*, 52, 53–68.

Harlan, J. R., 1975a, Our vanishing genetic resources. *Science*, 188, 617–22.

——, 1975b, *Crops and man*. Madison: American Society of Agronomy.

——, 1976, The plants and animals that nourish man. *Scientific American*, 235, 3, 88–97.

Harris, D. R., 1966, Recent plant invasions in the arid and semi-arid southwest of the United States. *Annals of the Association of American Geographers*, 56, 408–22.

——, (ed.), 1980, *Human ecology in savanna environments*. London: Academic Press.

Hartmann, H. C., 1990, Climate change impacts on Laurentian Great Lakes levels. *Climatic Change*, 17, 49–67.

Harvey, A. M. and Renwick, W. H., 1987, Holocene alluvial fan and terrace formation in the Bowland Fells, Northwest England. *Earth surface processes and landforms*, 12, 249–57.

Harvey, A. M., Oldfield, F., Baron, A. F. and Pearson, G. W., 1981, Dating of post-glacial landforms in the central Howgills. *Earth surface processes and landforms*, 6, 401–12.

Hawksworth, D. L., 1990, The long-term effects of air pollutants on lichen communities in Europe and North America. In G. M. Woodwell (ed.),

The earth in transition: patterns and processes of biotic impoverishment. Cambridge: Cambridge University Press, 45–64.

Hay, J., 1973, Salt cedar and salinity on the Upper Rio Grande. In M. T. Farvar and J. P. Milton (eds), *The careless technology.* London: Tom Stacey, 288–300.

Haynes, C. V., 1991, Geoarchaeological and palaeohydrological evidence for a Clovis-age drought in North America and its bearing on extinction. *Quaternary Research*, 35, 438–50.

Heinselman, M. L. and Wright, H. E., 1973, The ecological role of fire in natural conifer forests of western and northern America. *Quaternary research*, 3, 317–482.

Helldén, U., 1985, Land degradation and land productivity monitoring – needs for an integrated approach. In A. Hjört (ed.), *Land management and survival.* Uppsala: Scandinavian Institute of African Studies, 77–87.

Helliwell, D. R., 1974, The value of vegetation for conservation. II: M1 motorway area. *Journal of environmental management*, 2, 75–8.

Henderson-Sellers, A. and Blong, R. 1989, *The greenhouse effect: living in a warmer Australia.* Kensington, NSW: New South Wales University Press.

Henderson-Sellers, A. and Gornitz, V., 1984, Possible climatic impacts of land cover transformation with particular emphasis on tropical deforestation. *Climatic change*, 6, 231–57.

Henderson-Sellers, A. and Robinson, P. J., 1986, *Contemporary climatology.* London: Longman.

Hess, W. N. (ed.), 1974, *Weather and climate modification.* New York: Wiley.

Hewlett, J. D., Post, H. E. and Doss, R., 1984, Effect of clear-cut silviculture on dissolved ion export and water yield in the Piedmont. *Water resources research*, 20, 7, 1030–8.

Heywood, V. H., 1989, Patterns, extents and modes of invasions by terrestrial plants. In J. A. Drake (ed.), *Biological invasions: a global perspective.* Chichester: Wiley, 31–55.

Hickey, J. J. and Anderson, O. W., 1968, Chlorinated hydrocarbons and eggshell changes in raptorial and fish-eating birds. *Science*, 162, 271–2.

Hill, A. R., 1975, Ecosystems stability in relation to stresses caused by human activities. *Canadian geographer*, 19, 206–20.

Hillel, D., 1971, Artificial inducement of runoff as a potential source of water in arid lands. In W. G. McGinnies, B. J. H. Goldman and P. Paylore (eds), *Food, fiber and the arid lands.* Tucson: University of Arizona Press, 324–30.

Hills, T. L., 1965, Savannas: a review of a major research problem in tropical geography. *Canadian geographer*, 9, 216–28.

Hobbs, P. V. and Radke, L. F., 1992, Airborne studies of the smoke from the Kuwait oil fires. *Science*, 256, 987–91.

Holdgate, M. W., 1979, *A perspective on environmental pollution.* Cambridge: Cambridge University Press.

Holdgate, M. W. and Wace, N. M., 1961, The influence of man on the floras and faunas of southern islands. *Polar record*, 10, 473–93.

Holdgate, M. W., Kassas, M. and White, G. F., 1982, *The world environment 1972–1982.* Dublin: Tycooly.

Hollis, G. E., 1975, The effects of urbanization on floods of different recurrence interval. *Water resources research*, 11, 431–5.

——, 1978, The falling levels of the Caspian and Aral Seas. *Geographical journal*, 144, 62–80.

Hollis, G. E. and Luckett, J. K., 1976, The response of natural river channels to urbanization: two case studies from Southeast England. *Journal of hydrology*, 30, 351–63.

Holtz, W. G., 1983, The influence of vegetation on the swelling and shrinking of clays in the United States of America. *Géotechnique*, 33, 159–63.

Holzer, T. L., 1979, Faulting caused by groundwater extraction in South-central Arizona. *Journal of geophysical research*, 84, 603–12.

Hopkins, B., 1965, Observations on savanna burning in the Olikemeji forest reserve, Nigeria. *Journal of applied ecology*, 2, 367–81.

Hornbeck, J. W., 1981, Acid rain: facts and fallacies. *Journal of forestry*, 79, 438–43.

Hotes, F. L. and Pearson, E. A., 1977, Effects of irrigation on water quality. In E. B. Worthington (ed.), *Arid land irrigation in developing countries: environmental problems and effects*. Oxford: Pergamon, 127–58.

Houghton, J. T., Jenkins, G. J. and Ephraums, J. J., 1990, *Climate change: the IPCC Scientific Assessment*. Cambridge: Cambridge University Press.

Houghton, J. T., Callander, B. A. and Varney, S. K. (eds), 1992, *Climate change 1992: the supplementary report of the IPCC scientific assessment*. Cambridge: Cambridge University Press.

Howe, G. M., Slaymaker, H. O. and Harding, D. M., 1966, Flood hazard in mid-Wales. *Nature*, 212, 584–5.

——, —— and ——, 1967, Some aspects of the flood hydrology of the upper catchments of the Severn and Wye. *Transactions of the Institute of British Geographers*, 41, 33–58.

Hoyt, D. V. and Fröhlich, C., 1983, Atmospheric transmissions at Davos, Switzerland, 1909–1979. *Climatic change*, 5, 61–71.

Hudson, B. J., 1979, Coastal land reclamation with special reference to Hong Kong. *Reclamation review*, 2, 3–16.

Hudson, N., 1971, *Soil conservation*. London: Batsford.

——, 1987, Soil and water conservation in semi-arid areas. *FAO soils bulletin*, 55.

Hughes, M. K., Lepp, N. W. and Phipps, D. A., 1980, Aerial heavy metal pollution and terrestrial ecosystems. *Advances in ecological research*, 11, 217–327.

Hughes, R. J., Sullivan, M. E. and Yok, D., 1991, Human-induced erosion in a highlands catchment in Papua New Guinea: the prehistoric and contemporary records. *Zeitschrift für Geomorphologie, Supplement-band*, 83, 227–39.

Hull, S. K. and Gibbs, J. N., 1991, Ash dieback: a survey of non-woodland trees. *Forestry Commission Bulletin*, 93, 32.

Huntington, E., 1914, *The climatic factor as illustrated in arid America*. Carnegie Institution of Washington Publication, 192.

Hurd, L. E., Mellinger, M. W., Wold, L. L. and McNaughton, S. J., 1971, Stability and diversity at three trophic levels in terrestrial successional ecosystems. *Science*, 173, 1134–6.

Husar, R. B. and Husar, J. D., 1991, Sulfur. In B. L. Turner, W. C. Clark, R. W. Kates, J. F. Richards, J. T. Matthews and W. B. Meyer (eds), *The earth as transformed by human action*. Cambridge: Cambridge University Press, 409–21.

Hutchinson, G. E., 1973, Eutrophication. *American scientist*, 61, 269–79.

Hutchinson, G. L. and Mosier, A. R., 1979, Nitrous oxide emissions from an irrigated cornfield. *Science*, 205, 1125–7.

Hutchinson, T. C. and Havas, M., 1980, *Effects of acid precipitation on terrestrial ecosystems*. New York: Plenum.

Huybrechts, P., Letreguilly, A. and Rech, N., 1990, The Greenland ice sheet and greenhouse warming. *Palaeogeography, Palaeoclimatology, Palaeoecology*, 89, 399–412.

Idso, S. B., 1982, *Carbon dioxide: friend or foe?* Tempe: IBR Press.

——, 1983, Carbon dioxide and global temperature: what the data show. *Journal of environmental quality*, 12, 159–63.

——, 1989, *Carbon dioxide and global change: earth in transition.* Tempe, Arizona: IBR Press.

Idso, S. B. and Brazel, A. J., 1978, Climatological effects of atmospheric particulate pollution. *Nature*, 274, 781–2.

Ikawa-Smith, F., 1980, Current issues in Japanese archaeology. *American scientist*, 68, 134–45.

Illies, J., 1974, *Introduction to zoogeography.* London: Macmillan.

Imeson, A. C., 1971, Heather burning and soil erosion on the North Yorkshire Moors. *Journal of applied ecology*, 8, 537–41.

Innes, J. L., 1983, Lichenometric dating of debris-flow deposits in the Scottish Highlands. *Earth surface processes and landforms*, 8, 579–88.

——, 1987, *Air pollution and forestry.* Forestry Commission bulletin, 70.

——, 1992, Forest decline. *Progress in Physical Geography*, 16, 1–64.

Innes, J. L. and Boswell, R. C., 1990, Monitoring of forest condition in Great Britain 1989. *Forestry Commission Bulletin*, 94, 57.

Institute of Hydrology, 1991, *Institute of Hydrology Report 1990–91.* Wallingford: Institute of Hydrology.

Irving, W. M., 1985, Context and chronology of early man in the Americas. *Annual review of anthropology*, 14, 529–55.

Isaac, E., 1970, *Geography of domestication.* Englewood Cliffs: Prentice Hall.

Isachenko, A. G., 1974, On the so-called anthropogenic landscapes. *Soviet geography*, 15, 467–75.

——, 1975, Landscape as a subject of human impact. *Soviet geography*, 16, 631–43.

Isaksen, I. S. A. (ed.), 1988, *Tropospheric ozone: regional and global scale interactions.* Dordrecht: Reidel.

Ives, J. D. and Messerli, B., 1989, *The Himalayan dilemma: reconciling development and conservation.* London: Routledge.

Jacks, G. V. and Whyte, R. O., 1939, *The rape of the earth: a world survey of soil erosion.* London: Faber & Faber.

Jacobs, J., 1969, *The economy of cities.* New York: Random House.

——, 1975, Diversity, stability and maturity in ecosystems influenced by human activities. In W. H. Van Dobben and R. H. Lowe-McConnell (eds), *Unifying concepts in ecology.* The Hague: Junk, 187–207.

Jacobsen, T. and Adams, R. M., 1958, Salt and silt in ancient Mesopotamian agriculture. *Science*, 128, 1251–8.

Jarman, M. R., 1977, Early animal husbandry. In J. Hutchinson, J. G. G. Clark, E. M. Jope and R. Riley (eds), *The early history of agriculture.* Oxford: Oxford University Press, 85–97.

Jarvis, P. H., 1979, The ecology of plant and animal introductions. *Progress in physical geography*, 3, 187–214.

Jennings, J. N., 1952, *The origin of the Broads.* Royal Geographical Society research series, 2.

——, 1966, Man as a geological agent. *Australian journal of science*, 28, 150–6.

Jenny, H., 1941, *Factors of soil formation.* New York: McGraw-Hill.

Jickells, T. D., Carpenter, R. and Liss, P. S., 1991, Marine environment. In B. L. Turner, W. C. Clark, R. W. Kates, J. F. Richards, J. T. Matthews and W. B. Meyer (eds), *The earth as transformed by human action.* Cambridge: Cambridge University Press, 313–34.

Johannessen, C. L., 1963, Savannas of interior Honduras. *Ibero-Americana*, 46.

Johnson, A. I. (ed.), 1991, *Land subsidence.* International Association of Hydrological Sciences Publication 200.

Johnson, N. M., 1979, Acid rain: neutralization within the Hubbard Brook ecosystem and regional implications. *Science,* 204, 497–9.

Johnston, D. W., 1974, Decline of DDT residues in migratory songbirds. *Science,* 186, 841–2.

Johnston, D. W., Turner, J. and Kelly, J. M., 1982, The effects of acid rain on forest nutrient status. *Water resources research,* 18, 448–61.

Jones, D. K. C., 1983, Human occupance and the physical environment. In R. J. Johnston and J. C. Doornkamp (eds), *The changing geography of the United Kingdom.* London: Methuen, 327–61.

Jones, P. D., Raper, S. C. B., Bradley, R. S., Diaz, H. F., Kelly, P. M. and Wigley, T. M. L., 1986, Northern hemisphere surface air temperature variations, 1851–1984. *Journal of climate and applied meteorology,* 25, 161–79.

Jones, R., Benson-Evans, K. and Chambers, F. M., 1985, Human influence upon sedimentation in Llangorse Lake, Wales. *Earth surface processes and landforms,* 10, 227–35.

Judd, W. R., 1974, Seismic effects of reservoir impounding. *Engineering geology,* 8, 1–212.

Judson, S., 1968, Erosion rates near Rome, Italy. *Science,* 160, 1444–5.

Kadomura, H., 1983, Some aspects of large-scale land transformation due to urbanization and agricultural development in recent Japan. *Advances in space research,* 2, 8, 169–78.

Karnes, L. B., 1971, Reclamation of wet and overflow lands. In G.-H. Smith (ed.), *Conservation of natural resources.* New York: Wiley, 241–55.

Kates, R. W., Turner, B. L. I. and Clark, W. C., 1991, The great transformation. In B. L. Turner, W. C. Clark, R. W. Kates, J. F. Richards, J. T. Matthews and W. B. Meyer (eds), *The earth as transformed by human action.* Cambridge: Cambridge University Press, 1–17.

Keller, E. A., Channelisation: environmental, geomorphic and engineering aspects. In D. R. Coates (ed.), *Geomorphology and engineering.* Stroudsburg: Dowden, Hutchinson and Ross, 115–40.

Kellman, M., 1975, Evidence for late glacial age fire in a tropical montane savanna. *Journal of biogeography,* 2, 57–63.

Kellogg, W. W., 1977, *Effects of human activities on global climate.* WHO technical note, 156.

——, 1978, Global influence of mankind on the climate. In J. Gribbin (ed.), *Climatic change.* London: Cambridge University Press, 205–27.

——, 1982, Precipitation trends on a warmer earth. In R. A. Peck and J. R. Hummel (eds), *Interpretation of climate and photochemical models, ozone and temperature measurements.* New York: American Institute of Physics, 35–46.

Kent, M., 1982, Plant growth problems in colliery spoil reclamation. *Applied geography,* 2, 83–107.

Khalil, M. A. K. and Rasmussen, R. A., 1987, Atmospheric methane: trends over the last 10,000 years. *Atmospheric environment,* 21, 2445–52.

Kiersch, G. A., 1965, The Vaiont reservoir disaster. *Mineral information service,* 18, 129–38.

Kilgore, B. M. and Taylor, D., 1979, Fire history of a sequoia–mixed conifer forest. *Ecology,* 60, 129–42.

King, C. A. M., 1974, Coasts. In R. U. Cooke and J. C. Doornkamp (eds), *Geomorphology in environmental management.* Oxford: Clarendon Press, 188–222.

——, 1975, *Introduction to physical and biological oceanography*. London: Edward Arnold.

Kinsey, D. W. and Hopley, D., 1991, The significance of coral reefs as global carbon sinks – response to greenhouse. *Palaeogeography, Palaeoclimatology, Palaeoecology*, 89, 363–77.

Kirch, P. V., 1982, Advances in Polynesian prehistory: three decades in review. *Advances in world archaeology*, 2, 52–102.

Kittredge, J. H., 1948, *Forest influences*. New York: McGraw-Hill.

Klein, R. G., 1983, The stone age prehistory of southern Africa. *Annual review of anthropology*, 12, 25–48.

Knox, J. C., 1977, Human impacts on Wisconsin stream channels. *Annals of the Association of American Geographers*, 67, 323–42.

——, 1987, Historical valley floor sedimentation in the Upper Mississippi Valley. *Annals of the Association of American Geographers*, 77, 224–44.

Koide, M. and Goldberg, E. D., 1971, Atmospheric and fossil fuel combustion. *Journal of geophysical research*, 76, 6589–96.

Komar, P. D., 1976, *Beach processes and sedimentation*. Englewood Cliffs: Prentice Hall.

Komarov, B., 1978, *The destruction of nature in the Soviet Union*. London: Pluto Press.

Kotlyakov, V. M., 1991, The Aral Sea basin: a critical environmental zone. *Moscow Environment*, 33(1), 4–9, 36–8.

Kovda, V. A., 1980, *Land aridization and drought control*. Boulder: Westview Press.

Krantz, G. S., 1970, Human activities and megafaunal extinctions. *American scientist*, 58, 164–70.

Krug, E. C. and Frink, C. R., 1983, Acid rain on acid soil: a new perspective. *Science*, 221, 520–5.

Kühlmann, D. H., 1988, The sensitivity of coral reefs to environmental pollution. *Ambio*, 17, 13–21.

Kuo, C., 1986, Flooding in Taipeh, Taiwan and coastal drainage. In J. G. Titus (ed.), *Effects of changes in stratospheric ozone and global climate*. Washington DC: UNEP/USEPA, 37–46.

La Marche, V. C., Graybill, D. A., Fritts, H. C. and Rose, M. R., 1984, Increasing atmospheric carbon dioxide: tree ring evidence for growth enhancement in natural vegetation. *Science*, 225, 1019–21.

Labadz, J. C., Burt, T. P. and Potter, A. W. L., 1991, Sediment yield and delivery in the blanket peat moorlands of the southern Pennines. *Earth Surface Processes and Landforms*, 16, 255–71.

Lamb, H. H., 1977, *Climate: present, past and future. 2: Climatic history and the future*. London: Methuen.

Lambert, J. H., Jennings, J. N., Smith, C. T., Green, C. and Hutchinson, J. N., 1970, *The making of the Broads: a reconsideration of their origin in the light of new evidence*. Royal Geographical Society research series, 3.

Lamprey, H., 1975, The integrated project on arid lands. *Nature and resources*, 14, 2–11.

Landes, K. K., 1973, Mother nature as an oil polluter. *Bulletin of the American Association of Petroleum Geologists*, 57, 637–41.

Landsberg, H. E., 1970, Man-made climatic changes. *Science*, 170, 1265–8.

——, 1981, *The urban climate*. New York: Academic Press.

Langford, T. E., 1972, A comparative assessment of thermal effects in some British and North American rivers. In R. T. Oglesby, C. A. Carlson and J. A. McCann (eds), *River ecology and man*. New York: Academic Press, 318–51.

Langford, T. E. L., 1990, *Ecological effects of thermal discharges*. London: Elsevier Applied Science.

Lanly, J. P. and Clement, J., 1979, Present and future natural forest and plantation areas in the tropics. *Unasylva*, 31, 12–20.

Lanly, J. P., Singh, K. D. and Janz, K., 1991, FAO's 1990 reassessment of tropical forest cover. *Nature and Resources*, 27, 21–6.

Laporte, L. F., 1975, *Encounter with the earth*. San Francisco: Canfield Press.

La Roe, E. T., 1977, Dredging – ecological impacts. In J. R. Clarke (ed.), *Coastal ecosystem management*. New York: Wiley, 610–14.

Larson, F., 1940, The role of bison in maintaining the short grass plains. *Ecology*, 21, 113–21.

Lawson, D. E., 1986, Response of permafrost terrain to disturbance: a synthesis of observations from northern Alaska, USA. *Arctic and Alpine research*, 18, 1–17.

Le Houérou, H. N., 1977, Biological recovery versus desertization. *Economic geography*, 63, 413–20.

Lean, J. and Warrilow, D. A., 1989, Simulation of the regional climatic impact of Amazon deforestation. *Nature*, 342, 126–33.

Lee, D. O., 1992, Urban warming – an analysis of recent trends in London's heat island. *Weather*, 47, 50–6.

Lemon, P. C., 1968, Effects of fire on an African plateau grassland. *Ecology*, 49, 316–22.

Leopold, L. B., 1951, Rainfall frequency: an aspect of climatic variation. *Transactions of the American Geophysics Union*, 32, 347–57.

Leopold, L. B., Wolman, M. G. and Miller, J. P., 1964, *Fluvial processes in geomorphology*. San Francisco: Freeman.

Lewin, J., Bradley, S. B. and Macklin, M. G., 1983, Historical valley alluviation in mid-Wales. *Geological journal*, 18, 331–50.

Likens, G. E. and Bormann, F. H., 1974, Acid rain: a serious regional environmental problem. *Science*, 184, 1176–9.

Likens, G. E. and Butler, T. J., 1981, Recent acidification of precipitation in North America. *Atmospheric environment*, 15, 1103–9.

Likens, G. E., Wright, R. F., Galloway, J. N. and Butler, T. J., 1979, Acid rain. *Scientific American*, 241, 4, 39–47.

Lloyd, J. W., 1986, A review of aridity and groundwater. *Hydrological processes*, 1, 63–78.

Lockwood, J. G., 1979, *Causes of climate*. London: Arnold.

Lowe, P. D., 1983, Values and institutions in the history of British nature conservation. In A. Warren and F. B. Goldsmith (eds), *Conservation in perspective*. Chichester: Wiley, 329–52.

Lowe-McConnell, R. H., 1975, Freshwater life on the move. *Geographical magazine*, 47, 768–75.

Lugo, A. E., Cintron, G. and Goenaga, C., 1981, Mangrove ecosystems under stress. In G. W. Barrett and R. Rosenberg (eds), *Stress effects on natural ecosystems*. Chichester: John Wiley, 129–53.

Lugo, A. E., 1988, Estimating reductions in the diversity of tropical forest species. In E. O. Wilson (ed.), *Biodiversity*. Washington DC: National Academy Press.

Luke, R. H., 1962, *Bush fire control in Australia*. Melbourne: Hodder & Stoughton.

Lund, J. W. G., 1972, Eutrophication. *Proceedings of the Royal Society of London*, 180B, 371–82.

Lyell, C., 1835, *Principles of geology* (4th edn), vol. III. London: Murray (12th edn 1875).

Lynch, J. A., Rishel, G. B. and Corbett, E. S., 1984, Thermal alteration of streams draining clearcut watersheds: quantification and biological implications. *Hydrobiologia*, 111, 161–9.

Mabbutt, J. A., 1985, Desertification of the world's rangelands. *Desertification Control Bulletin*, 12, 1–11.

Macdonald, G. J., 1982, *The long-term impacts of increasing atmospheric carbon dioxide levels.* Cambridge, Mass.: Ballinger.

Macfarlane, M. J., 1976, *Laterite and landscape.* London: Academic Press.

Macklin, M. G. and Lewin, J., 1986, Terraced fills of Pleistocene and Holocene age in the Rheidol Valley, Wales. *Journal of Quaternary Science*, 1, 21–34.

Macklin, M. G., Passmore, D. G., Stevenson, A. C., Colwey, A. C., Edwards, D. N. and O'Brien, C. F., 1991, Holocene alluviation and land-use change on Callaly Moor, Northumberland, England. *Journal of Quaternary Science*, 6, 225–32.

McKnight, T. L., 1959, The feral horse in Anglo-America. *Geographical review*, 49, 506–25.

——, 1971, Australia's buffalo dilemma. *Annals of the Association of American Geographers*, 61, 759–73.

Mader, H. J., 1984, Animal habitat isolation by roads and agricultural fields. *Biological conservation*, 29, 81–96.

Maignien, R., 1966, *A review of research on laterite.* UNESCO, Natural resources research, 4.

Makkaveyev, N. I., 1972, The impact of water engineering projects on geomorphic processes in stream valleys. *Soviet geography*, 13, 387–93.

Maltby, E., 1986, *Waterlogged wealth. Why waste the world's wet places?* London: Earthscan.

Manabe, S. and Stouffer, R. J., 1980, Sensitivity of a global climate model to an increase of CO_2 concentration in the atmosphere. *Journal of atmospheric science*, 37, 99–118.

Manabe, S. and Wetherald, R. T., 1986, Reduction in summer soil wetness by an increase in atmospheric carbon dioxide. *Science*, 232, 626–8.

Manabe, S., Wetherald, R. T. and Stouffer, R. J., 1981, Summer dryness due to an increase of atmosperhic CO_2 concentration. *Climate change*, 3, 347–86.

Mandel, S., 1977, The overexploitation of groundwater resources in dry regions. In Y. Munklak and S. F. Singer (eds), *Arid zone development: potentialities and problems*, Cambridge, Mass.: Ballinger, 31–52.

Manners, I. R., 1978, Agricultural activities and environmental stress. In K. A. Hammond (ed.), *Sourcebook of the environment*, Chicago: University of Chicago Press, 263–94.

Manners, I. R. and Mikesell, M. W. (eds), 1974, *Perspectives on environment.* Washington DC: Association of American Geographers.

Mannion, A. M., 1991, *Global environmental change.* Harlow: Longman.

——, 1992, Acidification and eutrophication. In A. M. Mannion and S. R. Bowlby (eds), *Environmental issues in the 1990s*. Chichester: Wiley, 177–95.

Manshard, W., 1974, *Tropical agriculture.* London: Longman.

Mark, A. F. and McSweeney, G. D., 1990, Patterns of impoverishment in natural communities: case studies in Forest ecosystems – New Zealand. In G. M. Woodwell (ed.), *The earth in transition: patterns and processes of biotic impoverishment.* Cambridge: Cambridge University Press, 151–76.

Marks, P. L. and Bormann, F. H., 1972, Revegetation following forest cutting: mechanisms for return to steady-state nutrient cycling. *Science*, 176, 914–15.

Marquiss, M., Newton, I. and Ratcliffe, D. A., 1978, The decline of the raven, *Corvus corax*, in relation to afforestation in southern Scotland and northern England. *Journal of applied ecology*, 15, 129–44.

Marsh, G. P., 1864, *Man and nature.* New York: Scribner.

—— 1965, *Man and nature*, edited by D. Lowenthal. Cambridge, Mass.: Belknap Press.

Marshall, L. G., 1984, Who killed Cock Robin? An investigation of the extinction controversy. In P. S. Martin and R. G. Klein (eds), *Quaternary extinctions*. Tucson: University of Arizona Press.

Martens, L. A., 1968, Flood inundation and effects of urbanization in Metropolitan Charlotte, North Carolina. *United States geological survey water supply paper*, 1591–C.

Martin, P. S., 1967, Prehistoric overkill. In P. S. Martin and H. E. Wright (eds), *Pleistocene extinctions*. New Haven: Yale University Press, 75–120.

——, 1974, Palaeolithic players on the American stage: man's impact on the late Pleistocene megafauna. In J. D. Ives and R. G. Barry (eds), *Arctic and alpine environments*. London: Methuen.

——, 1982, The pattern and meaning of Holarctic mammoth extinction. In D. M. Hopkins, J. V. Matthews, C. S. Schweger and S. B. Young (eds), *Paleoecology of Beringia*. New York: Academic Press, 399–408.

Martin, P. S. and Klein, R. G., 1984, *Pleistocene extinctions*. Tucson: University of Arizona Press.

Martin, P. S. and Wright, H. E. (eds), 1967, *Pleistocene extinctions*. New Haven: Yale University Press, 75–120.

Martinez, J. D. 1971, Environmental significance of salt. *Bulletin of the American Association of Petroleum Geologists*, 55, 810–25.

Marx, J. L., 1975, Air pollution: effects on plants. *Science*, 187, 731–3.

Mather, A. S., 1983, Land deterioration in upland Britain. *Progress in physical geography*, 7, 2, 210–28.

Matley, I. M., 1966, The Marxist approach to the geographical environment. *Annals of the Association of American Geographers*, 56, 97–111.

Matthews, W. H., Kellogg, W. W. and Robinson, G. D. (eds), 1971, *Man's impact on the climate*. Cambridge, Mass.: MIT Press.

Maugh, T. H., 1979, The Dead Sea is alive and well . . . *Science*, 205, 178.

May, R. M., 1979, Fluctuations in abundance of tropical insects. *Nature*, 278, 505–7.

May, T., 1991, Südspanische matorrales als Kulturofolge-vegetation. *Geoökodynamik* 12, 87–107.

Mead, W. R., 1954, Ridge and furrow in Buckinghamshire. *Geographical journal*, 120, 34–42.

Meade, R. B., 1991, Reservoirs and earthquakes. *Engineering Geology*, 30, 245–62.

Meade, R. H. and Trimble, S. W., 1974, Changes in sediment loads in rivers of the Atlantic drainage of the United States since 1900. *Publication of the International Association of Hydrological Science*, 113, 99–104.

Mellanby, K., 1967, *Pesticides and pollution*. London: Fontana.

Mendelssohn, H., 1973, Ecological effects of chemical control of rodents and jackals in Israel. In H. T. Farvar and J. P. Milton (eds), *The careless technology*. London: Tom Stacey, 527–44.

Mercer, D. E. and Hamilton, L. S., 1984, Mangrove ecosystems: some economic and natural benefits. *Nature and resources*, 20, 14–19.

Mercer, J. H., 1978, West Antarctic ice sheet and CO_2 greenhouse effect: a threat of disaster. *Nature*, 271, 321–5.

Merryfield, D. L. and Moore, P. D., 1971, Prehistoric human activity and blanket peat initiation on Exmoor. *Nature*, 250, 439–41.

Meybeck, M., 1979, Concentration des eaux fluviales en éléments majeurs et apports en solution aux océans. *Revue de géographie physique et géologie dynamique*, 21a, 215–46.

Micklin, P. P., 1972, Dimensions of the Caspian Sea problem. *Soviet geography*, 13, 589–603.

Mieck, I., 1990, Reflections on a typology of historical pollution: complementary conceptions. In P. Brimblecombe and C. Pfister, (eds), *The silent countdown*. Berlin: Springer Verlag, 73–80.

Mikesell, M. W., 1969, The deforestation of Mount Lebanon. *Geographical review*, 59, 1–28.

Miller, R. S. and Botkin, D. B., 1974, Endangered species: models and predictions. *American scientist*, 62, 172–81.

Milliman, J. D., 1990, Fluvial sediment in coastal seas: flux and fate. *Nature and Resources*, 26, 12–22.

Milliman, J. D., Qin, Y. S., Ren, M. E. and Yoshiki Saita, 1987, Man's influence on erosion and transport of sediment by Asian rivers: the Yellow River (Huanghe) example. *Journal of geology*, 95, 751–62.

Milne, W. G., 1976, Induced seismicity. *Engineering geology*, 10, 83–388.

Ministry of Agriculture, Fisheries and Food, 1976, *Agriculture and water quality*. London: Her Majesty's Stationery Office.

Mintzer, I. M. and Miller, A. S., 1992, Stratospheric ozone depletion: can we save the sky? In *Green Globe Yearbook 1992*. Oxford: Oxford University Press, 83–91.

Mitchell, J. F. B., 1983, The seasonal response of a general circulation model to changes in CO_2 and sea temperatures. *Quarterly journal of the Royal Meteorological Society*, 109, 113–52.

Mitchell, J. F. B., Wilson, C. A. and Cunnington, W. M., 1987, On CO_2 climate sensitivity and model dependence of results. *Quarterly journal of the Royal Meteorological Society*, 113, 293–322.

Mooney, H. A. and Parsons, D. J., 1973, Structure and function of the California Chaparral – an example from San Dimas. *Ecological studies*, 7, 83–112.

Moore, D. M., 1983, Human impact on island vegetation. In W. Holzner, M. J. A. Werger and I. Ikusima (eds), *Man's impact on vegetation*. The Hague: Junk, 237–48.

Moore, N. W., Hooper, M. D. and Davis, B. N. K., 1967, Hedges, I. Introduction and reconnaissance studies. *Journal of applied ecology*, 4, 201–20.

Moore, P. D., 1973, Origin of blanket mires. *Nature*, 256, 267–9.

——, 1986, Unravelling human effects. *Nature*, 321, 204.

Moore, R. M., 1959, Ecological observations on plant communities grazed by sheep in Australia. In A. Keast, R. L. Crocker and C. S. Christian (eds), *Biogeography and ecology in Australia*. The Hague: Junk, 500–13.

Moore, T. R., 1979, Land use and erosion in the Machakos Hills. *Annals of the Association of American Geographers*, 69, 419–31.

Morgan, G. S. and Woods, C. A., 1986, Extinction and the zoogeography of West Indian land mammals. *Biological journal of the Linnean Society*, 28, 167–203.

Morgan, R. P. C., 1977, *Soil erosion in the United Kingdom: field studies in the Silsoe area, 1973–75*. National College of Agricultural Engineering, occasional paper, 4.

——, 1979, *Soil erosion*. London: Longman.

Morgan, W. B. and Moss, R. P., 1965, Savanna and forest in Western Nigeria. *Africa*, 35, 286–93.

Mote, V. L. and Zumbrunne, C., 1977, Anthropogenic environmental alteration of the Sea of Azov. *Soviet geography*, 18, 744–59.

Moyle, P. B., 1976, Fish introductions in California: history and impact on native fishes. *Biological conservation*, 9, 101–18.

Mrowka, J. P., 1974, Man's impact on stream regimen and quality. In I. R. Manners and M. W. Mikesell (eds), *Perspectives on environment*, Washington DC: Association of American Geographers.

Mudge, G. P., 1983, The incidence and significance of ingested lead pellet poisoning in British wildfowl. *Biological conservation*, 27, 333–72.

Murdoch, W. W., 1975, Diversity, complexity, stability and pest control. *Journal of applied ecology*, 12, 795–807.

Murozumi, M., Chow, T. J. and Paterson, C., 1969, Chemical concentrations of pollutant lead aerosols, terrestrial dusts and sea salt in Greenland and Antarctic snow strata. *Geochimica et cosmoschimica acta*, 33, 1247–94.

Murton, R. K., 1971, *Man and birds*. London: Collins.

Musk, L. F., 1991, The fog hazard. In A. H. Perry and L. J. Symons (eds), *Highway meteorology*. London: Spon, 91–130.

Myers, N., 1979, *The sinking ark: a new look at the problem of disappearing species*. Oxford: Pergamon Press.

——, 1983, Conversion rates in tropical moist forests. In F. B. Golley (ed.), *Tropical rain forest ecosytems*. Amsterdam: Elsevier Scientific, 289–300.

—— 1984, *The primary source: tropical forests and our future*. New York: Norton.

——, 1988, *Natural resource systems and human exploitation systems: physiobiotic and ecological linkages*. World Bank policy planning and research staff, environment department working paper, 12.

——, 1992, Future operational monitoring of tropical forests: an alert strategy. In J. P. Mallingreau, R. da Cunha and C. Justice (eds), *Proceedings World Forest Watch Conference*. Sao Jose dos Campos Brazil, 9–14.

Mylne, M. F. and Rowntree, P. R., 1992, Modelling the effects of albedo change associated with tropical deforestation. *Climatic Change*, 21, 317–43.

Nakano, T. and Matsuda, I., 1976, A note on land subsidence in Japan. *Geographical reports of Tokyo Metropolitan University*, 11, 147–62.

National Academy of Sciences, 1972, *The earth and human affairs*. San Francisco: Camfield Press.

Nature Conservancy Council, 1977, *Nature conservation and agriculture*. London: Her Majesty's Stationery Office.

——, 1984, *Nature conservation in Great Britain*. Shrewsbury: Nature Conservancy Council.

Nesterov, A. I., 1974, The development of anthropogenic landscape science at the Geography Faculty of Voronezh University. *Soviet geography*, 15, 463–6.

Newman, J. R., 1979, Effects of industrial pollution on wildlife. *Biological conservation*, 15, 181–90.

Newman, W. S. and Fairbridge, R. W., 1986, The management of sea-level rise. *Nature*, 320, 319–21.

Newton, J. G., 1976, Induced and natural sinkholes in Alabama: continuing problem along highway corridors. In F. R. Zwanig (ed.), *Subsidence over mines and caverns*. Washington DC: National Academy of Sciences, 9–16.

Nicholson, S. E., 1978, Climatic variations in the Sahel and other African regions during the past five centuries. *Journal of arid environments*, 1, 3–24.

——, 1988, Land surface atmosphere interaction: physical processes and surface changes and their impact. *Progress in physical geography*, 12, 36–65.

Nicod, J., 1986, Facteurs physico-chimiques de l'accumulation des formations travertineuses. *Méditerranée*, 10, 161–4.

Nikonov, A. A., 1977, Contemporary technogenic movements of the Earth's crust. *International geology review*, 19, 1245–58.

Nir, D., 1983, *Man, a geomorphological agent, an introduction to anthropic geomorphology.* Jerusalem: Keter.

Nisbet, E., 1990, Climate change and methane. *Nature*, 347, 23.

Nossin, J. J., 1972, Landsliding in the Crati basin, Calabria, Italy. *Geologie en Mijnbouw*, 51, 591–607.

Noy-Meir, I., 1974, Stability in arid ecosystems and effects of men on it. *Proceedings of the 12th International Congress of Ecology, Wageningen*, 220–5.

Nriagu, J. D., 1979, Global inventory of natural and anthropogenic emissions of trace metals in the atmosphere. *Nature*, 279, 409–11.

Nunn, P. D., 1991, *Human and natural impacts on Pacific island environments.* Occasional paper of the East–West Environment and Policy Institute, 13. Honolulu.

Nutalaya, P. and Ran, J. L., 1981, Bangkok: the sinking metropolis. *Episodes*, 4, 3–8.

Nye, P. H. and Greenland, D. J., 1964, Changes in the soil after clearing tropical forest. *Plant and Soil*, 21, 101–12.

Oberle, M., 1969, Forest fires: suppression policy has its ecological drawbacks. *Science*, 165, 568–71.

OECD (Organisation for Economic Co-operation and Development), 1991, *The state of the environment.* Paris: OECD.

Oke, T. R., 1978, *Boundary layer climates.* London: Methuen.

O'Sullivan, P. E., Coard, M. A. and Pickering, D. A., 1982, The use of laminated lake sediments in the estimation and calibration of erosion rates. *Publication of the International Association of Hydrological Science*, 137, 385–96.

Otterman, J., 1974, Baring high albedo soils by overgrazing: a hypothesised desertification mechanism. *Science*, 186, 531–3.

Overpeck, J. T., Rind, D. and Goldberg, R., 1990, Climate-induced changes in forest disturbance and vegetation. *Nature*, 343, 51–3.

Oxley, D. J., Fenton, M. B. and Carmody, G. R., 1974, The effects of roads on populations of small mammals. *Journal of applied ecology*, 11, 51–9.

Ozenda, P. and Borel, J. L., 1990, The possible responses of vegetation to a global climatic change. In M. M. Boer and R. S. de Groot (eds), *Landscape – ecological impact of climatic change.* Amsterdam: IOS Press, 221–49.

Pakiser, L. C., Eaton, J. P., Healy, J. H. and Raleigh, C. B., 1969, Earthquake prediction and control. *Science*, 166, 1467–74.

Panel on Weather and Climate Modification, 1966, *Weather and climate modification problems and prospects*, publication 1350. Washington DC: National Academy of Sciences.

Panofsky, H., 1976, Man's impact on climate. In T. A. Ferrar (ed.), *The urban costs of climate modification.* New York: Wiley.

Park, C. C., 1977, Man-induced changes in stream channel capacity. In K. J. Gregory (ed.), *River channel change.* Chichester: Wiley, 121–44.

——, 1987, *Acid rain: rhetoric and reality.* London: Methuen.

Park, R. A., Armentano, T. V. and Cloonan, C. L., 1986, Predicting the effects of sea level rise on coastal wetlands. In J. G. Titus (ed.), *Effects of changes in stratospheric ozone and global climate*, vol. 4, *Sea level rise.* Washington DC: UNEP/USEPA, 129–52.

Parkinson, C. L. and Kellogg, W. W., 1979, Arctic sea ice decay simulated for a CO_2-induced temperature rise. *Climatic change*, 2, 149–62.

Parsons, J. J., 1960, Fog drip from coastal stratus. *Weather*, 15, 58.

Passmore, J., 1974, *Man's responsibility for nature.* London: Duckworth.

Pawson, E., 1978, *The early industrial revolution: Britain in the eighteenth century.* London: Batsford.

Peck, A. J., 1975, Effects of land use on salt distribution in the soil. *Ecological studies*, 15, 77–90.

——, 1978, Salinization of non-irrigated soils and associated streams: a review. *Australian journal of soil research*. 16, 157–68.

——, 1983, Response of groundwaters to clearing in Western Australia. *Papers, international conference on groundwater and man*, 2, 327–35.

Peet, R. K., Glenn-Lewin, D. C. and Wolf, J. W., 1983, Prediction of man's impact on plant species diversity. In W. Holzner, M. J. A. Werger and I. Ikusima (eds), *Man's impact on vegetation*. The Hague: Junk, 41–54.

Peierls, B. L., Caraco, N. F., Pace, M. L. and Cole, J. J., 1991, Human influence on river nitrogen. *Nature*, 350, 386.

Pennington, W., 1974, *The history of British vegetation* (2nd edn). London: English University Press.

——, 1981, Records of a lake's life in time: the sediments. *Hydrobiologia*, 79, 197–219.

Pereira, H. C., 1973, *Land use and water resources in temperate and tropical climates*. Cambridge: Cambridge University Press.

Perla, R., 1978, Artificial release of avalanches in North America. *Arctic and Alpine research*, 10, 235–40.

Peters, R. L., 1988, The effect of global climatic change on natural communities. In E. O. Wilson (ed.), *Biodiversity*. Washington DC: National Academy Press, 450–61.

Petersen, K. K., 1981, *Oil shale, the environmental challenges*. Golden, Colorado: Colorado School of Mines.

Petts, G. E., 1979, Complex response of river channel morphology subsequent to reservoir construction. *Progress in physical geography*, 3, 329–62.

——, 1985, *Impounded rivers: perspectives for ecological management*. Chichester: Wiley.

Petts, G. E. and Lewin, J., 1979, Physical effects of reservoirs on river systems. In G. E. Hollis (ed.), *Man's impact on the hydrological cycle in the United Kingdom*, Norwich: Geobooks, 79–91.

Pimentel, D., 1976, Land degradation: effects on food and energy resources. *Science*, 194, 149–55.

Pitt, D., 1979, Throwing light on a black secret. *New scientist*, 81, 1022–5.

Pluhowski, E. J., 1970, Urbanization and its effects on the temperature of the streams on Long Island, New York. *United States geological survey professional paper*, 627–D.

Pollard, E. and Miller, A., 1968, Wind erosion in the East Anglian Fens. *Weather*, 23, 414–17.

Ponting, C., 1991, *A green history of the world*. London: Penguin.

Poore, M. E. D., 1976, The values of tropical moist forest ecosystems. *Unasylva*, 28, 127–43.

Pope, J. C., 1970, Plaggen soils in the Netherlands. *Geoderma*, 4, 229–55.

Porter, E., 1978, *Water management in England and Wales*. Cambridge: Cambridge University Press.

Potter, G. L., Ellsaesser, H. W., MacCracken, M. C. and Luther, F. M., 1975, Possible climatic impact of tropical deforestation. *Nature*, 258, 697–8.

Potter, G. L., Ellsaesser, H. W., MacCracken, M. C. and Ellis, J. C., 1981, Albedo change by man: test of climatic effects. *Nature*, 291, 47–9.

Preston, A., 1973, Heavy metals in British waters. *Nature*, 242, 95–7.

Price, M. and Reed, D. W., 1989, The influence of mains leakage and urban drainage on groundwater levels beneath conurbations in the United Kingdom. *Proceedings Institution of Civil Engineers*, 86(I), 31–9.

Prince, H. C., 1959, Parkland in the Chilterns. *Geographical review*, 49, 18–31.

—— 1962, Pits and ponds in Norfolk. *Erdkunde*, 16, 10–31.

——, 1964, The origin of pits and depressions in Norfolk. *Geography*, 49, 15–32.

——, 1979, Marl pits or dolines of the Dorset Chalklands? *Transactions of the Institute of British Geographers*, new series, 4, 116–17.

Proffitt, M. H., Margitan, J. J., Kelly, K. K., Loewenstein, M., Podolske, J. R. and Chan, K. R., 1990, Ozone loss in the Arctic polar vortex inferred from high-altitude aircraft measurements. *Nature*, 347, 31–3.

Prokopovich, N. P., 1972, Land subsidence and population growth. *24th International Geological Congress Proceedings*, 13, 44–54.

Prospero, J. M. and Nees, R. T., 1977, Dust concentration in the atmosphere of the equatorial North Atlantic; possible relationship to Sahelian drought. *Science*, 196, 1196–8.

Pyne, S. J., 1982, *Fire in America – a cultural history of wildland and rural fire*. Princeton: Princeton University Press.

Rackham, O., 1980, *Ancient woodland*. London: Arnold.

Radley, J., 1962, Peat erosion on the high moors of Derbyshire and west Yorkshire. *East Midlands Geographer*, 3(1), 40–50.

Radley, J. and Sims, C., 1967, Wind erosion in East Yorkshire. *Nature*, 216, 20–2.

Raison, R. J., 1979, Modification of the soil environment by vegetation fires with particular reference to nitrogen transformation: a review. *Plant and Soil*, 51, 73–108.

Raleigh, C. B., Healy, J. H. and Bredehoeft, J. D., 1976, An experiment in earthquake control at Rangeley, Colorado. *Science*, 191, 1230–7.

Ramanathan, V., 1988, The greenhouse theory of climate change: a test by an inadvertent global experiment. *Science*, 240, 293–9.

Ranwell, D. S. and Boar, R., 1986, *Coast dune management guide*. Monks Wood: Institute of Terrestrial Ecology.

Rapp, A., 1974, *A review of desertization in Africa – water, vegetation and man*. Secretariat for International Ecology, Stockholm, report no. 1.

Rapp, A., Murray-Rust, D. H., Christansson, C. and Berry, L., 1972, Soil erosion and sedimentation in four catchments near Dodoma, Tanzania. *Geografiska annaler*, 54A, 255–318.

Rapp, A., Le Houérou, H. N. and Lundholm, B., 1976, Can desert encroachment be stopped? *Ecological bulletin*, 24.

Rasid, H., 1979, The effects of regime regulation by the Gardiner Dam on downstream geomorphic processes in the South Saskatchewan River. *Canadian geographer*, 23, 140–58.

Rasool, S. I. and Schneider, S. H., 1971, Atmospheric carbon dioxide and aerosols: effects of large increases on global climate. *Science*, 173, 138–41.

Ratcliffe, D. A., 1974, Ecological effects of mineral exploitation in the United Kingdom and their significance to nature conservation. *Proceedings of the Royal Society of London*, 339A, 355–72.

Reck, R. A., 1975, Aerosols and polar temperature changes. *Science*, 188, 728–30.

Reclus, E., 1871, *The earth* (2 vols). London: Chapman & Hall.

——, 1873, *The ocean, atmosphere and life*. New York: Harper and Brothers.

Reed, C. A., 1970, Extinction of mammalian megafauna in the old world late Quaternary. *Bioscience*, 20, 284–8.

Reed, D. J., 1990, The impact of sea level rise on coastal salt marshes. *Progress in Physical Geography*, 14, 465–81.

Reed, L. A., 1980, Suspended-sediment discharge, in five streams near Harrisburg, Pennsylvania, before, during and after highway construction. *United States geological survey water supply paper*, 2072.

Renberg, I. and Hellberg, T., 1982, The pH history of lakes in SW Sweden, as calculated from the subfossil diatom flora of the sediments. *Ambio*, 11, 30–3.

Revelle, R. R. and Waggoner, P. E., 1983, Effect of a carbon dioxide-induced climatic change on water supplies in the western United States. In Carbon Dioxide Assessment Committee, *Changing climate*. Washington DC: National Academy Press, 419–32.

Rhoades, J. D., 1990, Soil salinity – causes and controls. In A. S. Goudie (ed.), *Techniques for Desert Reclamation*. Chichester: Wiley, 109–34.

Richards, J. F., 1991, Land transformation. In B. L. Turner, W. C. Clark, R. W. Kates, J. F. Richards, J. T. Matthews and W. B. Meyer (eds) *The earth as transformed by human action*. Cambridge: Cambridge University Press, 163–78.

Richards, K. S. and Wood, R., 1977, Urbanization, water redistribution, and their effect on channel progresses. In K. J. Gregory (ed.), *River channel changes*. Chichester: Wiley, 369–88.

Richards, P. W., 1952, *The tropical rainforest*. Cambridge: Cambridge University Press.

Richardson, J. A., 1976, Pit heap into pasture. In J. Lenihan and W. W. Fletcher (eds), *Reclamation*. Glasgow: Blackie, 60–93.

Richardson, S. J. and Smith, J., 1977, Peat wastage in the East Anglian Fens. *Journal of soil science*, 28, 485–9.

Richter, D. O. and Babbar, L. I., 1991, Soil diversity in the tropics. *Advances in Ecological Research*, 21, 315–89

Ripley, E. A., 1976, Drought in the Sahara: insufficient geophysical feedback? *Science*, 191, 100.

Ritchie, J. C., 1972, Sediment, fish and fish habitat. *Journal of soil and water conservation*, 27, 124–5.

Roberts, N., 1989, *The Holocene: an environmental history*. Oxford: Basil Blackwell.

Robin, G. de Q., 1986, Changing the sea level. In B. Bolin et al. (eds), *The greenhouse effect, climatic change and ecosystems*. Chichester: Wiley, 322–59.

Robinson, M., 1979, The effects of pre-afforestation ditching upon the water and sediment yields of a small upland catchment. Working paper 252, School of Geography, University of Leeds.

——, 1990, *Impact of improved land drainage on river flows*. Institute of Hydrology, Wallingford. Report 113.

Robinson, M. A. and Lambrick, G. H., 1984, Holocene alluviation and hydrology in the Upper Thames Basin. *Nature*, 308, 809–14.

Rodda, J. C., Downing, R. A. and Law, F. M., 1976, *Systematic hydrology*. London: Newnes-Butterworth.

Roots, C., 1976, *Animal invaders*. Newton Abbot: David & Charles.

Rose, R., 1970, Lichens as pollution indicators. *Your environment*, 5.

Rosepiler, M. J. and Reilinger, R., 1977, Land subsidence due to water withdrawal in the vicinity of Pecos, Texas. *Engineering geology*, 11, 295–304.

Routson, R. C., Wildung, R. E. and Bean, R. M., 1979, A review of the environmental impact of ground disposal of oil shale wastes. *Journal of environmental quality*, 8, 14–19.

Royal Commission on Environmental Pollution, 1972, *Pollution in some British estuaries and coastal waters*. London: Her Majesty's Stationery Office.

Royal Society Study Group, 1983, *The nitrogen cycle of the United Kingdom*. London: The Royal Society.

Rozanov, B. G., Targulian, V. and Orlov, D. S., 1991, Soils. In B. L.

Turner, W. C. Clark, R. W. Kates, J. F. Richards, J. T. Matthews and W. B. Meyer (eds), *The earth as transformed by human action*. Cambridge: Cambridge University Press, 203–14.

Russell, E. J., 1961, *The world of the soil*. London: Fontana.

Russell, J. S. and Isbell, R. F. (eds), 1986, *Australian soils: the human impact*. St Lucia: University of Queensland Press.

Ryder, M. L., 1966, The exploitation of animals by man. *Advancement of science*, 23, 9–18.

Ryding, S. O. and Rast, R. W., 1989, *The control of eutrophication of lakes and reservoirs*. Paris: UNESCO.

Sabadell, J. E., Risley, E. M., Jorgensen, H. T. and Thornton, B. S. 1982, *Desertification in the United States: Status and Issues*. Bureau of Land Management, Department of the Interior.

Sanchez, P. A. and Buol, S. W., 1975, Soils of the tropics and the world food crisis. *Science*, 188, 598–603.

Sanders, W. M., 1972, Nutrients. In R. T. Oglesby, C. A. Carlson and J. A. McCann (eds), *River ecology and man*. New York: Academic Press, 389–415.

Sarmiento, G. and Monasterio, M., 1975, A critical consideration of the environmental conditions associated with the occurrence of savanna ecosystems in tropical America. *Ecological studies*, 11, 233–50.

Sarre, P., 1978, The diffusion of Dutch elm disease. *Area*, 10, 81–5.

Sauer, C. O., 1938, Destructive exploitation in modern colonial expansion. *International Geographical Congress, Amsterdam*, vol. III, sect. IIIC, 494–9.

——, 1952, *Agricultural origins and dispersals*, Isaiah Bowman lecture series, 2. New York: American Geographical Society.

——, 1969, *Seeds, spades, hearths and herds*. Cambridge, Mass.: MIT Press.

Savage, M., 1991, Structural dynamics of a southwestern pine forest under chronic human influence. *Annals of the Association of American Geographers*, 81, 271–89.

Savini, J. and Kammerer, J. C., 1961, Urban growth and the water regime. *United States geological survey water supply paper*, 1591A.

Sawyer, J. S., 1971, Possible effects of human activity on world climate. *Weather*, 26, 252–62.

Schimper, A. F. W., 1903, *Plant-geography upon a physiological basis*. Oxford: Clarendon Press.

Schindler, D. W., 1974, Eutrophication and recovery in experimental lakes: implications for lake management. *Science*, 184, 897–9.

Schlesinger, M. E. and Mitchell, J. F. B., 1985, Model projections of equilibrium response to increased carbon dioxide. In M. C. MacCracken and F. M. Luther (eds), *Projecting the climatic effects of increasing carbon dioxide*. Washington DC: US Dept of Energy, 81–147.

Schmid, J. A., 1974, The environmental impact of urbanization. In I. R. Manners and M. W. Mikesell (eds), *Perspectives on environment*. Washington DC: Association of American Geographers.

——, 1975, Urban vegetation. Research paper, Department of Geography, University of Chicago, 161.

Schmieder, O., 1927a, The Pampa – a natural or culturally induced grassland? *University of California publications in geography*, 2, 255–70.

——, 1927b, Alteration of the Argentine Pampa in the colonial period. *University of California publications in geography*, 2, 303–21.

Schneider, S. H., 1984, On the empirical verification of model-predicted CO_2-induced climatic effects. *Geophysical monograph*, 29, 187–201.

Schneider, S. H. and Thompson, S. L., 1988, Simulating the effects of nuclear war. *Nature*, 333, 221–7.

Schumm, S. A., 1977, *The fluvial system*. New York: Wiley.

Schumm, S. A., Harvey, M. D. and Watson, C. C., 1984, *Incised channels: morphology, dynamics and control*. Littleton, Colorado: Water Resources Publications.

Schwarz, E. H. L., 1923, *The Kalahari or Thirsland redemption*. Cape Town and Oxford: Oxford University Press.

Schwarz, H. E., Emel, J., Dickens, W. J., Rogers, P. and Thompson, J., 1991, Water quality and flows. In B. L. Turner, W. C. Clark, R. W. Kates, J. F. Richards, J. T. Matthews and W. B. Meyer (eds), *The earth as transformed by human action*. Cambridge: Cambridge University Press, 253–70.

Scott, G. J., 1977, The role of fire in the creation and maintenance of savanna in the Montana of Peru. *Journal of biogeography*, 4, 143–67.

Sears, P. B., 1957, Man the newcomer: the living landscape and a new tenant. In L. H. Russwurm and E. Sommerville (eds), *Man's natural environment, a system approach*. North Scituate: Duxbury, 43–55.

Segall, P., 1989, Earthquakes triggered by fluid extraction. *Geology*, 17, 942–6.

Selby, M. J., 1979, Slopes and weathering. In K. J. Gregory and D. E. Walling (eds), *Man and environmental processes*. Folkestone: Dawson, 105–22.

Semtner, A. J., 1984, The climate response of the Arctic Ocean to Soviet river diversions. *Climatic change*, 6, 109–30.

Shaler, N. S., 1912, *Man and the earth*. New York: Duffield.

Sheail, J., 1971, *Rabbits and their history*. Newton Abbot: David & Charles.

Sherlock, R. L., 1992, *Man as a geological agent*. London: Witherby.

Sherratt, A., 1981, Plough and pastoralism: aspects of the secondary products revolution. In I. Hodder, G. Isaac and N. Hammond (eds), *Pattern of the past*. Cambridge, Cambridge University Press, 261–305.

Shiklomanov, I. A., 1985, Large scale water transfers. In J. C. Rodda (ed.), *Facets of hydrology II*. Chichester: Wiley, 345–87.

Shirahata, H., Elias, R. W., Patterson, C. C. and Koide, M., 1980, Chronological variations in concentrations and isotopic composition of anthropogenic atmospheric lead in sediments of a remote sub-alpine pond. *Geochimica et cosmochimica acta*, 44, 149–62.

Sinclair, A. R. E. and Fryxell, J. M., 1985, The Sahel of Africa: ecology of a disaster. *Canadian journal of zoology*, 63, 987–94.

Simmonds, N. W., 1976, *Evolution of crop plants*. London: Longman.

Simmons, I., 1979, *Biogeography: natural and cultural*. London: Arnold.

——, 1989, *Changing the face of the earth*. Oxford: Basil Blackwell.

Simon, J. L. and Kahn, H., 1984, *The resourceful earth – a response to global 2000*. Oxford: Basil Blackwell.

Simoons, F. J., 1974, Contemporary research themes in the cultural geography of domesticated animals. *Geophysical review*, 64, 557–76.

Smith, C. T., 1978, *An historical geography of western Europe before 1800* (2nd edn). London: Longman.

Smith, J. E. (ed.), 1968, *'Torrey Canyon' pollution and marine life*. Cambridge: Cambridge University Press.

Smith, K., 1975, *Principles of applied climatology*. London: McGraw-Hill.

Smith, L. B. and Hadley, C. H., 1926. The Japanese beetle. *Department Circular, US Department of Agriculture*, 363, 1–66.

Smith, N., 1976, *Man and water*. London: Davies.

Smith, T. M., Shugart, H. H., Bonan, G. B. and Smith, J. B., 1992, Modelling the potential response of vegetation to global climate change. *Advances in Ecological Research*, 22, 93–116.

Smith, W. H., 1974, Air pollution – effects on the structure and function of the temperate forest ecosystem. *Environmental pollution*, 6, 111–29.

Snelgrove, A. K., 1967, *Geohydrology of the Indus River, West Pakistan*. Hyderabad: Sind University Press.

So, C. L., 1971, Mass movements associated with the rainstorm of 1966 in Hong Kong. *Transactions of the Institute of British Geographers*, 53, 55–65.

Solomon, A. M. and West, D. C., 1986, Atmospheric carbon dioxide change: agent of future forest growth or decline. In J. G. Titus (ed.), *Effects of changes in stratospheric ozone and global climate*, vol. 3, *Climatic change*. Washington DC: UNEP/USEPA, 23–8.

Somerville, M., 1858, *Physical geography* (4th edn). London: Murray.

Sopper, W. E., 1975, Effects of timber harvesting and related management practices on water quality in forested watersheds. *Journal of environmental quality*, 4, 24–9.

Sorkin, A. J., 1982, *Economic aspects of natural hazards*. Lexington: Lexington Books.

Southwick, C. H., 1976, *Ecology and the quality of our environment* (2nd edn). New York: Van Nostrand.

Spanier, E. and Galil, B. S., 1991, Lessepsian migration: a continuous biogeographical process. *Endeavour*, 15, 102–6.

Sparks, B. W. and West, R. G., 1972, *The Ice Age in Britain*. London: Methuen.

Spate, O. H. K. and Learmonth, A. T. A., 1967, *India and Pakistan*. London: Methuen.

Speight, M. C. D., 1973, Outdoor recreation and its ecological effects. A bibliography and review. Discussion papers in conservation, 4, University College, London.

Spencer, J. E. and Hale, G. A., 1961, The origin, nature and distribution of agricultural terracing. *Pacific viewpoint*, 2, 1–40.

Spencer, J. E. and Thomas, W. L., 1978, *Introducing cultural geography* (2nd edn). New York: Wiley.

Spencer, T. and Douglas, I. 1985, The significance of environmental change: diversity, disturbance and tropical ecosystems. In I. Douglas and T. Spencer (eds), *Environmental change and tropical geomorphology*. London: Allen & Unwin, 13–33.

Sperling, C. H. B., Goudie, A. S., Stoddart, D. R. and Poole, G. C., 1979, Origin of the Dorset dolines. *Transactions of the Institute of British Geographers*, new series, 4, 121–4.

Speth, W. W., 1977, Carl Ortwin Sauer on destructive exploitation. *Biological Conservation*, 11, 145–60.

Steadman, D. W., Stafford, T. W., Donahue, D. J. and Jull, A. J. T., 1991, Chronology of Holocene vertebrate extinction in the Galápagos Islands. *Quaternary Research*, 36, 126–33.

Stensland, G. J. and Semonin, R. G., 1982, Another interpretation of the pH trend in the United States. *Bulletin of the American Meteorological Society*, 63, 11, 1277–84.

Stephens, J. C., 1956, Subsidence of organic soils in the Florida Everglades. *Proceedings of the Soil Science Society of America*, 20, 77–80.

Sternberg, H. O'R., 1968, Man and environmental change in South America. *Monographiae biologicae (New Guinea)*, 18, 413–45.

Stevenson, A. C., Jones, V. J. and Battarbee, R. W., 1990, The cause of peat erosion: a palaeolimnological approach. *New Phytologist*, 114, 727–35.

Stewart, O. C., 1956, Fire as the first great force employed by man. In W. L. Thomas (ed.), *Man's role in changing the face of the earth*. Chicago: University of Chicago Press, 115–33.

Stocking, M., 1984, Erosion and soil productivity: a review. *FAO soil conservation programme land and water development division consultants' working paper*, 1.

Stoddart, D. R., 1968, Catastrophic human interference with coral atoll ecosystems. *Geography*, 53, 25–40.

——, 1971, Coral reefs and islands and catastrophic storms. In J. A. Steers (ed.), *Applied coastal geomorphology*. London: Macmillan, 154–97.

——, 1990, Coral reefs and islands and predicted sea-level rise. *Progress in Physical Geography*, 14, 521–36.

Stott, P. A., 1978, Tropical rain forest in recent ecological thought: The reassessment of a non-renewable resource. *Progress in Physical Geography*, 2, 80–98.

Strahler, A. N. and Strahler, A. H., 1973, *Environmental geoscience: interaction between natural systems and man*. Santa Barbara: Hamilton.

Strandberg, C. H., 1971, Water pollution. In G. H. Smith (ed.), *Conservation of natural resources* (4th edn), New York: Wiley, 189–219.

Sturrock, F. and Cathie, J., 1980, Farm modernisation and the countryside. Occasional paper 12, Department of Land Economy, University of Cambridge.

Sugden, D. E., 1991, The stepped response of ice sheets to climatic change. In C. Harris and B. Stonehouse (eds), *Antarctica and global climatic change*. London: Bellhaven Press, 107–14.

Swank, W. T. and Douglas, J. E., 1974, Streamflow greatly reduced by converting deciduous hardwood stands to pine. *Science*, 18, 857–9.

Swanston, D. N. and Swanson, F. J., 1976, Timber harvesting, mass erosion and steepland forest geomorphology in the Pacific north-west. In D. R. Coates (ed.), *Geomorphology and engineering*. Stroudberg: Dowden, Hutchinson & Ross, 199–221.

Swift, J., 1974, *The other Eden. A new approach to man, nature and society*. London: Dent.

Swift, L. W. and Messer, J. B., 1971, Forest cuttings raise temperatures of small streams in the southern Appalachians. *Journal of soil and water conservation*, 26, 111–16.

Swift, M. J. and Sanchez, P. A., 1984, Biological management of tropical soil fertility for sustained productivity. *Nature and resources*, 20, 2–10.

Tajikistan Academy of Sciences, 1975, *Induced seismicity of the Nurek reservoir*. Dushanbe: Tajik Academy.

Tallis, J. H., 1965, Studies on southern Pennine peats, IV: evidence of recent erosion. *Journal of Ecology*, 53, 509–20.

Tallis, J. H., 1985, Erosion of blanket peat in the southern Pennines: new light on an old problem. In R. H. Johnson (ed.), *The geomorphology of north-west England*. Manchester: Manchester University Press, 313–36.

Taylor, C., 1975, *Fields in the English landscape*. London: Dent.

Taylor, J. A., 1985, Bracken encroachment rates in Britain. *Soil use and management*, 1, 53–6.

Thirgood, J. V., 1981, *Man and the Mediterranean forest – a history of resource depletion*. London: Academic Press.

Thomas, R. H., Sanderson, T. J. O. and Rose, K. E., 1979, Effect of climatic warming on the West Antarctic ice sheet. *Nature*, 277, 355–8.

Thomas, W. L. (ed.), 1956, *Man's role in changing the face of the Earth*. Chicago: University of Chicago Press.

Thompson, J. R., 1970, Soil erosion in the Detroit metropolitan area. *Journal of Soil and Water Conservation*, 25, 8–10.

Thompson, R., 1978, Atmospheric contamination – a review of the air pollution problem. *Geographical Papers, Department of Geography, University of Reading*, 68.

Thornthwaite, C. W., 1956, Modification of the rural microclimates. In W. L. Thomas (ed.), *Man's role in changing the face of the Earth*. Chicago: University of Chicago Press, 567–83.

Titus, J. G. (ed.), 1986, *Effects of changes in stratospheric ozone and global climate*, vol. 4, *Sea level rise*. Washington DC: UNEP/USEPA.

Titus, J. G. and Seidel, S., 1986, Overview of the effects of changing the atmosphere. In J. G. Titus (ed.), *Effects of changes in stratospheric ozone and global climate*. Washington DC: UNEP/USEPA, 3–19.

Tivy, J., 1971, *Biogeography. A study of plants in the ecosphere*. Edinburgh: Oliver and Boyd.

Tivy, J. and O'Hare, G., 1981, *Human impact on the ecosystem*. Edinburgh: Oliver and Boyd.

Todd, D. K., 1959, *Groundwater hydrology*. New York: Wiley.

——, 1964, Groundwater. In Ven Te Chow (ed.), *Handbook of applied hydrology*, sect. 13. New York: McGraw-Hill.

Tomaselli, R., 1977, Degradation of the Mediterranean maquis. *UNESCO, MAB technical note*, 2, 33–72.

Tomlinson, T. E., 1970, Trends in nitrate concentration in English rivers in relation to fertilizer use. *Water treatment and examination*, 19, 277–89.

Trimble, S. W., 1974, *Man-induced soil erosion on the southern Piedmont*. Soil Conservation Society of America.

——, 1976, Modern stream and valley sedimentation in the Driftless Area, Wisconsin, USA. *23rd International Geographical Congress*, sect. 1, 228–31.

——, 1988, The impact of organisms on overall erosion rates within catchments in temperate regions. In H. A. Viles (ed.), *Biogeomorphology*. Oxford: Basil Blackwell, 83–142.

Trimble, S. W. and Lund, S. W., 1982, Soil conservation and the reduction of erosion and sedimentation in the Coon Creek Basin, Wisconsin. *US geological survey professional paper*, 1234.

Troels-Smith, J., 1956, Neolithic period in Switzerland and Denmark. *Science*, 124, 876–9.

Turco, R. P., Toon, O. B., Ackermann, T. P., Pollack, J. B. and Sagan, C., 1983, Nuclear winter: global consequences of multiple nuclear explosions. *Science*, 222, 1283–92.

Turner, B. L., Kasperson, R. E., Meyer, W. B., Dow, K. M., Golding, D., Kasperson, J. X., Mitchell, R. C. and Ratick, S. J., 1990, Two types of global environmental change: definitional and spatial-scale issues in their human dimensions. *Global Environmental Change*, 1, 14–22.

Tyldesley, J. A. and Bahn, P. G., 1983, The use of plants in the European palaeolithic: a review of the evidence. *Quaternary Science Reviews*, 2, 53–81.

UNEP, 1989, *Environmental data report*. Oxford and Cambridge, Mass.: Basil Blackwell.

——, 1991, *United Nations Environment Programme environmental data report*, 3rd edn. Oxford: Basil Blackwell.

US Bureau of Entomology and Plant Quarantine, 1941, *Insect pest survey bulletin*, 21, 801–2.

US Department of Energy, 1985, *Glaciers, ice sheets, and sea level: effect of a CO_2-induced climatic change*. Washington DC: US Dept of Energy.

Usher, M. B., 1973, *Biological management and conservation*. London: Chapman & Hall.

Vale, T. R., 1974, Sagebrush conversion projects: an element of contemporary environmental change in the western United States. *Biological conservation*, 6, 272–84.

Vale, T. R. and Vale, G. R., 1976, Suburban bird population in west-central California. *Journal of biogeography*, 3, 157–65.

Vandermeulen, J. H. and Hrudey, S. E. (eds), 1987, *Oil in freshwater: chemistry, biology, countermeasure technology*. New York: Pergamon.

Vankat, J. L., 1977, Fire and man in Sequoia National Park. *Annals of the Association of American Geographers*, 67, 17–27.

Vaudour, J., 1986, Travertins holocènes et pression anthropique. *Mediterranée*, 10, 168–73.

Veblen, T. T. and Stewart, G. H., 1982, The effects of introduced wild animals on New Zealand forests. *Annals of the Association of American Geographers*, 72, 372–97.

Vendrov, S. L., 1965, A forecast of changes in natural conditions in the northern Ob'basin in case of construction of the lower Ob'Hydro Project. *Soviet geography*, 6, 3–18.

Vice, R. B., Guy, H. P. and Ferguson, G. E., 1969, Sediment movement in an area of suburban highway construction, Scott Run Basin, Fairfax, County, Virginia, 1961–64. *United States geological survey water supply paper*, 1591–E.

Viessman, W., Knapp, J. W., Lewis, G. L. and Harbaugh, T. E., 1977, *Introduction to hydrology* (2nd edn). New York: IEP.

Viets, F. G., 1971, Water quality in relation to farm use of fertilizer. *Bioscience*, 21, 460–7.

Vine, H., 1968, Developments in the study of soils and shifting agriculture in tropical Africa. In R. P. Moss (ed.), *The soil resources of tropical Africa*. Cambridge: Cambridge University Press, 89–119.

Vita-Finzi, C., 1969, *The Mediterranean valleys*. Cambridge: Cambridge University Press.

Vitousek, P. M., Gosz, J. R., Gruer, C. C., Melillo, J. M., Reiners, W. A. and Todd, R. L., 1979, Nitrate losses from disturbed ecosystems. *Science*, 204, 469–73.

Vogl, R. J., 1974, Effects of fires on grasslands. In T. T. Kozlowski and C. C. Ahlgren (eds), *Fire and ecosystems*. New York: Academic Press, 139–94.

——, 1977, Fire: a destructive menace or a rational process. In J. Cairns, K. L. Dickson and E. E. Herricks (eds), *Recovery and restoration of damaged ecosystems*. Charlottesville: University Press of Virginia, 261–89.

Von Broembsen, S. L., 1989, Invasions of natural ecosystems by plant pathogens. In J. A. Drake (ed.), *Biological invasions: a global perspective*. Chichester: Wiley, 77–83.

Wagner, R. H., 1974, *Environment and man*. New York: Norton.

Waldichuk, M., 1979, Review of the problems. *Philosophical transactions of the Royal Society*, 286B, 399–429.

Wall, G. and Wright, C., 1977, The environmental impact of outdoor recreation. *Department of Geography publication series, University of Waterloo, Canada*, 11.

Walling, D. E. and Gregory, K. J., 1970, The measurement of the effects of building construction on drainage basin dynamics. *Journal of hydrology*, 11, 129–44.

Wallwork, K. L., 1956, Subsidence in the mid-Cheshire industrial area. *Geographical journal*, 122, 40–53.

——, 1960, Some problems of subsidence and land use in the mid-Cheshire industrial area. *Geographical journal*, 126, 191–9.

——, 1974, *Derelict land*. Newton Abbot: David & Charles.

Walker, H. J., 1988, *Artificial structures and shorelines*. Dordrecht: Kluwer.

Walker, H. J., Coleman, J. M., Roberts, H. H. and Tye, R. S., 1987, Wetland loss in Louisiana. *Geografiska annaler*, 69A, 189–200.

Walling, D. E. and Quine, T. A., 1991, *Recent rates of soil loss from areas of arable cultivation in the UK*. IAHS Publication 203, 123–31.

Walsh, R. P., Hudson, R. N. and Howells, K. A., 1982, Changes in the magnitude-frequency of flooding and heavy rainfalls in the Swansea Valley since 1875. *Cambria*, 9, 2, 36–60.

Wang, W. C., Yung, Y. L., Lacis, A. A., Mo, T. and Hanson, J. E., 1976, Greenhouse effects due to man-made perturbations of trace gases. *Science*, 194, 685–90.

Ward, R. C., 1978, *Floods – a geographical perspective*. London: Macmillan.

Ward, S. D., 1979, Limestone pavements – a biologist's view. *Earth science conservation*, 16, 16–18.

Warren, A. and Maizels, J. K., 1976, *Ecological change and desertification*. London: University College.

Warren, A. and Maizels, J. K., 1977, Ecological change and desertification. In United Nations, *Desertification: its causes and consequences*. Oxford: Pergamon, 171–260.

Warrick, R. A., 1988, Carbon dioxide, climatic change and agriculture. *Geographical journal*, 154, 221–33.

Watson, A., 1976, The origin and distribution of closed depressions in south-west Lancashire and north-west Cheshire. Unpublished BA dissertation, University of Oxford.

Weare, B. C., Temkin, R. L. and Snell, C. M., 1974, Aerosols and climate: some further considerations. *Science*, 186, 827–8.

Weaver, H., 1974, Effects of fire on temperate forests: western United States. In T. T. Kozlowski and C. C. Ahlgren (eds), *Fire and ecosystems*. New York: Academic Press, 279–319.

Weaver, J. E., 1954, *North American prairie*. Lincoln: Johnsen.

Webb, T. and Wigley, T. M. L., 1985, What past climates can indicate about a warmer world. In M. C. MacCracken and F. M. Luther (eds), *Projecting the climatic effects of increasing carbon dioxide*, Washington DC: US Dept of Energy, 237–57.

Wein, R. W. and Maclean, D. A. (eds), 1983, *The role of fire in northern circumpolar ecosystems*. Chichester: Wiley.

Weisrock, A., 1986, Variations climatiques et periodes de sedimentation carbonatée a l'Holocène-l'age des depôts. *Mediterrannée*, 10, 165–7.

Wellburn, A., 1988, *Air pollution and acid rain: the biological impact*. London: Longman.

Wells, P. V., 1965, Scarp woodlands, transported grass soils and concept of grassland climate in the Great Plains region. *Science*, 148, 246–9.

Wertine, T. A., 1973, Pyrotechnology: man's first industrial uses of fire. *American scientist*, 61, 670–82.

Westhoff, V., 1983, Man's attitude towards vegetation. In W. Holzner, M. J. A. Werger and I. Ikusima (eds), *Man's impact on vegetation*. The Hague: Junk, 7–24.

Westing, A. and Pfeiffer, E. W., 1972, The cratering of Indochina. *Scientific American*, 226, 5, 21–9.

Whitaker, J. R., 1940, World view of destruction and conservation of natural resources. *Annals of the Association of American Geographers*, 30, 143–62.

Whittaker, E., 1961, Temperatures in heath fires. *Journal of ecology*, 49, 709–15.

Whyte, A. V. T., 1977, Guidelines for field studies in environmental perception. *UNESCO, MAB technical note 5*.

Wigley, T. M. L., 1983, The pre-industrial carbon dioxide level. *Climatic change*, 5, 315–20.

Wigley, T. M. L., Jones, P. D. and Kelly, P. M., 1980, Scenario for a warm, high-CO_2 world. *Nature*, 283, 17–21.

Wilken, G. C., 1972, Microclimate management by traditional farmers. *Geographical review*, 62, 544–60.

Wilkinson, W. B. and Brassington, F. C., 1991, Rising groundwater levels – an international problem. In R. A. Downing and W. B. Wilkinson (eds), *Applied groundwater hydrology – a British perspective.* Oxford: Clarendon Press, 35–53.

Williams, G. P., 1978, The case of the shrinking channels – the North Platte and Platte rivers in Nebraska. *United States geological survey circular,* 781.

Williams, M., 1970, *The draining of the Somerset levels.* Cambridge: Cambridge University Press.

——, 1988, The death and rebirth of the American forest: clearing and reversion in the United States, 1900–1980. In J. F. Richards and R. P. Tucker (eds), *World deforestation in the twentieth century.* Durham, NC and London: Duke University Press, 211–29.

——, 1989, *Americans and their Forests,* Cambridge: Cambridge University Press.

——, 1990, *Wetlands: a threatened landscape.* Oxford: Basil Blackwell.

Williams, R. S. and Moore, J. G., 1973, Iceland chills lava flow. *Geotimes.* 18, 14–17.

Williamson, D. R., 1983, The application of salt and water balances to quantify causes of the dryland salinity problem in Victoria. *Proceedings of the Royal Society of Victoria,* 95, 103–11.

Wilshire, H. G., 1980, Human causes of accelerated wind erosion in California's deserts. In D. R. Coates and J. D. Vitek (eds), *Geomorphic thresholds.* Stroudsburg: Dowden, Hutchinson & Ross, 415–33.

Wilshire, H. G., Nakata, J. K. and Hallet, B., 1981, Field observations of the December 1977 wind storm, San Joaquin Valley, California. In T. L. Péwé (ed.), *Desert dust: origin, characteristics and effects on man.* Geological Society of America, 233–51.

Wilson, A. T., 1978, Pioneer agriculture explosion and CO_2 levels in the atmosphere. *Nature,* 273, 40–1.

Wilson, E. O., 1988, *Biodiversity.* Washington DC: National Academy Press, 58–70.

Wilson, K. V., 1967, A preliminary study of the effect of urbanization on floods in Jackson, Mississippi. *United States geological survey professional paper,* 575–D, 259–61.

Winkler, E. M., 1970, The importance of air pollution in the corrosion of stone and metals. *Engineering geology,* 4, 327–34.

Winstanley, D., 1973, Rainfall patterns and general atmospheric circulation. *Nature,* 245, 190–4.

Wolman, M. G., 1967, A cycle of sedimentation and erosion in urban river channels. *Geografiska annaler,* 49A, 385–95.

Wolman, M. G. and Schick, A. P., 1967, Effects of construction on fluvial sediment, urban and suburban areas of Maryland. *Water resources research,* 3, 451–64.

Wong, C. S., 1978, Atmospheric input on carbon dioxide from burning wood. *Science,* 200, 197–9.

Wood, C. M., Lee, M., Linker, J. A. and Saunders, P. J. W., 1974, *The geography of pollution, a study of greater Manchester.* Manchester: Manchester University Press.

Woodroffe, C. D., 1990, The impact of sea-level rise on mangrove shorelines. *Progress in Physical Geography,* 14, 483–520.

Woodwell, G. M., 1978, The carbon dioxide question. *Scientific American,* 238, 1, 34–43.

——, 1990, *The earth in transition: patterns and processes of biotic impoverishment.* Cambridge: Cambridge University Press.

Woodwell, G. M., Whittaker, R. H., Reiners, W. A., Likens, G. E., Delwiche, C. C. and Borkin, D. B., 1978, The biota and world carbon budget. *Science,* 199, 141–6.

Wooster, W. S., 1969, The ocean and man. *Scientific American*, 221, 218–23.

World Resources Institute, 1988, *World resources 1988–9*. New York: Basic Books.

——, 1992, *World Resources 1990–91*. New York and Oxford: Oxford University Press.

Worthington, E. B. (ed.), 1977, *Arid land irrigation in developing countries: environmental problems and effects*. Oxford: Pergamon.

Wrigley, E. A., 1965, Changes in the philosophy of geography. In R. J. Chorley and P. Haggett (eds), *Frontiers in geographical teaching*, London: Methuen.

Wright, H. E., 1974, Landscape development, forest fires and wilderness management. *Science*, 186, 487–95.

Wright, L. W. and Wanstall, P. J., 1977, The vegetation of Mediterranean France: a review. Occasional paper 9, Department of Geography, Queen Mary College, University of London.

Wymer, J., 1982, *The palaeolithic age*. London: Croom Helm.

Yaalon, D. H. and Yaron, B., 1966, Framework for man-made soil changes – an outline of metapedogenesis. *Soil science*, 102, 272–7.

Yi-fu Tuan, 1971, Man and nature. *Commission on College Geography resource paper*, 10.

Yorke, T. H. and Herb, W. J., 1978, Effects of urbanization on stream-flow and sediment transport in the Rock Creek and Anacostia River basins, Montgomery County, Maryland, 1962–74. *United States Geological survey professional paper*, 1003.

Young, J. E., 1992, *Mining the earth*. Worldwatch Paper 109, 1–53.

Zaret, T. M. and Paine, R. T., 1973, Species introduction in a tropical lake. *Science*, 182, 449–55.

Zimina, R. P., Polevaya, Z. A. and Yelkin, K. F., 1972, The plowing up of the virgin lands and the Bobac Marmot resources of Central Kazakhstan. *Soviet geography*, 13, 246–56.

Index

Note: Page numbers in italics refer to figures and plates. Author references are included only where their work is quoted directly in the text (or in tables and figures) or discussed in some detail (i. e. more than a sentence or two). References to particular places, species, etc. will be found most easily by looking for the specific entry rather than searching under more general headings (e. g. Grimes Graves, rather than East Anglia or Britain; Africanized honey bee rather than animals or insects).

Index compiled by Ann Barham